IFMBE Proceedings 130

Series Editor

The IFMBE Proceedings Book Series is an official publication of *the International Federation for Medical and Biological Engineering* (IFMBE). The series gathers the proceedings of various international conferences, which are either organized or endorsed by the Federation. Books published in this series report on cutting-edge findings and provide an informative survey on the most challenging topics and advances in the fields of medicine, biology, clinical engineering, and biophysics.

The series aims at disseminating high quality scientific information, encouraging both basic and applied research, and promoting world-wide collaboration between researchers and practitioners in the field of Medical and Biological Engineering.

Topics include, but are not limited to:

- Diagnostic Imaging, Image Processing, Biomedical Signal Processing
- Modeling and Simulation, Biomechanics
- Biomaterials, Cellular and Tissue Engineering
- Information and Communication in Medicine, Telemedicine and e-Health
- Instrumentation and Clinical Engineering
- Surgery, Minimal Invasive Interventions, Endoscopy and Image Guided Therapy
- Audiology, Ophthalmology, Emergency and Dental Medicine Applications
- Radiology, Radiation Oncology and Biological Effects of Radiation
- Drug Delivery and Pharmaceutical Engineering
- Neuroengineering, and Artificial Intelligence in Healthcare

IFMBE proceedings are indexed by SCOPUS, EI Compendex, Japanese Science and Technology Agency (JST), SCImago. They are also submitted for consideration by WoS.

Proposals can be submitted by contacting the Springer responsible editor shown on the series webpage (see "Contacts"), or by getting in touch with the series editor Ratko Magjarevic.

Simona Vlad · Nicolae Marius Roman
Editors

9th International Conference on Advancements of Medicine and Health Care Through Technology

Proceedings of MEDITECH 2024, September 30 – October 2, 2024, Cluj-Napoca, Romania

Springer

Editors
Simona Vlad
Faculty of Electrical Engineering
Technical University of Cluj-Napoca
Cluj-Napoca, Romania

Nicolae Marius Roman
Faculty of Electrical Engineering
Technical University of Cluj-Napoca
Cluj-Napoca, Romania

ISSN 1680-0737 ISSN 1433-9277 (electronic)
IFMBE Proceedings
ISBN 978-3-031-95670-6 ISBN 978-3-031-95671-3 (eBook)
https://doi.org/10.1007/978-3-031-95671-3

This Springer imprint is published by the registered company Springer Nature Switzerland AG
The registered company address is: Gewerbestrasse 11, 6330 Cham, Switzerland

Preface

Advancements of Medicine and Health Care through Technology and Promoting Medical Technology in Health Care by Biomedical Engineering Societies

Ratko Magjarević
President, IFMBE
University of Zagreb Faculty of Electrical Engineering and Computing, Zagreb, Croatia

It is a rare pleasure to be able to follow the development and growth of a truly scientific and professional series of conferences in biomedical engineering: I am one of the lucky ones and I am glad to have been invited to write this foreword to the Proceedings of the 9th International Conference on Advancements of Medicine and Health Care through Technology—MediTech 2024 which was held from September 30 to October 2, 2024, in Cluj-Napoca in Romania.

Like the conferences in previous editions, researchers and professionals from the fields of biomedical and clinical engineering, medical physics, medicine and health care gathered together to exchange their knowledge and experience, discuss innovative ideas and strengthen their communication networks.

In this conference, participants were able to hear more on the new trends in digital health, which is an important extension of eHealth due to advancement in digital technology and extensive use in health care: big data, extended health records, extensive use of telemedicine especially after the pandemics, data acquisition from numerous wearable devices, machine learning, artificial intelligence, just to name a few. It is essential for patients and for the healthy to make use of benefits brought by newly developed technologies. International scientific organizations such as the International Federation for Medical and Biological Engineering (IFMBE) unify the world-wide Medical and Biological Engineering Community in promoting health and quality of life through the advancement of research, development, application and management of medical devices and technology. Engineering jobs are directly present in medicine and health care, primarily through research, development and manufacturing of medical products, devices and systems, but is increasingly encountered in clinical settings.

Since its founding in 1959, IFMBE has in sixty+ years grown from a group of enthusiastic researchers, engineers and physicians into the world largest society based association in the field of biomedical engineering.

The International Federation for Medical and Biological Engineering is primarily a federation of national and transnational societies. In 2024, the number of affiliated societies has grown to more than seventy. Among the aims of IFMBE is building a strong international and regional network of BME societies that would enable and support projects of interest for medicine and health care. Also, IFMBE is actively included in international collaboration between other international organizations and associations aiming to improve health care and wellness and plan important actions for global society in the future.

The IFMBE is a non-governmental organization (NGO) for the United Nations and the World Health Organization (WHO), where it is uniquely positioned to influence the delivery of health care to the world through biomedical and clinical engineering. In the forthcoming years, IFMBE aims to support WHO to provide guidance documents on innovation, assessment, nomenclature, and management of medical devices which would be suitable for all countries affiliated to WHO. IFMBE is supporting activities on development of training materials, lately specially on the provision of services for emergencies, disasters and that are supportive toward the achievement of universal health coverage. IFMBE may support or propose resolutions to the World Health Assembly and is involved in WHO projects and in preparing publications.

IFMBE has a close association with the International Organization of Medical Physics (IOMP). Together, these two international bodies have established the umbrella organization International Union for Physical and Engineering Sciences in Medicine (IUPESM).

The principal objective of IUPESM is to contribute to the advancement of physical and engineering sciences in medicine for the benefit and wellbeing of humanity. These two organizations coordinate the World Congress on Medical Physics and Biomedical Engineering every three years, the next in Adelaide, Australia, in 2025.

International Union for Physical and Engineering Sciences in Medicine (IUPESM) is a member of International Science Council (ISC) which works under the auspices of UNESCO (United Nations Educational, Scientific and Cultural Organization). The International Science Council (ISC) is a non-governmental organization with a unique global membership that brings together over 200 international scientific unions and associations as well as national and regional scientific organizations including academies and research councils.

As a member of World Federation of Engineering Organizations (WFEO), IFMBE is involved with other engineering professional organizations in projects aiming of achievement of UN's Sustainable Development Goals as the only member to foster achievement of the SDG3, good health and well-being.

Participation and contributions of a large number of volunteers from all national and transnational affiliates to IFMBE, like the Romanian national biomedical engineering society *Societatea Națională de Inginerie Medicală și Tehnologie Biologică* (SNIMTB), are essential for achieving our mission. We hope that this new volume in the IFMBE Proceedings Series will contribute to the visibility of the Romanian BME community and IFMBE activities and open a space to even closer collaboration among BME researchers, professionals and students.

Prof. Kang-Ping Lin
Vice-Pesident, IFMBE
DHC, Universitatea Tehnică din Cluj-Napoca, Romania
Distinguished Professor, Chung Yuan Christian University, Taiwan

This volume presents the Proceedings of the 9th International Conference on Advancements of Medicine and Health Care through Technology, MediTech 2024, held from September 30 to October 2 in Cluj-Napoca Romania. As with its predecessors, MediTech 2024 continues the tradition of advancing the intersection of medicine, health care, and technology, bringing together researchers and professionals from diverse disciplines, including biomedical engineering, clinical engineering, medical physics and medicine.

It is with deep humility and immense gratitude that I address you, both as a participant and as an award recipient at this prestigious conference. Being honored with the opportunity to contribute to this gathering of distinguished scholars and healthcare professionals is an experience I will cherish throughout my career. The drive for innovation and our shared commitment to improving health care are themes that resonate deeply with me, and it is these values that make conferences like MediTech 2024 so vital.

MediTech 2024 brings together a wide range of knowledge, with high-quality papers reflecting excellence in fields of biomedical engineering, biomedical signal processing, telemedicine, health technology assessment and more. After a rigorous double-blind review process, a large number of submissions are accepted for publication. These are the selfless contributions of the Scientific Advisory Board, and we thank them for their expertise and efforts in evaluating these submissions.

This conference has brought forward critical advancements in not only in biomedical engineering but also in healthcare technology linking to medicine, further reinforcing the significance of our shared commitment to science and technology for the benefit of humanity. The presence of such remarkable experts from across the globe highlights the conference's essential role in fostering global collaboration, and I look forward to the continued exchange of ideas and breakthroughs that will shape the future of health care.

I would like to express my deepest gratitude to the organizers of MediTech 2024, the reviewers, authors and all those who made this event a success. Your collective contributions enrich our field and provide invaluable insight into the progress of biomedical engineering.

I extend my heartfelt congratulations to all the participants and award recipients at MediTech 2024. Your contributions continue to inspire us all to push the boundaries of knowledge and technology for the betterment of society. The papers presented at this conference will undoubtedly contribute to the body of knowledge in biomedical engineering and healthcare technology, influencing the future of the field for years to come. I look forward to continued collaboration among all of us as we advance the field of biomedical engineering and explore new ways to improve healthcare worldwide. Let us work together to ensure that technology and innovation serve humanity, paving the way for a healthier, more sustainable future.

Prof. Nicolae Marius Roman
Technical University of Cluj Napoca
President, SNIMTB Chair of MediTech Conference
Honorary Life Member, IFMBE

The MediTech 2024 continues series of MediTech conferences organized by the "Romanian National Society for Medical Engineering and Biological Technology" in collaboration with Technical University of Cluj-Napoca, "Iuliu Haţieganu" University of Medicine and Pharmacy of Cluj-Napoca and "Dr. Constantin Papilian" Military Emergency Hospital Cluj-Napoca, Romania, under the auspices of "International Federation for Medical and Biological Engineering" every two years, in even years, in Cluj-Napoca Romania.

It is worth noting that at this edition of the MediTech 2024 conference, the acceptance rate of articles submitted for publication, following reviews, is 88%.

We were honored and happy to have with us on *Prof. Ratko Magjarević*, IFMBE President, and *Prof. Kang-Ping Lin,* IFMBE Vice-Pesident, not only as guests but also as the keynote speaker of the conference.

We would like to especially thank the decision-making colleagues who supported the organizational effort:

Prof. Vasile Ţopa–Rector, Technical University of Cluj-Napoca

Prof. Dr. Mihaela Felicia Băciuţ–Vice-Rector, "Iuliu Haţieganu" University of Medicine and Pharmacy of Cluj-Napoca

Col. Dr. Doina Baltaru – Manager, "Dr. Constantin Papilian" Military Emergency Hospital Cluj-Napoca

Also, we would like to thank all the participants, invited speakers, academics, researchers, medical specialists, sponsors, media partners, members of the scientific and organizing committee for their contribution and dedication to the success of the conference.

Organization

Organizers

Romanian National Society for Medical Engineering and Biological Technology
Technical University of Cluj-Napoca, Romania
"Iuliu Haţieganu" University of Medicine and Pharmacy Cluj-Napoca, Romania
"Dr. Constantin Papilian" Military Emergency Hospital, Cluj-Napoca, Romania

Conference Chair

Nicolae Marius Roman — Romanian National Society for Medical Engineering and Biological Technology

Honorary Chair

Radu Vasile Ciupa — Romanian National Society for Medical Engineering and Biological Technology

Scientific Advisory Committee

Ancuta-Coca Abrudan, Romania
Laura Bacali, Romania
Doina Baltaru, Romania
Mihaela Baritz, Romania
Radu Ciorap, Romania
Eva Dulf, Romania
Paul Farago, Romania
Stefan Gergely, Romania
Zoltan German-Sallo, Romania
Laura Grindei, Romania
Rodica Holonec, Romania
Alexandru Ianosi-Andreeva-Dimitrova, Romania
Ioan Jivet, Romania
Mihaela-Ruxandra Lascu, Romania
Mircea Leabu, Romania
Kang-Ping Lin, Taiwan

Angela Lungu, Romania
Eugen Lupu, Romania
Dan S. Mandru, Romania
Winfried Mayr, Austria
Bogdan Micu, Romania
Dan D. Micu, Romania
L. Dan Milici, Romania
Marius Muji, Romania
Mihai S. Munteanu, Romania
Vlad Muresan, Romania
Anca I. Nicu, Romania
Maria Olt, Romania
Gabriel Oltean, Romania
Tudor Palade, Romania
Alessandro Pepino, Italy
Dorin Petreus, Romania
Doina Pisla, Romania
Dan V. Rafiroiu, Romania
Nicolae Marius Roman, Romania
Corneliu Rusu, Romania
Adrian Samuila, Romania
Mircea S. Serbanescu, Romania
Doru Ursutiu, Romania
Radu C. Vlad, Romania
Simona Vlad, Romania

Local Organizing Committee

Ancuta-Coca Abrudan
Mihaela F. Baciut
Doina Baltaru
Adrian I. Bojita
Levente Czumbil
Rodica Holonec
Angela Lungu
Dan S. Mandru
Bogdan Micu
Dan D. Micu
Anca I. Nicu
Razvan Nicu
Maria Olt
Sever Pasca
Doina Pisla
Nicolae Marius Roman
Adrian Samuila
Simona Vlad

Speakers at the Opening Ceremony

Ratko Magjarević — President of the International Federation for Medical and Biological Engineering (IFMBE)

Kang-Ping Lin	Vice-president of the International Federation for Medical and Biological Engineering (IFMBE)
Mihaela Băciuţ	Vice-rector of the "Iuliu Haţieganu" University of Medicine and Pharmacy, Cluj-Napoca, Romania
Gabriel Oltean	Vice-rector of the Technical University of Cluj-Napoca, Romania
Dan Doru Micu	President of the Senate Research Commission, Technical University of Cluj-Napoca, Romania
Nicolae Marius Roman	President of the Romanian National Society for Medical Engineering and Biological Technology, Romania

Keynote Speakers

Ratko Magjarević, President at International Federation for Medical and Biological Engineering (IFMBE); University of Zagreb, Faculty of Electrical Engineering and Computing, Department for Electronic Systems and Information Processing, Zagreb, Croatia
Key drivers in medical technology and their implications to health care

Kang-Ping Lin, Vice-president at International Federation for Medical and Biological Engineering (IFMBE); Chung-Yuan Christian University, Taiwan
How does ICT support the patient safety and medical quality?

Invited Speakers

Helmut Hutten, Institute of Biomedical Imaging, University of Technology, Graz, Austria

Inherent Ambivalence Risks in Biomedical Engineering Progress: From Technical Aids to Cyborgs, from Signal Processing to Artificial Intelligence, from Gene Analysis to Genome Editing

Magdalena Stoeva, Department of Diagnostic Imaging, Medical University of Plovdiv, Bulgaria; Secretary General IOMP, Secretary General IUPESM

Africa's Readiness for Artificial Intelligence in Radiotherapy–Technological Aspects and the Way Forward

Winfried Mayr, Center of Medical Physics and Biomedical Engineering, Medical University of Vienna, Austria

Functional electrical stimulation is a versatile, highly effective and efficient tool for rehabilitation. Why is it not better accessible for patients in need?

Cecilia Cristea, Group of Biosensors and Bioelectrochemistry, Analytical Chemistry Department; Faculty of Pharmacy, "Iuliu Haţieganu" University of Medicine and Pharmacy Cluj-Napoca, Romania

Point of Use Electrochemical Systems for Fast and On-site Detection

Andreea Catalina Forțu, Antoanela Cristina Teoroc, Bucharest-Jilava Penitentiary, Romania

Ana-Maria Culai, National Administration of Penitentiaries, Romania
Addressing the problem of drug use in prisons. Methods and perspectives

Dan Lucian Dumitrașcu, 2nd Medical Clinic, Cluj County Clinical Emergency Hospital; University of Medicine and Pharmacy "Iuliu Haţieganu" Cluj-Napoca, Romania
13C Urease breath test to prevent gastric cancer

Mircea Leabu, "Victor Babeş" National Institute of Pathology, Bucharest, Romania
Biomedical engineering a tool to apply the Noumenal principle "Everything is possible" into Phenomenal

Alessandro Pepino, Department of Electrical Engineering and Information Technology, University of Naples "Federico II", Italy
The assessment of the impact of Simulation and Blended learning in university course, a case study

Irina David, Ioana Țîrcomnicu-Unga,
INAS SA – Romania
Simulation Software Solutions in Biomedicine: Enhancing Innovation, Precision, and Accuracy

Bogdan Micu, Fifth Surgical Clinic, Municipal Clinical Hospital; University of Medicine and Pharmacy "Iuliu Hatieganu" Cluj-Napoca, Romania
Implantable ports in medical practice

Cristian Dinu,
Mihaela Băciuț, Simion Bran, Florin Onişor, Sergiu Văcăraş, Gabriel Armencea, Ileana Mitre, Sebastian Stoia, Tiberiu Tamaş, Rareş Mocan, Avram Manea, Grigore Băciuț
University of Medicine and Pharmacy, Maxillofacial Surgery and Implantology Department, Cluj-Napoca, Romania
The Evolution in the 21st Century of Personalised Parotid Surgery

Horațiu Rotaru, Department of Cranio-Maxillofacial Surgery, "Iuliu Hatieganu" University of Medicine and Pharmacy, Cluj-Napoca, Romania
Custom reconstructions of post-ablative cranio-maxillofacial defects

Tony Ward, Department of Electronic Engineering, University of York, England
A general design approach to engineering programmes designed to produce employment ready graduates with a Medical Engineering degree example described

Doru Ursutiu[1,2], Horia Modran[1], Cornel Samoila[1,3]
[1] Center for Valorization and Transfer of Competences CVTC–"Transilvania" University of Brasov, Romania
[2] Romanian Academy of Scientists **AOSR**
[3] Academy of Technical Sciences of Romania **ASTR**
Novel trends and technologies in Engineering and Biomedical Engineering Education

Mihaela Ioana Baritz, Product design Mechatronics and Environment Department, "Transilvania" University from Brasov, Romania
Integrated multimodular structures for monitoring the biomechanics of the human locomotor system

Laurențiu-Dan Milici, Ștefan cel Mare University of Suceava, Romania
Stress Monitoring Device for IT Professionals

Thordur Helgason, Science and Technology at Reykjavik University, Iceland; President of the Icelandic Society for Biomedical Engineering and Medical Physics (HTFI)
Patellar tendon test, evoked potentials and brain connectivity

Chairmen

Radu Ciorap
Laura Grindei
Rodica Holonec
Angela Lungu
Eugen Lupu
Dan S. Mandru
Mihai S. Munteanu
Anca Iulia Nicu
Andra Păstrăv
Nicolae Marius Roman
Adrian Samuila
Simona Vlad

Sponsors

Farmec SA
Cluj-Napoca, Romania

INAS SA
Craiova, Romania

Tehno Industrial SA
Bucharest, Romania

Media Partners

Contents

Medical Devices, Measurements and Instrumentation; Biomedical Signal Processing

Biometric Medication Storage System

Rodica Holonec(✉), Laura Grindei, and Laszlo Rapolti

Department of Electrotechnics and Measurements, Technical University of Cluj-Napoca, Cluj-Napoca, Romania

rodica.holonec@ethm.utcluj.ro

Abstract. In response to the increasing need for secure and efficient management of prescription drugs, particularly controlled substances like opioids and sedatives, this paper proposes a biometric medication storage system. The system leverages fingerprint and facial recognition technologies to provide stringent access control, ensuring that only authorized personnel can access sensitive medications. Unlike traditional security methods, biometric systems offer higher security levels as biometric traits are unique, difficult to forge, and cannot be easily lost or forgotten. The integration of these technologies enhances not only security but also operational efficiency, reducing authentication time and simplifying access in critical situations. The system's design includes a dual-mode biometric authentication process, where users are first identified by their fingerprint, followed by facial recognition to verify identity before access is granted. The implementation includes two subsystems: the first is composed of an Arduino board connected to a fingerprint sensor and a servomotor, which conditions the operation of the second subsystem. The second subsystem is based on a Raspberry Pi as the central processing unit, along with a camera and a solenoid lock mechanism. This biometric approach ensures secure, traceable, and efficient medication management, meeting both safety and legal requirements while optimizing the storage and distribution processes of controlled substances.

Keywords: Secure Medication Management · Fingerprint Recognition · Face Recognition · Access Control · Raspberry Pi

1 Introduction

In the context of technological advancements in the medical field and the growing need for safe and efficient management of prescription drugs, implementing a biometric medication storage system is essential. Special medications, such as opioids, sedatives, and other psychotropic drugs, require careful management due to their potential for abuse and associated risks. A biometric storage system ensures strict access control while enhancing the traceability and monitoring of these drugs.

Compared to traditional security methods like keys or passwords, biometric technology offers a higher level of security since biometric characteristics cannot be lost, forgotten, or easily forged. Implementing such a system ensures that only authorized

S. Vlad and N. M. Roman (Eds.): MEDITECH 2024, IFMBE Proceedings, 130, pp. 3–12, 2025.
https://doi.org/10.1007/978-3-031-95671-3_1

and qualified personnel have access to these substances, reducing the risk of theft, unauthorized use, or administration errors. Besides security and traceability, biometric systems also contribute to operational efficiency. By eliminating the need to manage keys or access cards, authentication time is significantly reduced, allowing medical staff to quickly and securely access necessary medicines, which is crucial in emergencies. Additionally, advanced systems can provide data analytics to optimize storage and distribution processes, further improving the management of medical resources. For example, in a hospital setting, an automated medication dispensing system could incorporate biometric access controls to ensure that only authorized healthcare professionals can retrieve controlled substances. In [1], the authors describe an automated medication dispensing device designed to help patients take medication on time, especially those with physical disabilities. Biometric security is built into this device to ensure only authorized individuals can access medication. The system stores medication information on a server so the patient can access it at any time. This measure increases security and ensures that only legitimate users can manage or access medications dispensed by the device. In other cases, smart pill dispensers use IoT technology to improve drug delivery, incorporating biometric security measures such as facial recognition and biometric verification to ensure that drugs are given to the right person [2].

Special containers for controlled drugs ensure safety and strict access control. These containers are designed to meet legal and safety requirements for storing controlled substances and can include features such as rugged construction, high-security locking systems, biometric access, monitoring systems, temperature control, and audit trails.

Fingerprint-based access solutions, known for their scalability, are increasingly adopted across various industries, including healthcare, education, government, and commercial enterprises. Their advanced security features make them a valuable investment for enhancing security and streamlining access control processes.

In healthcare, facial recognition access systems are also gaining popularity for regulating access to controlled substances by authorized personnel.

In practical single-mode biometric systems utilizing fingerprint or facial recognition, performance, accessibility, and vulnerability to bypass are critical considerations. In [3] the authors present an Internet of Things (IoT) based smart safe, system that uses fingerprint and face recognition. They demonstrated that the integration of facial recognition with fingerprint authentication provides an additional layer of security, ensuring that only authorized staff can access the safe.

2 Fingerprint and Facial Recognition in Biometric Security

2.1 Fingerprint Recognition

Fingerprint-based access systems are an important component of biometric technologies used to secure access to various spaces and information systems. These systems are based on the unique characteristics of individual fingerprints, providing a high level of security and mitigating the risks associated with easily compromised access data, such as passwords. They operate by capturing the user's fingerprint image and comparing it to a database of authorized fingerprints. This process involves two main stages: enrollment and authentication. During enrollment, the user's fingerprint is captured and stored in a secure database. In the authentication phase, the captured fingerprint is compared with those stored in the database, granting access only if a match is found [4].

Fingerprint-based systems have many advantages, such as high security, since fingerprints are unique and difficult to duplicate; convenience, as users do not have to remember passwords or access cards, which reduces the risk of loss or theft; and speed, as login only takes a few seconds. However, these systems also have disadvantages, such as high implementation and maintenance costs, especially for large organizations; potential problems recognizing damaged, dirty or worn fingerprints; and data privacy and security concerns related to storage and management. of biometric information [5].

Fingerprint-based access systems are widely used in various sectors, including healthcare, for managing access to secure areas like hospitals, safeguarding patient data in electronic health records, and securing medical devices. Additionally, these systems are employed in government institutions, mobile devices, and financial transactions, offering a blend of security, convenience, and speed. However, it's crucial to implement these systems carefully to protect the security and privacy of biometric data, especially in sensitive fields like healthcare [6].

One of the most widely used biometric methods to confirm a person's identity is fingerprint recognition. Contemporary fingerprint recognition systems use a combination of hardware technologies and complex algorithms to collect, process and compare fingerprints to existing databases [7]. Because of their reliability and affordability, optical systems are among the most widely used of these technologies. Optical technologies were the first methods of fingerprint capture that used light to obtain a detailed image of the fingerprint surface. In an optical sensor, the user's finger is placed on a glass surface illuminated by a light source. A camera or CMOS sensor captures the light reflected by the ridges and valleys of the fingerprint, and the captured image is then processed and analyzed. The ability to obtain high-resolution images, durability and simplicity of design are some of the advantages of optical technologies. However, they also have disadvantages, such as the sensitivity of the sensor to dirt and lighting, as well as difficulties in capturing fingerprints from fingers that are dry or damaged. Modern optical technologies use special filters to reduce image noise and improve capture clarity. Some systems use advanced algorithms to compensate for pressure variations or surface imperfections.

The fingerprint recognition process identifies a user by capturing the fingerprint image and applying sophisticated algorithms to analyze its unique characteristics. This process involves several stages: preprocessing, where the image is enhanced to reduce noise and improve clarity; feature extraction, where key points such as ridge bifurcations and terminations are identified, along with other features like ridge density and orientation; and matching, where these extracted features are compared against a database of stored fingerprints. Convolutional Neural Networks (CNNs) are increasingly used in machine learning and deep learning to enhance the accuracy of fingerprint recognition [8].

2.2 Facial Recognition Techniques

Facial recognition systems have gained importance in image processing and biometric applications, especially in healthcare and research, where they offer a more accessible alternative to other biometric methods like fingerprints, iris scans, or signatures, particularly when dealing with uncooperative or incapacitated individuals. These systems are increasingly utilized not only in urban security for surveillance and crime prevention but also in healthcare settings to verify patient identities, ensuring that medical records and treatments are accurately matched to the correct individual, thereby enhancing both security and patient care.

Facial recognition technology presents a complex challenge in real-world applications due to varying factors like lighting, occlusion, and imaging conditions. It involves a combination of face detection and recognition techniques within image analysis [9]. Detection identifies the location of faces in an image, while recognition classifies these images based on known structured properties, which are typically used in computer vision applications. These "standard images" share characteristics like consistent resolution and similar alignment of facial features.

Finding faces in an image is the first step in the recognition process, which starts with obtaining an image from a camera. Extracting face features, including the lips, nose, eyebrows, and eyes, is the next stage. Based on the findings of the recognition process, the person is identified in the last stage. While appearance-based approaches use learning and training techniques to locate faces, knowledge-based methods are based on human comprehension of facial traits. Faces are identified when they are detected by comparing them to a database of common face pictures. It is possible to process these photos in two or three dimensions. While 3D approaches use three-dimensional facial data and employ techniques like point correspondence measurement, average face calculation, and geometric measurement, 2D approaches use flat photos and classify them using various training or learning techniques.

Facial recognition technology encompasses a range of algorithms, each employing unique strategies for face identification and verification [10].

One feature descriptor that is frequently used in computer vision, especially for facial recognition applications, is the Histogram of Oriented Gradients (HOG). The gradient direction histogram for each cell in an image is calculated by the HOG algorithm, which then normalizes the output to increase accuracy. The image is divided into small, connected sections known as cells. Concatenated vectors of these histograms make up the final descriptor, which is very useful for tasks like face detection because it captures the fundamental appearance and structure of the picture. Because HOG is resilient to changes in lighting and posture, it is especially well-suited for facial recognition applications. It draws attention to an image's local edge structures, which are essential for differentiating between faces. HOG can be used as a pre-processing step in facial recognition applications, converting the raw picture data into a format that machine learning algorithms or matching processes can examine more successfully.

Facial recognition techniques encompass local, holistic, and hybrid approaches, utilizing either individual facial features or the entire face in the recognition process [11]. The practical implementation of these techniques often involves Python, particularly through the OpenCV library. A robust and efficient face recognition system must address challenges such as illumination variations, occlusions, and the complexities of real-world conditions. Additionally, key aspects such as feature extraction, coding, and the use of deep learning models are critical for achieving accurate recognition.

Implementing a facial recognition algorithm in Python typically involves using various libraries to simplify the recognition process. For instance, Haar cascades are commonly used in face detection by leveraging pre-trained classifiers available in the OpenCV library. These classifiers employ a series of increasingly complex stages of Haar-like features to efficiently and accurately identify facial regions in an image, quickly eliminating non-face areas in a hierarchical manner.

In facial recognition, the Histogram of Oriented Gradients (HOG) is a feature descriptor often employed for face detection. It operates by dividing an image into small, connected regions called cells, then computing the histogram of gradient directions or edge orientations for the pixels within each cell. These histograms are combined into a feature vector that effectively encodes local shape information, emphasizing the direction and intensity of edges. HOG is robust to variations in illumination and position, making it a powerful method for face detection and recognition under various conditions.

HOG can be implemented using Python's OpenCV library, where HOG descriptors are extracted from images and then fed into machine learning models for classification. The HOG method is renowned for its efficiency and accuracy in various face recognition tasks, particularly when combined with other techniques such as Support Vector Machines (SVM) for classification in the comparison module [12].

2.3 Integrating Fingerprint and Facial Biometrics

Biometric identification systems are essential to modern security solutions, with facial recognition and fingerprint identification being two of the most commonly employed methods. Each utilizes unique physical attributes-fingerprints and facial features-to authenticate individuals.

Both methods are highly accurate but they have certain limitations. Fingerprint identification can be compromised by dirt, moisture, or other external factors, while face recognition might struggle in poor lighting or with facial obstructions. Despite these challenges, continuous advancements in algorithms have improved the reliability of face recognition, making it increasingly suitable for diverse applications.

From the convenience and scalability point of view, the fingerprint identification requires physical contact, which can be inconvenient in environments where hygiene is crucial or when large numbers of people need to be processed quickly. On the other hand, face recognition is contactless, allowing for faster and more scalable identification, particularly in crowded or high-traffic areas. This makes face recognition particularly effective for real-time surveillance and access control in public spaces.

Both technologies can be integrated into existing systems, but face recognition is often easier to deploy in environments where devices already have built-in cameras. This can reduce the cost and complexity of implementation, especially in large-scale applications.

Unimodal biometric systems measure and analyze biological traits based on a single source or distinct individual biological characteristic. The most popular type of biometric systems uses fingerprints, but when a fingerprint is placed in front of the sensor, it can be spoofing or presented as a presentation attack. Nassim Ammour et al. [13] suggested an improved biometric system based on a multimodal approach utilizing two different kinds of biological traits in order to address this problem. The experimental findings show how well that the suggested strategy can strengthen the system's resistance to different attacks.

Using a multimodal biometric system that combines both face recognition and fingerprint identification can significantly enhance security. This approach leverages the strengths of each method, reducing the likelihood of errors and providing a more secure and reliable authentication process. In cases where one method may be less effective due to environmental conditions or other factors, the other can serve as a backup, ensuring consistent and accurate identification.

In conclusion, integrating both face recognition and fingerprint identification within a security framework offers numerous advantages, including enhanced accuracy, reliability, and scalability. Our proposed system that uses a multimodal approach not only improves overall security but also provides a more user-friendly and hygienic solution, suitable for the medication storage container.

3 System Design

In the design and implementation of our biometric system (Fig. 1), we employed a combination of fingerprint identification (first subsystem) and facial recognition (second subsystem) for controlling access to the medication storage container. The process begins with user identification via fingerprint; if the fingerprint matches an entry in the database, a servomotor is activated to open a shutter in front of the camera. This enables the second subsystem to proceed with facial recognition, using the image captured by the camera. Access is granted only when both the fingerprint and facial recognition criteria are successfully met.

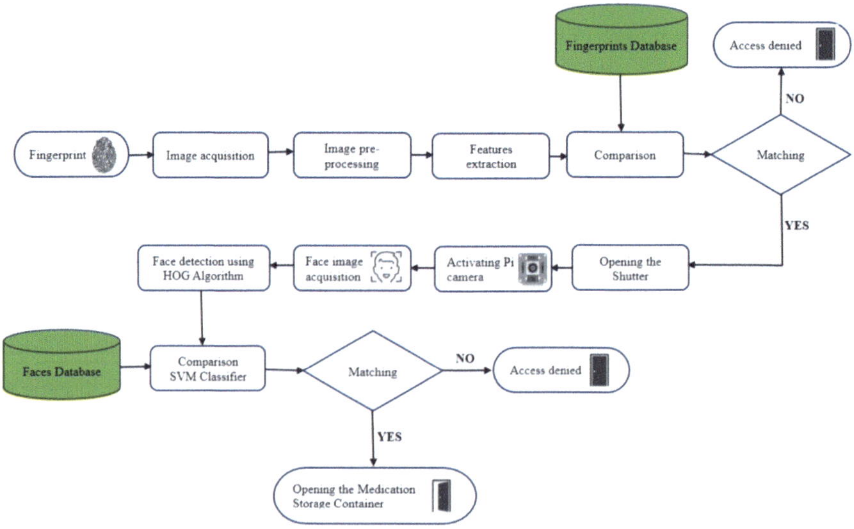

Fig. 1 Logical flowchart process for controlling access to the medication storage container

To access the contents of the medication container, the first subsystem involves following steps 1 through 4, and then for the second subsystem, steps 5 through 9 are followed as outlined below:

- Step 1: The user places their finger on the fingerprint scanner.
- Step 2: The user's fingerprint is acquired and preprocessed.
- Step 3: The fingerprint features are extracted and compared with those in the database for matching.
- Step 4: If the features match, the user is considered known, and the servomotor is activated to move the shutter.

The fingerprint recognition subsystem (Fig. 2) uses an AS608 fingerprint sensor, two LEDs (an Access Approved LED and an Access Denied LED), and an MG955 servomotor, all connected to an Arduino UNO board.

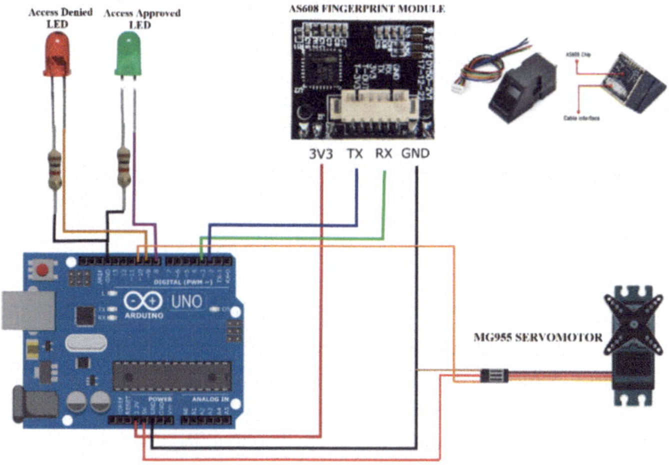

Fig. 2 The fingerprint recognition subsystem. The circuit diagram

The AS608 is an optical fingerprint sensor that captures and identifies fingerprints, transmitting the processed data to a microcontroller via serial communication. It can store up to 120 fingerprint records in its internal memory, making the authentication process efficient and straightforward. The operation of the AS608 involves two primary steps:

- Fingerprint Enrollment: This process involves placing, lifting, and reapplying the finger on the sensor to capture a digital image of the fingerprint. The image is then preprocessed to remove noise and extract unique features, such as bifurcations and ridges. These features are stored as a digital template for future authentication.
- Real-Time Authentication: In this step, the image acquisition and processing are repeated, and the captured fingerprint is compared with the stored templates. If a match is found, the face detection subsystem can be activated.

The setup involves uploading the necessary code and using the Adafruit Fingerprint Sensor library for integration and functionality.

- Step 5: Once the shutter is opened, the camera detects activity.
- Step 6: Images are captured by the camera in real-time and processed with the HOG algorithm for face detection.
- Step 7: The images are compared with those in the database (SVM Classifier).
- Step 8: If the images match, the user's identity is verified, otherwise the container remains closed
- Step 9: A command is sent to the Raspberry Pi, which activates the relay to open the door, allowing the user access to the medication in the container.

The face recognition subsystem architecture is presented in Fig. 3.

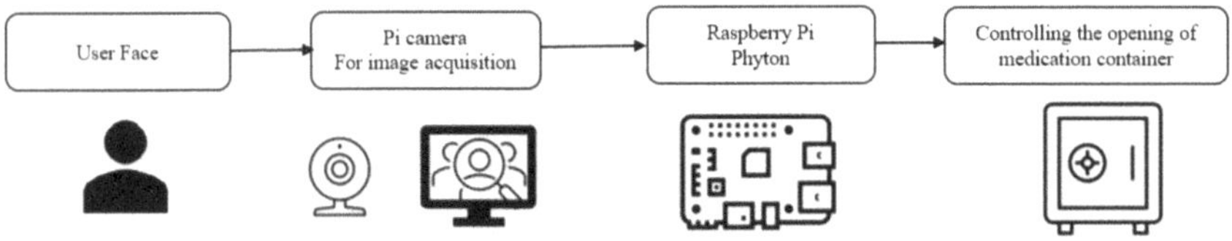

Fig. 3 The face recognition subsystem architecture

The Raspberry Pi 3B serves as the primary processing unit, running the operating system, facial recognition software, and controlling the input/output operations for connected components. Python is used to develop the software necessary for facial recognition, relay control, and solenoid lock management.

The PI Webcam captures images or video streams, and in this application, it is specifically used to capture the image of the person attempting to access the medication container. The webcam is connected to the Raspberry Pi via a flexible cable that is inserted into the camera port located between the Ethernet and HDMI ports.

The Solenoid Lock functions as an electronic locking mechanism, which can be controlled to lock or unlock the container door. It is operated by the Raspberry Pi through a relay, and is powered by an appropriate power supply. The relay allows the Raspberry Pi to control high-voltage devices, such as the solenoid lock, using a low-voltage signal. It connects to the GPIO pins of the Raspberry Pi and the solenoid lock circuit.

A second pair of red and green LEDs are used as visual indicators, where the red LED signifies that the door is locked, and the green LED indicates it is unlocked. These LEDs are connected to the Raspberry Pi through resistors to regulate the current flow.

Figure 4 depicts the implementation of the proposed system:

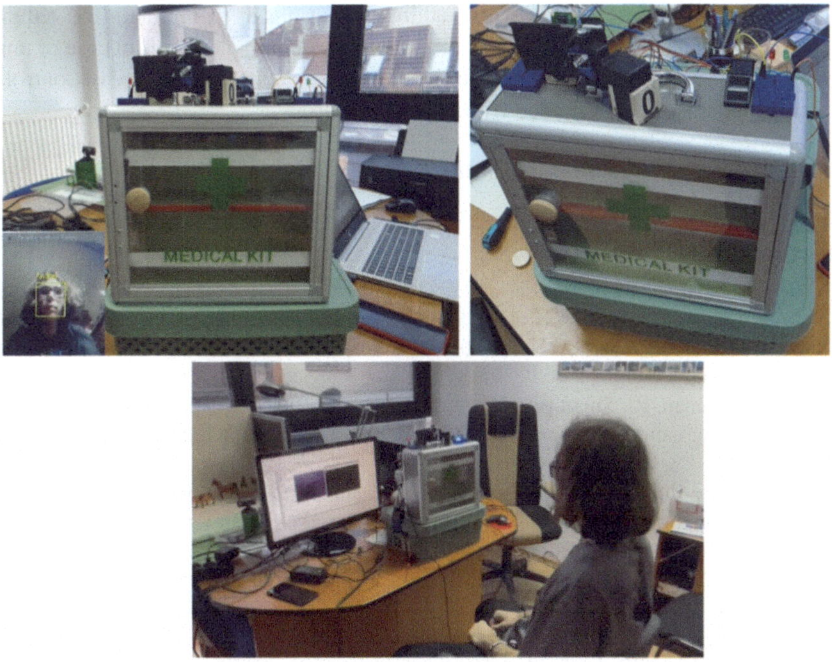

Fig. 4 Biometric medication storage system implementation

4 Conclusions

The implementation of a biometric medication storage system offers a significant improvement in the security and management of controlled substances, such as opioids and sedatives. By integrating fingerprint recognition with facial recognition, the proposed system ensures that access is strictly limited to authorized personnel, thereby reducing the risks of theft, unauthorized use, and administration errors. Additionally, this system enhances operational efficiency by providing quick and secure access to medications, which is crucial in emergency situations.

However, the biometric medication storage system has some limitations, such as potential delays in processing due to the limited computational power of the Arduino and Raspberry Pi, restricted storage capacity for biometric data and security vulnerabilities associated with local data storage. Furthermore, it relies on hardware components that may be prone to mechanical failure and is dependent on a stable power supply, which could hinder access during outages. A backup access solution, in case one or both subsystems fail, could include a manual override, such as a physical key, a keypad for PIN entry, a backup power supply to prevent downtime, or alternative user authentication methods like RFID cards or mobile-based authentication, ensuring continuous access to medication.

Future enhancements could involve adding a thermostat to maintain the required temperature for sensitive medications and an alarm system to alert staff in case of unauthorized access attempts. These upgrades would further secure the storage of controlled substances and ensure compliance with safety standards.

Conflict of Interest. The authors declare that they have no conflict of interest.

References

1. Mizanur Rahman, M., Aktar, R., Dey, S.K.: Design and implementation of a low-cost automated medicine dispenser. In: International Conference on Advancement in Electrical and Electronic Engineering, 24–26 February, Gazipur, Bangladesh (2022)
2. Tiwari, P., Raman, R.: Smart pill dispensers and IoT: advancements in medication administration. In: International Conference on Emerging Systems and Intelligent Computing (ESIC) (2024)
3. Mangata, B.B., Ilunga, D., N'kashama, P.B.C., Muamba, D.K.: Comparative studies between a facial recognition system and a fingerprint recognition system for access control. Int. J. Inform. Syst. Comput. Sci. **6**(2), 69–75 (2022)
4. Maltoni, D., Maio, D., Jain, A.K., Prabhakar, S.: Handbook of Fingerprint Recognition, 2nd edn. Springer: ISBN: 978-1848822542 (2009)
5. Jain, A.K., Ross, A., Prabhakar, S.: An introduction to biometric recognition. IEEE Trans. Circuits Syst. Video Technol. **14**, 4–20 (2004)
6. Sharmin Jahan; Mozammel Chowdhury; Rafiqul Islam, Robust fingerprint verification for enhancing security in healthcare system. International Conference on Image and Vision Computing New Zealand (IVCNZ), 4–6 Dec, (2017)
7. Sunitha, S.V., Vaishnavi, C.M., Trupti, R., Shilpa, R.: Biometric door access system using arduino. In: IEEE Fifth International Conference on Advances in Electronics, Computers and Communications (ICAECC), 07–08 September, Bengaluru, India (2023)
8. Militello, C., Rundo, L., Vitabile, S., Conti, V.: Fingerprint Classification Based on Deep Learning Approaches: Experimental Findings and Comparisons, Symmetry. MDPI (2021)
9. Shavetov, S., Sivtsov, V.: Access control system based on face recognition. In: 7th International Conference on Control, Decision and Information Technologies (CoDIT'20) | Prague, Czech Republic / June 29–July 2 (2020)
10. Singh, S., Prakash, U.M.: Facial recognition automation system using machine learning. In: 2nd International Conference on Advance Computing and Innovative Technologies in Engineering (ICACITE), pp. 28–29, Greater Noida, India (2022)
11. Abdelbar, A., Haytham, H., Ziad, M., Amr, M., Haytham, M., Ashraf, N., Kareem Moussa, M., Darweesh, S.: Face recognition using python and OpenCV. Rev. Int. J. Sci. Res. Publ. **14**(1), 70–94 (2024)
12. Savio, M.M.D., Deepa, T., Bonasu, A., Anurag, T.S.: Image processing for face recognition using HAAR, HOG and SVM algorithms. J. Phys. Conf. Ser. **1964**, 9 (2021)
13. Ammour, N., Bazi, Y., Alajlan, N.: Multimodal approach for enhancing biometric authentication, journal of. Imaging. **9**(9), 168., MDPI (2023)

Conceptual Model of an Eye Movement Simulation Procedure for Motor Dysfunction Prediction

Alexandra Maria Lazar(✉), Barbu Cristian Braun, Mihaela Ioana Baritz, Angela Repanovici, Daniela Mariana Barbu, Anca Ioana Tataru, and Mirela Gabriela Apostoaie

Product Design, Mechatronics and Environment Department, Transylvania University, Brasov, Romania
alexandra.lazar@unitbv.ro

Abstract. The visual process is sustained by complex interactions between the functioning of the ocular globes, the central unit (brain), and the surrounding environment. This process is supported by the recording of movement parameters, which can provide information about the state of normality, attention, memory, refractive visual dysfunctions, the condition of the oculomotor system, or emotional state. Over time, various procedures and ocular monitoring systems have been developed, and analyses of the types of eye movements have been conducted. However, significantly fewer studies have focused on a crucial aspect: the prediction of eye movement evolution for the assessment of normality or the presence of visual dysfunctions. In this paper, the authors present an analysis of the process of predicting potential ocular dysfunctions through the simulation of eye movements. The method used is based on the knowledge of simulating eye movements using the LabVIEW programming environment and the ability to correlate these simulations with video-recorded movements. In conclusion, the processing of the obtained data and the conceptualization of predictive analysis for the evolution of eye movements reveal the system's ability to reasonably estimate ocular behavior. This can contribute to the creation of a predictive kinematic model for the evolution of the visual system, with implications in learning processes, detection of dysfunctions, or behavioral manifestations.

Keywords: eye · eye movements · simulation · LabVIEW · prediction

1 Introduction

1.1 General Aspects Regarding Eye Movement Studies

Due to the dynamic evolution of electronic devices with which humans interact from an informational perspective, interest in tracking the eye movements of these users has significantly increased. The general explanation is that the information obtained from the analysis of eye movements can reveal important aspects of how human subjects interact with these devices (primarily computers), with the surrounding environment, or how they adapt their behavior to various external or internal changes.

S. Vlad and N. M. Roman (Eds.): MEDITECH 2024, IFMBE Proceedings, 130, pp. 13–22, 2025.
https://doi.org/10.1007/978-3-031-95671-3_2

Eye-tracking techniques, predominantly used in human-computer interaction, have proven effective in fields such as occupational ergonomics, learning environments, real-time data transmission and processing, and even in behavioral psychology. In terms of the benefits resulting from research utilizing these techniques, improvements have been observed in learning environments and outcomes [1–4], adaptation of workspaces to user needs [5–7], and broader applicability in the medical field [8, 9]. As highlighted in some studies, traditional eye movement analysis models [10] are based on various sequences of assumptions and empirical data, and they fail to fully simulate the entire range of oculomotor movements when the eyes are directed towards a fixed or moving target [11]. Moreover, eye movements can be influenced by the surrounding environment (natural, social, economic, cultural), and as such, a vast number of initial parameters must be identified, understood, and quantified to generalize eye-tracking methodologies across various domains [12].

Research on eye movements is not limited to recording trajectories, starting points, or endpoints, but also seeks to develop methodologies comprised of procedures and advanced technology tailored to the specific requirements of various fields where the visual system interacts with the external environment. Therefore, these methodologies fail to be unified, generalized, or conceptualized due to the fact that the human visual system itself is unique, personalized, and the applications are related to these characteristics.

1.2 Typology of Eye Movements

As defined in various specialized studies, eye movements can be classified into four basic categories based on several parameters: saccadic movements, smooth pursuit movements, vergence movements, and vestibulo-ocular movements [13].

Saccadic movements are primarily reflexive but can become voluntary when tracking a moving target is required. If the target shifts rapidly from one position to another, a saccadic movement occurs after a delay of approximately 200 ms. This delay is due to the decision-making process within the visual cortex, which evaluates how much the eye needs to move to relocate the target in the foveal region. This decision is then translated into a command for the extraocular muscles to adjust the fixation axis of the eyes toward the target's new position, considering the correct distance and direction. Saccadic movements occur continuously as long as the target remains in motion, both when all target positions are perceived and when some positions are missed [13, 14]. Thus, as highlighted in various applied studies [15], saccadic movements, accompanied by the fixation process, depend on the task being performed, the human subject's focus, and, not least, the frequency with which that action is carried out [16].

Smooth pursuit eye movements are significantly slower and aim to maintain the target's motion stimulus on the foveal region of each retina. Since this movement occurs consciously (whether or not a moving target is being tracked), it falls into the category of voluntary movements, which may be time-limited or continuous. However, they can also be categorized as involuntary movements when the human subject's attention and interest are divided across multiple tasks. Unlike the rapid movements characteristic of saccades, the visual behavior during smooth pursuit involves tracking a stimulus along a linear trajectory. As highlighted in the conclusions of several experiments, smooth pursuit movements can focus on a stimulus, typically reaching speeds of approximately 30°/s [16, 17].

In the same context, two types of eye tracking movements have been defined: open-loop and closed-loop tracking. The open-loop variant refers to the behavior during the initial 100 milliseconds of rapid eye movements that follow the initial appearance of a stimulus, without any correction. In contrast, closed-loop tracking reflects corrected and adjusted tracking that aligns more closely with the stimulus speed. By analyzing the dynamics of smooth pursuit movements, researchers have concluded that the speed variation falls within the range of 0.1 to 100 degrees per second, with a latency of approximately 100–125 milliseconds [17–20].

Vergence movements represent the eyes' ability to focus on different objects within the 3D surrounding space. This process involves the coordinated movement of both eyes in opposite directions to direct them toward the same point of stimulation (fixation) within the visual field. The two forms of vergence are: convergence (where the eyes are directed closer together) and divergence (where the eyes are directed farther apart). Convergence allows the visual system to focus on nearby objects, while divergence enables focusing on more distant objects. For a clear image to be formed on the retinas of both eyes, it is crucial that the accommodation process (the adjustment of the lens's refractive power during fixation) functions correctly, so that together with vergence movement, the accommodation reflex remains functional. Both conjugate and disjunctive movements can be either fast or slow, with their characteristics determined by whether the eyes are changing their point of fixation or maintaining focus on a target. This process is defined by Hering's law, which states that the movements of the two eyes are equal and symmetrical (synergy) [20–22].

Complex vestibulo-ocular movements, known as the vestibulo-ocular reflex (VOR), play a crucial role in stabilizing the position of the eyes relative to external stimuli by compensating for head and body movements (including respiration and blood circulation). These reflexive responses prevent the images formed on the retinas from shifting as the head position changes, thereby maintaining the stability of the visual field.

The effectiveness of vestibulo-ocular movements can be assessed by fixating on an object while moving the head from side to side. During this process, the eyes automatically compensate by moving the same distance but in the opposite direction, thereby keeping the object's image relatively stable on the retina (particularly in the foveal region). The vestibular system detects brief, transient changes in head position and generates rapid corrective eye movements to ensure that the visual cortex consistently receives fused images to create a "single 3D image."

2 Conceptual Model for Simulating Eye Movements

In this paper, the authors aim to define a model for evaluating eye movements through the concept of simulating these movements to identify elements indicative of oculomotor system dysfunctions. Simultaneously, using computer simulation, video recordings of eye movements are captured and analyzed in terms of trajectory shape, duration, speed, and acceleration. These movements are compared to identify specific "patterns" of how various dysfunctions influence ocular dynamics.

Considering all previously discussed types of eye movements (monocular/binocular), the authors propose a system for evaluation through simulation and stimulation of these movements while tracking a target moving along different trajectory patterns (linear-horizontal, linear-plane, and sinusoidal, as shown in Fig. 1a, b, c). An application has been developed for simulating eye movements using the LabVIEW programming environment, which can be used to assess and categorize specific forms of ocular dysfunction by analyzing parameters such as speed, acceleration, direction, and trajectory of the ocular globes.

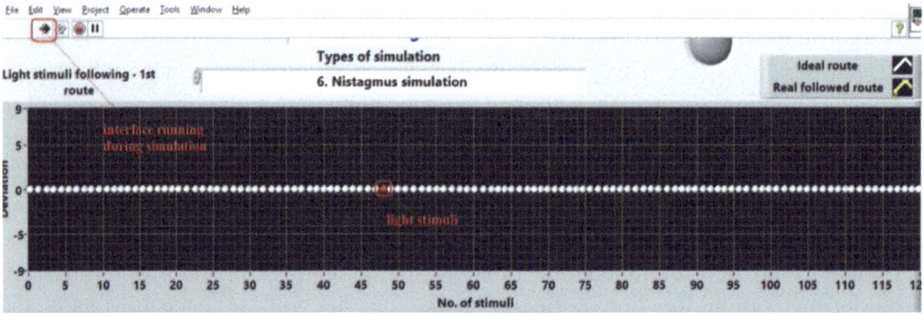

linear

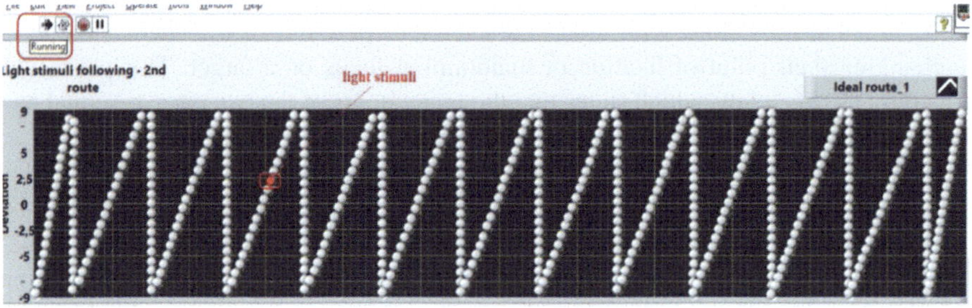

in the form of saw teeth

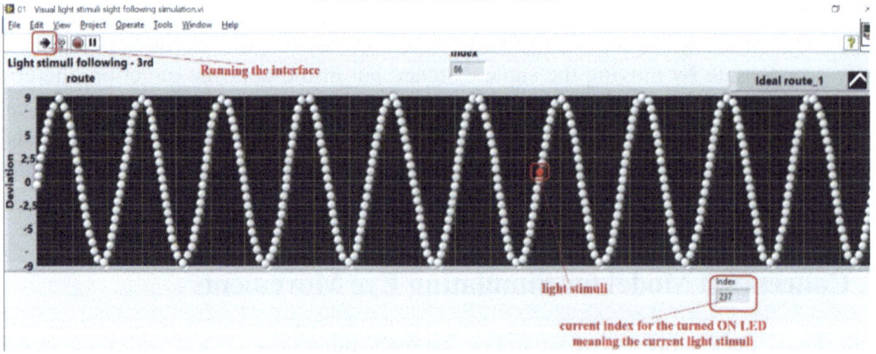

in the form of sinusoid

Fig. 1. Eye tracking movements' stimulation trajectory

Eye movement stimulation phase involves: training and learning the subjects from the sample using the LED panel but also running the Labview application with a virtual marker. This stage is followed by recording, evaluating and comparing the eye movements of the subjects in the sample using the previously performed simulations in the same application Labview.

The subjects in the sample for this evaluation were recorded using a video system to enable comparison of the movements stimulated by the LED panel, in prepara-tion for the next step, which involved tracking the movement of the virtual mark-er across the three categories of trajectories (Fig. 2). Compared to the system with LED panels and an integrated video camera (Fig. 2a), the simulation and stimulation variant using the LabVIEW application offers the advantage of directly tracking the movement of the virtual marker along the same viewing direction as the camera axis (Fig. 2b). This alignment allows for accurate synchronization with the image displayed on the screen.

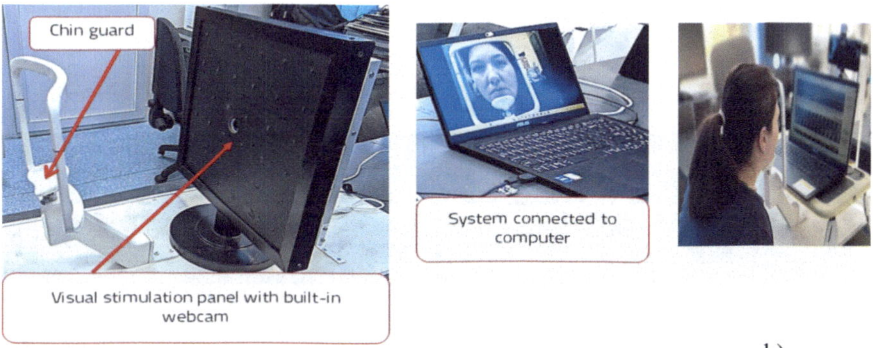

a) b)

Fig. 2. Stand for simulating and stimulating eye movements with Led panel (a) and Labview application on the laptop (b)

Thus, in this LabVIEW application, types of visual dysfunctions are defined (see Fig. 3) using specific parameters. These are then compared with the simulation of a normal visual system performed using the same application.

Tracking each of the three eye movement trajectories allows simultaneous recording with a video camera to correlate with the simulated variants, which are then processed using the open-source application CvMob [23]. The simulation variants defined in the LabVIEW application are derived by correlating values obtained from multiple studies in the field (a scientometric study was previously developed to synthesize these evaluations), with values derived from large subject samples [24–26].

A suggestive example is illustrated in Fig. 4, showing the manifestation of simulated nystagmus dysfunction across the three trajectories based on data from the literature. This type of simulation forms the foundation of the conceptual model for analyzing the typologies of visual dysfunctions observed during eye movement, established according to specific requirements (such as observing the visual system at various stages, assessing particular characteristics, behavioral evaluation, long-term monitoring, and under specific conditions, etc.).

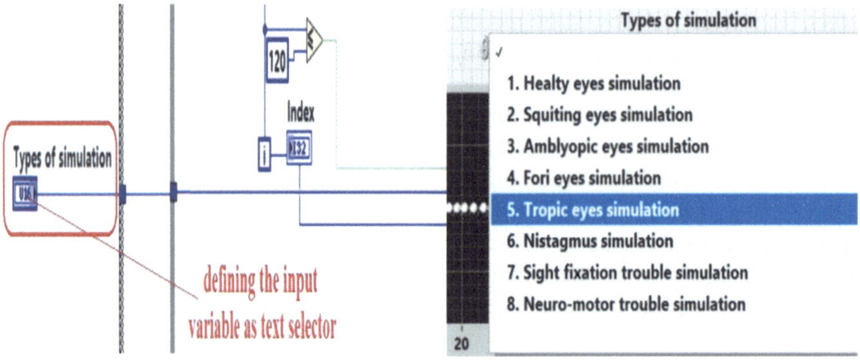

Fig. 3. Dialog box for setting the type of simulated pathology of the visual function by de-fine as a text selector type variable in the graphical programming window

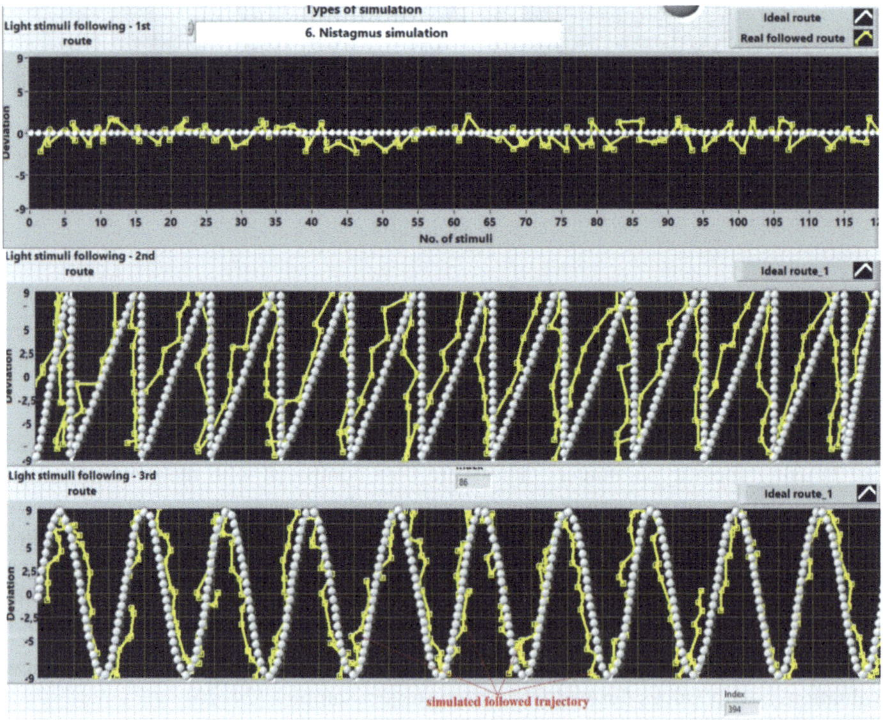

Fig. 4. Example of simulation for Nystagmus dysfunction across three trajectory types

In relation to this movement, patterns of eye movements can be generated to identify normal or abnormal manifestations, or to monitor certain forms of psychophysiological behavior.

These simulations were generated in the LabVIEW application for a set of 8 possible eye movement cases that may manifest at the level of the visual system. These cases are: healthy eyes, squinting eye, amblyopic eyes, fori eyes, tropic eyes, nistagmus (Fig. 4), sight fixation trouble, neuro-notor trouble.

3 Results

For the eight simulations conducted, the amplitude values of eye movements for each eye, obtained along the X and Y axes, were calculated and displayed by the LabVIEW application. For example, the nystagmus simulation is illustrated in Fig. 5. These values were subsequently summarized in Table 1.

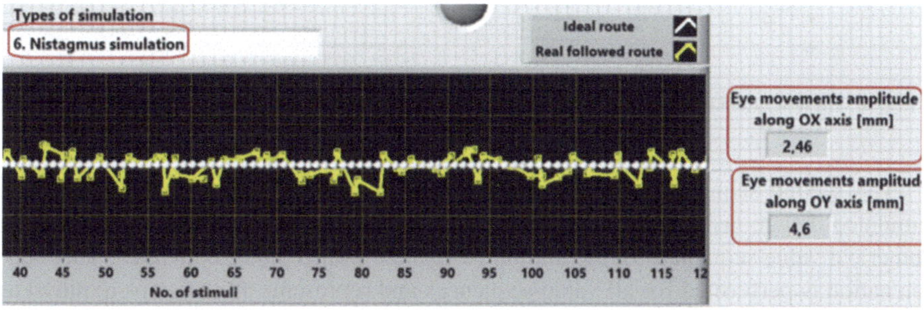

Fig. 5. Displaying the amplitude values of eye movements on the Ox and Oy axes

Table 1. Simulated values of eye movement amplitudes on the Ox and Oy axes

Ocular dysfunction	Eye movements amplitude on Ox axis [mm]			Eye movements amplitude on Oy axis [mm]		
	Line	Linear plan	Sine	Line	Linear plan	Sine
Healthy eyes	1.19	0.2	1.36	0.29	0.15	1.15
Squinting eye	1.9	0.8	3.57	0.44	0.25	0.96
Amblyopic eyes	2.18	0.85	6.49	0.52	0.59	4.12
Fori eyes	7.46	1.25	2.49	1	1.14	7.55
Tropic eyes	2.46	1.38	3.71	0.26	1.29	7.35
Nistagmus	2.46	2.48	4.04	4.6	2.24	5.15
Sight fixation trouble	3.99	4.08	15.79	5.99	3.44	56.2
Neuro-notor trouble	6.95	3.36	8.3	12.7	1.19	5.19

The displacements on the Ox and Oy axis are evaluated in millimeters and represent the deviations from the ideal trajectory of the eye movements in the simulation version.

For example, the first sequence of the program subroutine for determining and displaying the simulated deviations from straight-line trajectory tracking highlights the process of setting constants to simulate current fixation deviations at different stages of stimulus tracking. This includes simulation constants and random number functions for generating variations in the simulated deviations (see Fig. 6).

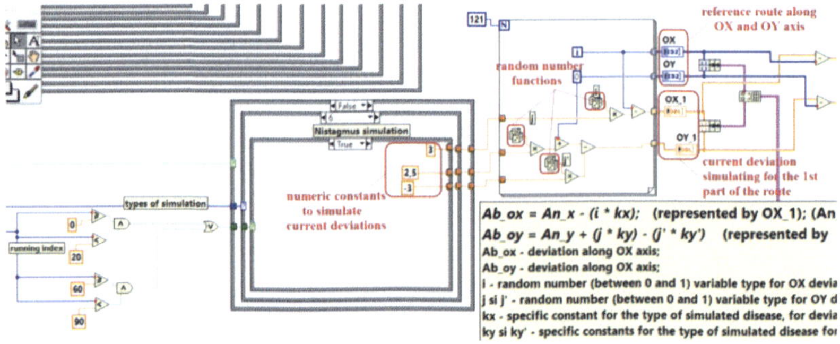

Fig. 6. First sequence of the subroutine for determining and displaying simulated deviations

The second sequence of the program subroutine for determining and displaying simulated deviations from straight-line trajectory tracking involves concatenating the numerical value sequences of the simulated fixation points along both axes. This concatenation is performed for all value sequences and represents all the steps required to traverse the entire trajectory (see Fig. 7).

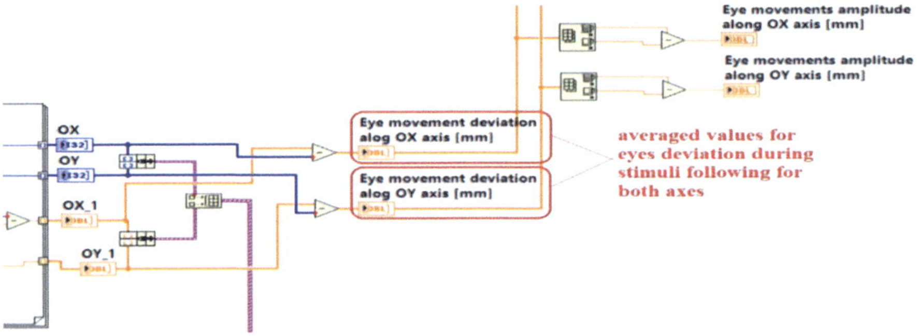

Fig. 7. The second sequence of the subroutine for determining and displaying simulated deviations

From the obtained values, reference numerical values are subtracted to determine the simulated deviations. As a result of these operations, two numerical value vectors are generated. To calculate the average deviations of the stimulus tracking along the entire path, the minimum deviation value is subtracted from the maximum deviation value for both the X (Ox) and Y (Oy) axes.

4 Conclusions

The conceptual model developed through this research provides specialists with an initial framework for predicting various forms of visual motor dysfunctions. This conceptual model leverages the capabilities of the LabVIEW programming environment and can be enhanced with additional subroutines for evaluating and simulating other important parameters within the visual system.

Prediting of potential ocular motor dysfunction is supplemented by video recordings of eye movements from subjects simulated by the same digital application. These video sequences are then processed in relation to the defined simulation. This entire mechanism can be subsequently automated and used for rapid evaluation and recording of visual dysfunction progression.

Acknowledgements. In these experiments, we have developed the investigations with equipment from Applied Optometry Laboratory and L11 laboratory from Research Institute of Transylvania University. The research is part of work of Alexandra Maria Lazar (Cateanu) PhD student.

Declaration. None of the authors have any conflicts of interest and/or incompatibilities with any of the organizers of the event/conference.

References

1. Toker, D., Conati, C., Steichen, B., Carenini, G.: Individual user characteristics and information visualization: connecting the dots through eye tracking. In: Proceedings of the SIGCHI Conference on Human Factors in Computing Systems (2013)
2. Ben Khedher, A., Jraidi, I., Frasson, C.: Static and dynamic eye movement metrics for students' performance assessment. Smart Learn. Environ. **5**, 14 (2018)
3. Susac, A., Bubic, A., Kaponja, J., Planinic, M., Palmovic, M.: Eye movements reveal students' strategies in simple equation solving. Int. J. Sci. Math. Educ. **12**(3), 555–577 (2014)
4. Ali, Q., Heldal, I., Helgesen, C.G., Dæhlen, A.: Serious games for vision training exercises with eye-tracking technologies: lessons from developing a prototype. Information. **13**, 569 (2022)
5. Haslbeck, A., Zimmermann, M., Rothkirch, I., Lange, C.: Eye-tracking in human factors. Ergon. Res. (2015). https://doi.org/10.13140/RG.2.2.16094.84806
6. Putri, D., Boucheix, J.-M.: Eye-tracking and its application in ergonomics research. Proceed. APCHI-ERGOFUTURE. **2010**, 581–585 (2010)
7. https://ergomania.eu/eye-tracking-in-practice/, last accessed 15 March 2024
8. https://www.ces.tech/articles/2021/january/four-eye-tracking-solutions-in-healthcare.aspx, last accessed 15 March 2024
9. Zammarchi, G., Conversano, C.: Application of eye tracking technology in medicine: a bibliometric analysis. Vision (Basel). **5**(4), 56 (2021)
10. Stockman, G., Shapiro, L.G.: Computer Vision. Prentice Hall (2001)
11. Xiaoming, W., Xinbo, Z., Jinchang, R.: A new type of eye movement model based on recurrent neural networks for simulating the gaze behavior of human reading. Complexity. **2019**, Article ID 8641074 (2019)
12. Holmqvist, K., Nyström, M., Andersson, R., van de Weijer, J.: Eyetracking. A Comprehensive Guide to Methods and Measures. Oxford University Press, Oxford (2011)

13. Purves, D., Augustine, G.J., Fitzpatrick, D., et al.: Neuroscience, 2nd edn. Sinauer Associates; Types of Eye Movements and Their Functions, Sunderland, MA (2001)
14. Land, M., Mennie, N., Rusted, J.: The roles of vision and eye movements in the control of activities of daily living. Perception. **28**, 1311–1328 (1999)
15. Foulsham, T.: Eye movements and their functions in everyday tasks. Eye. **29**, 196–199 (2015)
16. Tatler, B.W., Hayhoe, M., Land, M., Ballard, D.H.: Eye guidance in natural vision: reinterpreting salience. J. Vis. **11**, 5 (2011)
17. Erkelens, C.J.: Coordination of smooth pursuit and saccades. Vis. Res. **46**(1–2), 163–170 (2006)
18. https://imotions.com/blog/learning/best-practice/types-of-eye-movements/. Accessed Sept 2024
19. https://connect.tobii.com/s/article/types-of-eye-movements?language=en_US. Accessed Sept 2024
20. http://schorlab.berkeley.edu/passpro/oculomotor/html/chapter_15.html. Accessed Sept 2024
21. Gonçalves, V.P., Coelho Scharlach, R.: Oculomotor evaluation in adults: a study of the effect of age and visual alterations, Audiology Communication Research, ISSN 2317-6431, (2016)
22. Mahanama, B., Jayawardana, Y., Rengarajan, S., Jayawardena, G., Chukoskie, L., Snider, J., Jayarathna, S.: Eye movement and pupil measures: a review. Front. Comput. Sci. **3**, 733531 (2022)
23. https://sourceforge.net/projects/cvmob/. Accessed Sept 2024
24. Meo, M., Del Punta, J.A., Sánchez, I., de Luis García, R., Gasaneo, G., Martin, R.: A dynamical method to objectively assess infantile nystagmus based on eye tracking. A pilot study. J. Opt. **16**(3), 221–228., ISSN 1888-4296, (2023)
25. Kang, J.-J., Choi, S., Na, S., Sun-Young, O.: Recording and interpretation of ocular movements: spontaneous and induced nystagmus. Ann. Clin. Neurophysiol. **25**(1), 10–18 (2023)
26. Shichinohe, N., Barnes, G., Akao, T., Kurkin, S., Fukushima, J., Kase, M., Leigh, R.J., Belton, T., Fukushima, K.: Oscillatory eye movements resembling pendular nystagmus in normal Juvenile Macaques. Investig. Ophthalmol. Vis. Sci. **52**, 3458–3467 (2011)

Adaptable Mechanism for Walking Aids

Istvan Ver(✉), Alina-Elena Ver, and Silviu-Dan Mandru

Technical University of Cluj-Napoca, Memorandumului Street 28, Cluj-Napoca, Romania
veristvan69@yahoo.ro

Abstract. Health has a major impact on everyday life. Walking disorders affect a large number of people, who thus become dependent on the use of a walking device or walking aid. The walking devices are addressed not only to the elderly, to people suffering from various locomotor disorders, but also to people in the phase of post-operative recovery, who for a period of time must use such a device. Their use also implies certain recommendations such as avoiding walking on stairs with bilateral subaxillary crutches, or the walking frame, the need for a fixed support surface next to the antebrachial crutch when lifting / sitting down, the need for a handrail or help from a person to climbing / descending stairs with the help of antebrachial crutch, etc. This paper presents an adaptable mechanism suitable to any walking device, such as a cane, axillary crutch, antebrachial crutch (Canadian) or walking frame, having the role of making the functions of these walking devices more efficient, so that it provides a propulsion force when lifting and ensures a controlled sitting, with support only in the walking device. Thanks to the possibility of changing the length of the walking device, it makes it possible to go up / down stairs, with only support in the walking device, providing an increase in user independence and mobility.

Keywords: walking devices · adaptable mechanism · transfer · mobility

1 Introduction

Although studies show that the proportion of people with walking disorders are usually elderly [1], there is also a worrying increase in the last 30 years in the percentage of users of walking devices, [2] noting for example that over 72% of crutch users are active people under 65 years, [3]. Walking devices or walking aids supplement strength, increase the support surface, contribute to lower pain in the lower limbs by reducing the joint load, thus helping walking, [1]. They have also been shown to improve stability, increasing patients' sense of security and reducing the fear of falling, [4–7]. The use of walking aids is also motivated by the following aspects: walking assistance by increasing and redistributing the supporting force, reducing pain in the lower limbs, improving balance, producing a small propulsion force when standing upright, etc.

The choice of the walking aid is made following the evaluation of the patient and in accordance with his pathology. For example, in patients with post-stroke hemiparesis, studies of the use of walking devices show that between 32% and 76% of them use at least one walking device.

S. Vlad and N. M. Roman (Eds.): MEDITECH 2024, IFMBE Proceedings, 130, pp. 23–30, 2025.
https://doi.org/10.1007/978-3-031-95671-3_3

Walking devices include: canes, crutches, walking frames, in different construction variants, depending on the pathology and the needs of the patient. A unilateral cane provides offloading to the opposite side at the hip of 20-25%, forearm crutches 40-45% and bilateral subaxillary crutches 80%, [8]. The antebrachial crutch is one of the most used walking devices, and Table 1 shows its main characteristics.

Table 1. Main characteristics of antebrachial crutch

Image	Transfers techniques
	– crutches with antebrachial support, also called Canadian or Lofstrand crutches, are easier to use, being useful for patients who require a certain degree of unloading and bilateral support, [8]

The use of walking devices implies that the patient is trained to use them. In Table 2, according to [8, 9] the transfer techniques, with reference to the antebrachial crutch walking device, are presented.

2 The Main Objectives

The study presented in this paper focuses on achieving the following objectives, considered relevant in assisting walking and transfer:

- restoring the ability to move
- increasing muscle strength
- re-education of coordination, control and balance
- increasing independence, autonomy and mobility.

3 Description of the Technical Solution

According to [11], the functional role of this gas lifting piston (telescope/damper) mechanism, hereinafter referred to as piston, applied to walking devices is to assist the transfer from sitting to standing position, providing a force of lift propulsion, as it supplements the force required for lifting (the patient leaning on the walker) and to provide a controlled transfer from standing to sitting, the patient just rests with his own weight on the walker, which shortens and while supporting the patient's weight. It can also assist by shortening or lengthening and going up and down the steps (Fig. 1).

Table 2. Transfer techniques with antebrachial crutch

Techniques	Image
When setting/picking up – make sure the seat is stable; – position the walking device within reach; – the patient moves his pelvis towards the front edge of the chair; – the healthy (or healthier) lower limb will have the role of propulsion, so that it will be located posterior to the sick one; – both plants will be with their entire surface on the ground; – the patient will lean forward, then lean with one hand on the walking device and the other on the handle of the chair or other fixed surface and initiate the lifting movement; – the aim is to maintain balance during the transfer; – then the plug is made on the walking device. – walk back until the healthy leg reaches the chair; – keep the sick leg forward – keeping your weight on your right leg, take the crutches un- der your arm(s). (if there are 2 crutches, both are transferred to one hand). – with crutch support, lean the trunk forward and bend the good knee; – with the other hand supported by the armrest of the chair, slowly start the movement of the chair – the crutches are placed next to the chair.	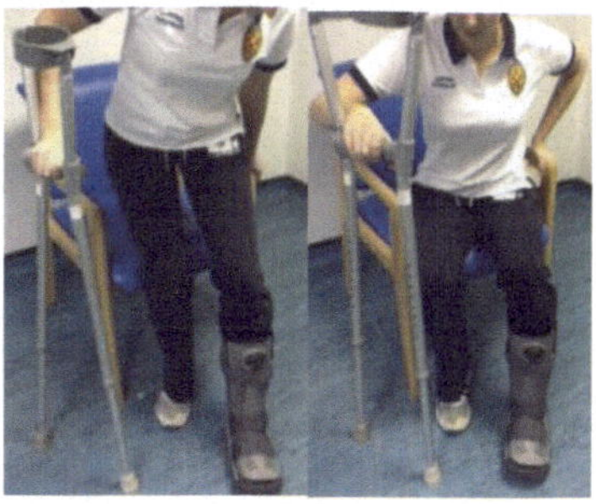
When going up / down stairs – when using a crutch, it is recommended that climbing/descending stairs should only be done where there are handrails or with the help of another person, [10]	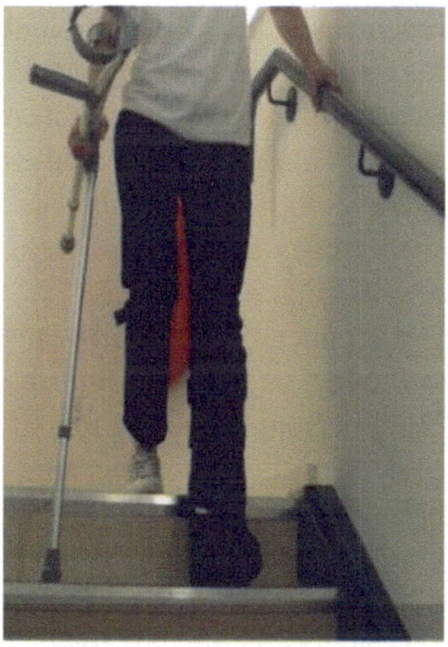

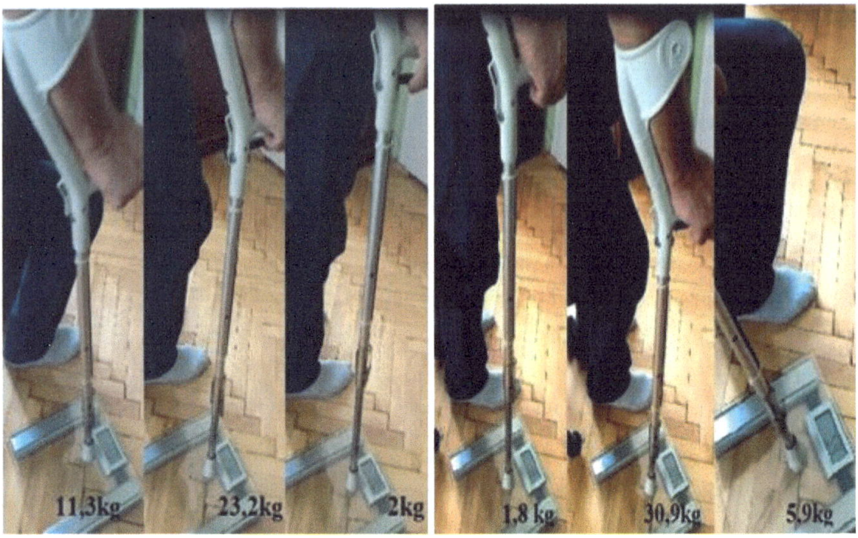

a) b)

Fig. 1. (a) Lifting from sitting in orthostatism; (b) Controlled placement from orthostatism to sitting.

Activating the device for lifting/sitting down, or when going up/down the steps, requires the actuation of a lever that unlocks the telescopic part of the walking device which can be lengthened or shortened depending on the needs of the patient, with the help of the lifting piston (telescope), as it is shown in Fig. 2.

Next, according to Figs. 2 and 3, the mode of action of the piston mechanism that is the subject of the present work, is presented, with an example on the crutch with antebrachial support, with the mention that the location mode and the resulting action is similar for the other walking devices, with the difference that in the case of the walking frame, we will have two such mechanisms, which will act independently two by two on the four legs of the walking frame (front/back).

The telescopic tube of the crutch can be composed of two elements I and II, or to provide a greater degree of adjustment of its height, it can be composed of three elements I, II and III (Fig. 3).

The parameters of the lifting piston are chosen as follows: the force is chosen according to the patient's weight and the required degree of assistance (the remaining force that must be supplied), and the stroke is determined according to the patient's height. Element (I) of the telescopic tube is positioned in continuation of the forearm support (1), being the element with the largest diameter of the telescopic tube; inside it is fixed and centered with the help of rings, a lifting piston (telescope/damper) with gas (20), hereafter referred to as piston, composed of working cylinder and rod, also on this element (I), at the end lower is fixed the locking/unlocking mechanism (B), which includes two flaps (12), (13), two springs (14), (15) and two rods (16), (17).

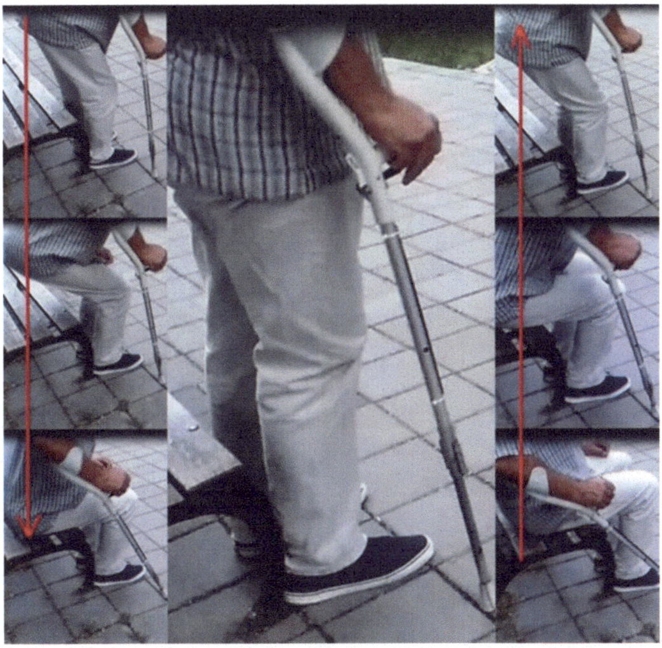

Fig. 2. How the proposed device (prototype) is used for sitting down or lifting

The element (II), of the telescopic tube, smaller in diameter than the element (I), thus allowing it to slide inside the element (I) between the holes (M1) and (M2), (M1) at the bottom, (M2) in the upper one of the element (II), between these positions defining the stroke of the piston rod (20), which is established individually according to the anthropometric dimensions of the user), has the end that enters the element (I), full, to allow the piston rod , to act upon it, during use, as follows: on the forearm support (1), an actuation mechanism (A) is fixed, containing a lever (4) and a bowden cable (5); when the lever (4) of the mechanism (A) is actuated by the user, the bowden cable (5) is actuated, which, upon exiting the mechanism (A), enters a bowden stop (6), fixed on the element (I) with the help a fixing clip, then the cable (5) descends along the element (I), entering a new bowden stop and when exiting it, the cable (5) will be fixed in the center of the cable fixing element where are contained on one side and on the other, the upper ends of the bowden cables, which will lead to the actuation of the locking/unlocking mechanism (B), thus, the flaps (12), and (13), will compress the springs (14), (15), extracting the rods (16), (17) from the holes (M1) for lifting (the crutch is in the shortened position), allowing the expansion of the piston (20), which will lead to the elongation of the crutch, the two rods (16), (17), remaining in contact with the body of the element (II), under the pressure of the springs (14), (15) through the flaps (12), (13), until they reach the holes (M2), the crutch locking in this elongated position, and for shortening, the mechanism will act similarly from (M1) to (M2); ensuring against twisting during the sliding of elements (I) and (II), is achieved through a guide channel, provided on the body of the two elements. If element III is needed, it is inserted into element II from the top of the

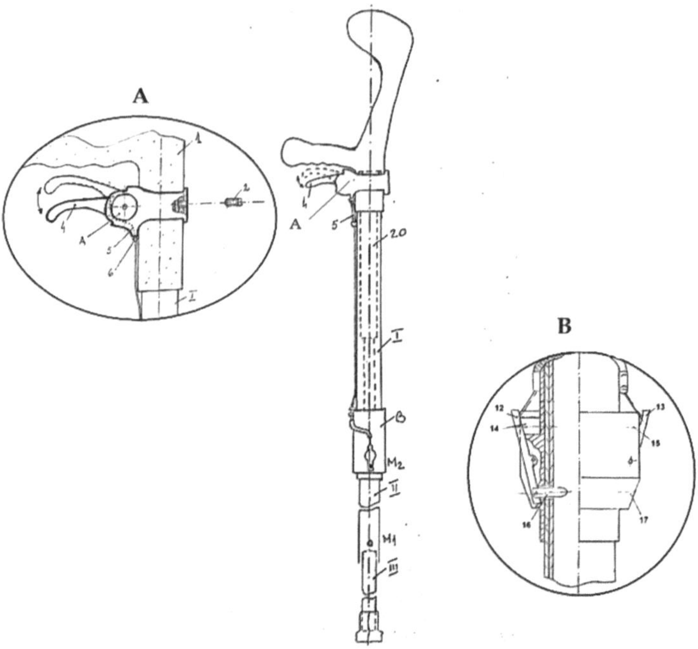

Fig. 3. Device components with details: (a) actuation mechanism, (b) lock/unlock mechanism

crutch, the guide channel helping to position the fixing holes correctly. On element III or II (if we have only two elements) the tip of the crutch is fixed, which can be replaced in case of wear.

When going up/down the steps:

- if the crutch with antebrachial support is used, she will be lengthened or shortened alternatively, supporting the patient in going up/down the stairs (Fig. 4).

  If the walking frame is used (Fig. 5), two actuation mechanisms with lever A, with two actuation cables each, will be used to achieve the independent actuation of the legs of the walking frame, two by two: when climbing the front legs of the walking frame will be shortened, and the rear legs when descending. If you want to extend/shorten all four legs at once, actuate both levers simultaneously.

In the case of axillary crutches (Fig. 6), the application of the present invention involves mounting the actuation mechanism with lever A, under the handle of the walking device. So, when the patient wants to get up from sitting upright with the crutch in shortened positions, he leans on the handle of the antebrachial support of the crutch, leaving part of his weight on the crutch, unlocking with the help of the lever of mechanism A, the mechanism of locking/unlocking B, the crutch thus providing him with a propulsive force that supplements his remaining strength and helps him rise to orthostatism (standing). When he wants to sit down, he will unlock mechanism B in the same way, only by leaning and leaving his weight on the crutch it will shorten, favoring a controlled sitting down.

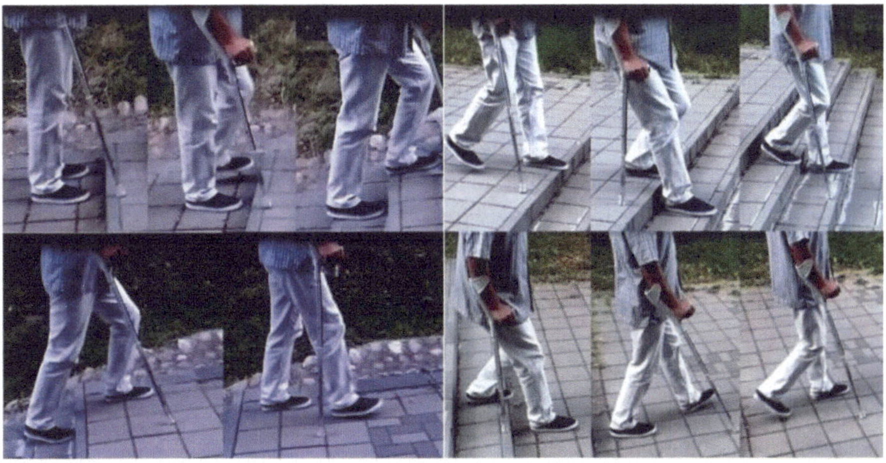

a) b)

Fig. 4. (a) Going up steps; (b) Descending steps.

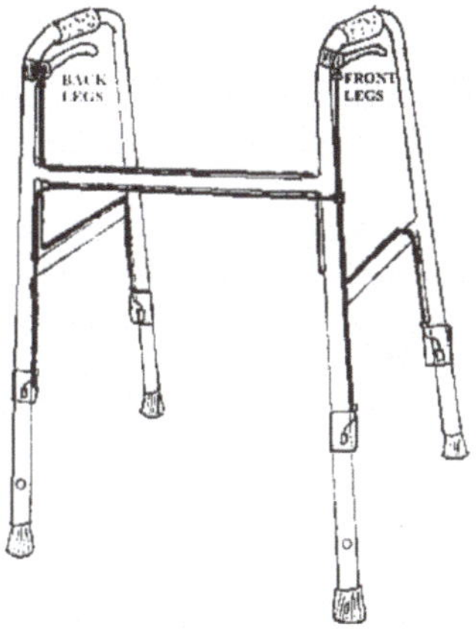

Fig. 5. Walking frame.

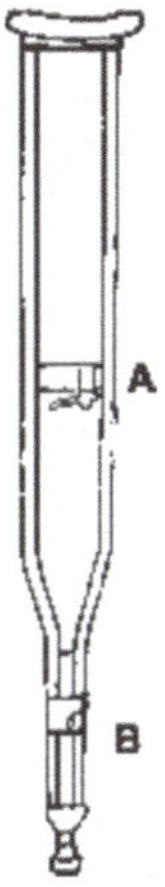

Fig. 6. Axillary crutches.

4 Conclusion

The advantages of the device with a piston mechanism attached to walking aids, which is the subject of the present work, are:

- the constructive solution proposed in the invention it does not necessarily require the design of new walking devices, it can be made an adaptation of existing ones;
- the unlocking time of the mechanism when the lever is actuated, which has the effect of lengthening or shortening the telescopic part of the walking device, is almost instantaneous;

- transfers from standing to sitting and from sitting to standing are faster, so the patient's scapulohumeral joint is not overloaded;
- provides stability when sitting down or standing up from other chair, bed, bench, toilet, etc., as well as for going up/down steps, or bending, only with support in the walking device, not like in the classic version, whose implementation was presented in Table 2, so it was increasing autonomy, self-esteem, because the patient not being dependent on another person;
- offers support and only supplements the remaining strength of the patient;
- provide low costs.
- The designed device thus succeeds in making existing walking devices more efficient, giving autonomy and mobility to the user, increasing his independence in daily activities, ensuring an increase in the quality of his life.

Conflict of Interest Statement. The authors have no conflict of interest to declare.

References

1. Irszay, L., Popa, A.: Occupational Therapy. "Iuliu Hațieganu" University Medical Publishing House, Cluj-Napoca (2016)
2. Gell, N.M., Wallace, R.B., LaCroix, A.Z., Mroz, T.M., Patel, K.: Mobility device use among older adults and incidence of falls and worry about falling. J. Am. Geriatr. Soc. **63**(5), 853–859 (2015)
3. Kaye H.S., , Kang T., LaPlante M.P., Device Use in the United States, Disability Statistics Center Institute for Health and Aging University of California, San Francisco, (2000)
4. Allet, L., Leemann, B., Guyen, E.: Effect of different walking aids on walking capacity of patients with poststroke hemiparesis. Arch Phys Med Rehabil. **90**, 1408–1413 (2009)
5. Laufer, Y.: The use of walking aids in the rehabilitation of stroke patients. Rev Clin Gerontol. **14**, 137–144 (2004)
6. Milczarek, J.J., Kirby, R.L., Harrison, E.R., MacLeod, D.A.: Standard and four-footed canes: their effect on the standing balance of patients with hemiparesis. Arch Phys Med Rehabil. **74**(3), 281–285 (1993)
7. Ashton-Miller, J.A., Yeh, M.W., Richardson, J.K., Galloway, T.: A cane reduces loss of balance in patients with peripheral neuropathy: results from a challenging unipedal balance test. Arch. Phys. Med. Rehabil. **77**, 446–452 (1996)
8. Irszay, L., Popa, A.: Orthotics and Prostheses. University Medical Publishing House "Iuliu Hațieganu", Cluj-Napoca (2016)
9. https://myhealth.alberta.ca/alberta/pages/use-crutches.aspx, last accessed 15 June 2021
10. Virtual fracture clinic Using your crutches page, https://www.fracturecare.co.uk/general-advice/using-your-crutches/, last accessed 3 July 2024
11. Ver, I., Ver, A., Ungur, R.A.: Walking device with a mechanism for assisting the transfer from the sitting position to the orthostatic position and from the orthostatic position to the sitting position. Patent Application OSIM A/00199 (2015)

Neonatal Incubator: An Arduino-Based Prototype with Multi-sensor Integration and Real-Time Data Streaming

Ioana Ariton, Laura-Nicoleta Ivanciu(✉), and Emilia Șipoș

Bases of Electronics Department, Technical University of Cluj-Napoca, 26-28 G. Baritiu Street, 400027 Cluj-Napoca, Romania
laura.ivanciu@bel.utcluj.ro

Abstract. With the rise in premature births and the high mortality rate of newborns within their first day, the need for incubators has become critical. This underscores the necessity of such a device to continuously track their daily progress and monitor the parameters such as temperature of the baby and of the environment from incubator, humidity, gas, pulse, sound and movement. This paper presents the implementation of an Arduino-based neonatal incubator prototype, capable of measuring various data: baby temperature, environment temperature and humidity, baby pulse, baby movement, and to detect whether the baby is crying, or the presence of gas or smoke. The acquired data is transmitted in real-time, processed and displayed, for a facile interpretation. The prototype proves to be accurate and robust, successfully passing a series of tests, aimed at demonstrating the correct functioning of the sensors, both separately and combined. The proposed solution can be used in neonatal hospital units, to ensure the best possible conditions for the baby's growth. The sensor system and data visualization facilities keep the medical team updated regarding the baby's progress, and issues regarding the baby's health are promptly identified.

Keywords: Incubator · Multi-Sensor Integration · Arduino · Data Streamer

1 Introduction

World Health Organization reports show that 2.3 million children died in the first 20 days of life in 2022, with approximately 6500 newborn daily deaths [1]. While most deaths occur in low and medium-income countries, prematurely born babies have fewer chances of survival, even when medical care is available. In this context, neonatal incubators play a crucial role in the care of premature and critically ill infants, by providing a stable environment that helps regulate their temperature, humidity, and oxygen levels. Moreover, affordable incubators are a possible solution for underdeveloped countries or regions where proper medical care is scarce.

Advanced prototypes of neonatal incubators are designed to provide a controlled and stable environment that supports the delicate physiological needs of newborns,

S. Vlad and N. M. Roman (Eds.): MEDITECH 2024, IFMBE Proceedings, 130, pp. 31–41, 2025.
https://doi.org/10.1007/978-3-031-95671-3_4

including optimal temperature, humidity, and oxygen levels. Moreover, remote real-time data acquisition and interpretation is a key aspect in modern incubators.

The incubator proposed in [2] is capable of monitoring real-time vital signs such as body temperature, humidity, pulse rate, oxygen concentration, air quality, and the ambient sound level, and transmitting the data through a mobile application. It features a microcontroller to store temporal sensor data. Specific sensors are used for different measurements: temperature and humidity sensors monitor environmental conditions, a pulse oximeter tracks oxygen levels and pulse rate, and a gas sensor detects harmful gases. Additionally, a microphone sensor monitors abnormal sound levels around the incubator. The collected sensor data is sent via Wi-Fi to a cloud-based platform for storage, analysis, and retrieval. A similar approach is used in [3], where an IoT-based health monitoring system tracks vital health parameters, using low-cost low-power sensors. Remote monitoring and instant alerts contribute to round-the-clock newborn care.

The newborn's vital signs are easily read when the baby is wearing a bodysuit, equipped with sensors. The direct contact between baby and wires/circuitry is avoided, this solution promises portability and comfort, while providing accurate and reliable sensor data [4].

The mathematical model of a neonatal incubator and neuro-control system is presented in [5]. The interaction between the newborn and incubator elements (walls, air, heater/cooler, humidifier, mattress) is modelled as a heat and mass transfer. The neuro-controller is built over a gradient-based neural network control algorithm, and the entire implementation is carried out on a Raspberry Pi 4B, using Python.

To control parameters such as temperature, humidity, amount of fresh air inside the incubator, various control methods are employed—multivariable PID control [6], adaptive PD control [7], fuzzy logic control [8] or fuzzy-PID hybrid control [9].

An artificial womb that incorporates the neonatal incubator principles is developed in [10]. The artificial womb and placenta are made of highly biocompatible materials, to provide infants with a safe and comfortable environment.

By ensuring a safer, more efficient, and supportive environment for neonates, these innovations hold the potential to significantly improve health outcomes and reduce the risks associated with preterm birth and other neonatal conditions. As research and development in this field continue to progress, neonatal incubator prototypes promise to transform neonatal care practices, offering new hope for the youngest and most vulnerable of patients.

This paper describes the implementation and testing of a prototype for an Arduino-based neonatal incubator, with multi-sensor integration and real-time data streaming. The novelty of the proposed solution consists of integrating movement and gas detection sensors, as well as data acquisition, visualization and interpretation features: all the values measured by the multi-sensor network are sent, interpreted and displayed, thus ensuring that the newborn is safe and monitored.

The remainder of the papers is organized as follows: Sect. 2 describes the materials and methods involved in the implementation of the incubator, the results are presented and discussed in Sect. 3, while Sect. 4 concludes the paper.

2 Materials and Methods

2.1 Overview of the Prototype

The neonatal incubator is built around an Arduino board, along with several sensors that continuously monitor: the temperature and humidity inside the incubator, the baby's temperature and pulse, whether the baby is crying/moving or not, whether there are any gases inside the incubator. Data collected from the sensors is sent to Excel by means of Data Streamer (a tool from the MS365 package), displayed and interpreted (Fig. 1).

Fig. 1. Block diagram of the proposed neonatal incubator prototype.

The medical staff is expected to interact with the babies by attaching three to six sensors to their skin and using the display monitor to visualize the parameters (Fig. 2). The incubator is meant for hospitals use only, under the supervision of the medical team.

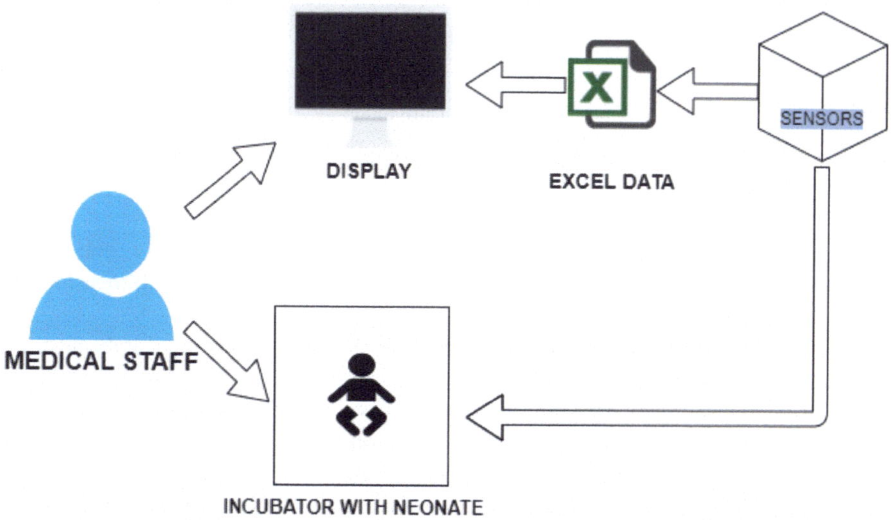

Fig. 2. Interaction between medical staff, incubator, sensor network and display.

2.2 Arduino UNO R3 and Sensors

The "brain" of the incubator is the Arduino UNO R3 board, which contains an ATmega328P microcontroller [11]. The technical characteristics of the sensors that are integrated in the neonatal incubator are shown in Table 1.

Table 1. Technical specifications of the sensors inside the neonatal incubator

Sensor name and function	Parameter	Value
DHT22 temperature and humidity digital sensor [13]	Supply	3…5 V DC
	Temperature range and accuracy	−40 °C… +80 °C, ±0.5 °C
	Humidity range and accuracy	0%...100% RH, ±2% RH
DHT11 temperature and humidity digital sensor [13]	Supply	3…5 V DC
	Temperature range and accuracy	0 °C…60 °C, ±2 °C
	Humidity range and accuracy	20%-95% RH, ±5% RH
Pulse sensor [14]	Supply	3.3…5 V DC
	Current consumption	4 mA
MQ2 gas sensor [12]	Supply	5 V DC
	Current consumption	150 mA
	Operating temperature	−20 °C… +50 °C
Sound sensor module with microphone [15]	Supply	3.3 V or 5 V DC
	Current consumption	20 mA
ADXL335 3-Axis accelerometer sensor [16]	Supply	2…3.6 V DC
	Current in measure mode	40 μA
	Current in standby mode	0.1 μA

There are two temperature and humidity sensors, DHT22 and DHT11, one for the temperature and humidity inside the incubator (DHT22), the other one to measure the baby's temperature.

The pulse sensor must be in direct touch with the skin and ambient light must be removed. The front-facing LED of the sensor is placed over a vein in the human body. The LED generates light, which falls directly on the vein. The veins will only have blood flow inside them when the heart pumps. If circulation is observed, the surrounding light sensor will take up light reflected by the blood.

The baby's movements are monitored with the ADXL 3-Axis accelerometer sensor, while the sound sensor with microphone detects whether the baby is crying or not. Finally, the MQ2 gas sensor detects gas leaks, as a safeguard preventing fire or toxicity. The gas sensor is highly sensitive, and is able to detect LPG (liquefied petroleum gas), carbon monoxide, isobutane, methane, alcohol, hydrogen, and smoke [12].

The complete schematic of the prototype is depicted in Fig. 3, where the connections between sensors and the Arduino UNO R3 board, via jumper wires and breadboard, are visible. To ensure the maintenance of optimal ambient temperature within the incubator, a closed-loop control system incorporating a 5 V DC fan was developed. The control system dynamically adjusts the fan's rotational speed based on the detected temperature: if the temperature slightly exceeds the desired threshold, the fan operates at a reduced speed; if the temperature surpasses this threshold significantly, the fan operates at maximum speed.

The DC fan requires more current than the Arduino board can provide (maximum 40 mA at 5 V), so a 2 N3904 BJT is included, thus increasing the current up to 0.2 A. A 1 N4007 diode protects the transistor from the electromotive force generated by the fan, when the transistor is off. The resistor in the base is determined as:

$$R = \frac{5 - V_{BE,on}}{I_B} = \frac{5 - 0.7}{0.016} = 268.75\ \Omega \tag{1}$$

where I_B is the base current, and 5 V is the voltage from the Arduino pin.

Finally, the LEDs placed on the breadboard will turn on if gas is detected, the baby has not been moving for a certain period, and when the baby is crying. A buzzer generates a sound when gas is detected, since this is the most dangerous case and must receive immediate attention from the medical staff. In the software implementation, the characteristics for various gases (LPG, CO and smoke) help approximate the sensor's output to the actual gas concentration, using a linear equation derived from the logarithmic representation. For example, the code for defining the LPG curve is:

$$LPGcurve3=\{2.3, 0.21, -0.47\}; \tag{2}$$

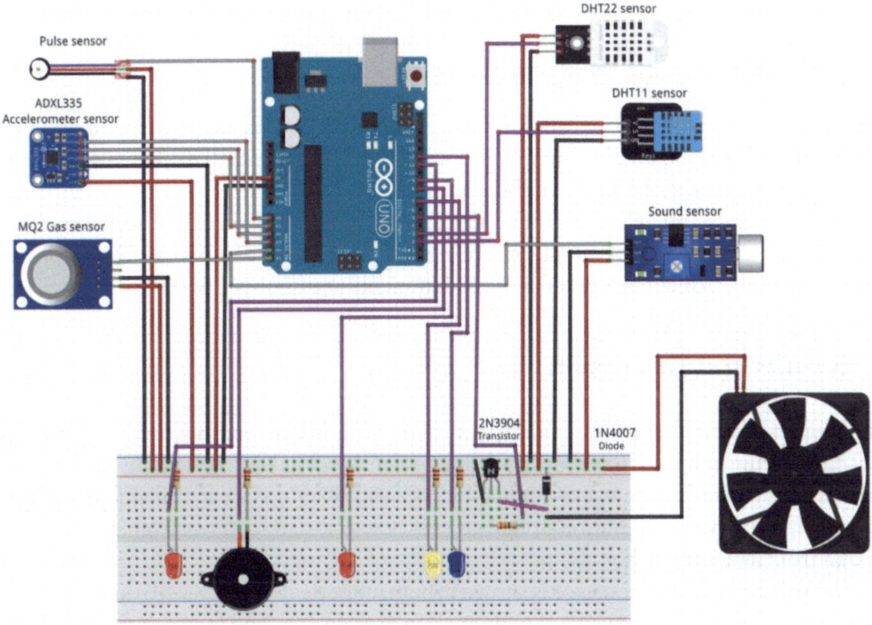

Fig. 3. Complete schematic of the proposed neonatal incubator prototype.

where the three values are determined as:

$$2.3 = log_{10}200 \tag{3}$$

$$0.21 = log_{10}1.65 \tag{4}$$

$$-0.47 = \frac{log_{10}\frac{0.27}{1.65}}{log_{10}\frac{10000}{200}} \tag{5}$$

2.3 Sensor Data Transmission Via Data Streamer

Sending sensor data through a serial port in an Excel spreadsheet for monitoring all the parameters was a must. By transmitting data in real time, plots of all the values and parameters from the sensors to track the newborn's health status can be visualized. For this purpose, Data Streamer was used, which enables two-way data transfer: sending data from Excel to the microcontroller and streaming live data from the microcontroller to Excel.

The interface includes three worksheets: *Data In, Data Out*, and *Settings*. Sensor data is sent to the *Data In* worksheet, where it will be periodically updated to monitor changes. Data in the *Data Out* worksheet is converted to CSV format and can be sent to an application or a connected device. The *Settings* worksheet is where the parameters for both *Data In* and *Data Out* can be modified.

2.4 Complete Neonatal Incubator Prototype

All the sensors and the necessary additional circuitry are connected to the Arduino UNO R3 board, and specific code snippets were written in the Arduino IDE, to ensure correct and accurate data acquisition from the sensors.

In the final setup, all the components are placed in a special transparent case (Fig. 4), ensuring optimal functionality and precise data acquisition. Additionally, *from/to* labeling is mandatory, for proper identification of jumper wires.

3 Results and Discussion

The correct functioning of the complete system, with all the sensors, was tested using real-life circumstances. However, no babies were involved in the testing process, all measurements were made using adult subjects, who were not located inside the plastic case. Additionally, the operation of the gas sensor was tested with gas from a lighter and smoke from an extinguished flame.

Data obtained from the sensor is shown in both numerical (Figs. 5 and 6) and graphical depictions (Fig. 7), for convenient and quick interpretation. The temperature inside the incubator and the baby's temperature are measured in °C, the humidity in %, while the gas detection measurements show values in parts per million (ppm). For the movement and

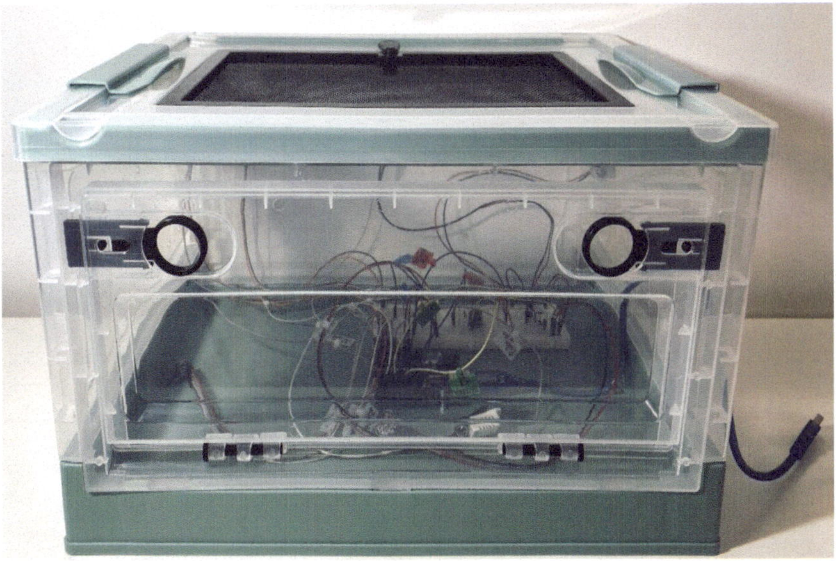

Fig. 4. Neonatal incubator prototype – circuit and plastic case.

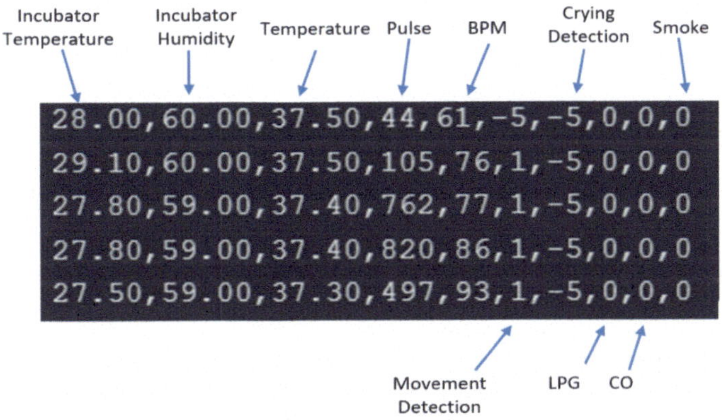

Fig. 5. Data from the sensors – numerical display on the serial monitor.

TIME	Incubator Temperature	Incubator Humidity	Temperature	Pulse	BPM	Movement Detection	Crying Detection	LPG	CO	SMOKE
14:51:39.59	27.2	60	36.2	497	70	-5	-5	0	0	0
14:51:41.16	27.2	60	36.8	502	61	-5	-5	0	0	0
14:51:42.73	27.2	60	36.8	516	58	-5	-5	0	0	0
14:51:44.29	27.2	60	37.3	505	58	-5	-5	0	0	0
14:51:45.86	27.3	60	37.3	456	60	-5	-5	0	0	0
14:51:47.44	27.3	60	37.8	516	62	-5	-5	0	0	0

Fig. 6. Data from the sensors – numerical display in Data Streamer.

Fig. 7. Graphical display of sensor data: incubator humidity and temperature, baby temperature, baby movement detection, baby crying detection, baby BPM and pulse, gas detection.

crying detection, the value 1 represents absence of said parameter, while −5 represents presence.

Any values that exceed pre-imposed limits will turn on the LEDs on the breadboard, while the presence of gas will also trigger the buzzer, for an audio alarm. The threshold values for which the LEDs light up can easily be changed from the Arduino IDE. The graphical representation of measured values enables the medical staff to monitor not only status, but also historical developments in the state of health of the newborn.

A comparison between the proposed neonatal incubator and other incubators available on the market is shown in Fig. 8 ([17–20]). While still being in the prototype phase, the proposed solution checks the important landmarks of a reliable incubator, with new features like movement and gas detection. The data acquisition and display facilities set it apart from its industrial competitors. However, extensive testing and validation from medical professionals is mandatory, before sending it into mass production.

Features \ Incubator type		Incubator prototype	Transport Incubator-TI-2100B	mOm Essential Incubator	Open Incubator Comen BQ80	Closed Incubator Dräger Babyleo TN500
Temperature of the environment	Measurement	X	X	X	X	X
	Control	X (not entirely)	X	X	X	X
Humidity of the environment	Measurement	X	X			X
	Control		X			X
Body Temperature		X	X	X	X	X
Pulse and BPM		X			X	
Movement Detection		X				
Sound Detection		X				X
Gas Detection		X				
Illuminating lamp			X			X
Additional power supply			X	X		
Hemoglobin monitoring			X			
Oxygen Concentration			X		X	
Audio stimulation						X

Fig. 8. Comparison between proposed neonatal incubator prototype and existing industrial solutions.

4 Conclusions and Future Work

With the rise in premature births and the high mortality rate of newborns within their first day, the need for incubators has become critical, often determining the difference between life and death. The proposed neonatal incubator prototype is outfitted with an array of sensors, including temperature and humidity sensors to monitor the baby's temperature and the incubator's internal environment, with the capability to control a fan for cooling. It also features a gas sensor that triggers alarms if the air becomes polluted, a pulse sensor to track the baby's heartbeat, a sound sensor to detect crying, and a motion sensor to monitor the newborn's activity. Together, these sensors and the data they provide create an optimal system for the care of premature babies.

This incubator prototype represents an advancement in neonatal care, integrating multiple sensors to provide a safe, controlled, and responsive environment for newborns. The graphs generated by the Data Streamer make reading and interpreting real-time data significantly easier and more user-friendly. Furthermore, access to historical data allows the medical staff to track the baby's progress over time and provide optimal care services.

Future work includes adding an uninterrupted power supply system, a heater to increase temperature as needed, and testing the incubator using medical personnel expertise. Since the incubator is designed to be used strictly under medical supervision, professionals in the field must be employed in the development and testing process.

Conflicts of Interest. The authors have no conflicts of interest to declare that are relevant to the content of this article.

References

1. World Health Organization: WHO: Newborn mortality. World Health Organization (2024b, March 14). https://www.who.int/news-room/fact-sheets/detail/levels-and-trends-in-child-mortality-report-2021, last accessed 14 July 2024
2. Mbanuzue, C.E., Hassan, O.E., Owolabi, F.M.: Real-time monitoring of neonatal Incubator parameters via Iot-integrated device: design, implementation, and evaluation. Int. J. Res. Innovat. Appl. Sci. **9**(4), 127–143 (2024)
3. Abdullah, M., Ching, A.H.K.: Internet of things (IoT) based neonatal Incubator for heart rate monitoring and caregiver motion detection. Evolut. Electr. Electron. Eng. **5**(1), 10–17 (2024)
4. Medehal, A., et al.: Portable smart neonatal Incubator with improvised alarm system. In: 2021 IEEE International Conference on Industry 4.0, Artificial Intelligence, and Communications Technology (IAICT). IEEE (2021)
5. Frolov, S.V., et al.: Building a simulation model of a neonatal Incubator with a neural controller. In: 2024 XXVII International Conference on Soft Computing and Measurements (SCM). IEEE (2024)
6. Woldeamanuel, Hailemichael, L., Ramaveerapathiran, A.: Design of Multivariable PID control scheme for humidity and temperature control of neonatal Incubator. IEEE Access. **12**, 6051–6062 (2024)
7. Kholiq, A., et al.: Development of adaptive PD control for infant Incubator using fuzzy logic. J. Robot. Control. **5**(3), 756–765 (2024)
8. Pramudia, H., et al.: A fuzzy logic-based temperature control system for baby incubators. Indonesian J. Electr. Electromed. Eng. Med. Inform. **5**, 4 (2023)
9. Alimuddin, A., et al.: Development and performance study of temperature and humidity regulator in baby incubator using fuzzy-PID hybrid controller. Energies. **14**(20), 6505 (2021)
10. Harish, M., Neela, S.P., Anushiya, K.: Modification of artificial womb in combination with modern neonatal Incubator principles. In: 2024 Tenth International Conference on Bio Signals, Images, and Instrumentation (ICBSII). IEEE (2024)
11. Arduino UNO – JavaTpoint: https://www.javatpoint.com/arduino-uno, last accessed 14 July 2024
12. MQ2 Sensor datasheet: https://www.pololu.com/file/0J309/MQ2.pdf, last accessed 14 July 2024
13. DHT 11, DHT 22 Sensor datasheet: https://www.mouser.com/datasheet/2/737/dht-932870.pdf, last accessed 14 July 2024
14. Pulse sensor datasheet: https://lastminuteengineers.com/pulse-sensor-arduino-tutorial/?utm_content=cmp-true, last accessed 14 July 2024
15. Microphone sound sensor module: https://littlebirdelectronics.com.au/products/microphone-sound-sensor-module-for-arduino, last accessed 14 July 2024
16. ADXL 335 3-axes accelerometer datasheet : https://www.analog.com/media/en/technical-documentation/data-sheets/adxl335.pdf, last accessed 14 July 2024

17. Transport Incubator - TI-2100B - Ningbo David Medical Device - PDF Catalogs | Technical Documentation. https://pdf.medicalexpo.com/pdf/ningbo-david-medical-device/transport-incubator-ti-2100b/69526-252552.html#open1031544, last accessed 14 July 2024
18. mOm Essential Incubator, UK: Medical Device Network. (2022, July 13). https://www.medicaldevice-network.com/projects/mom-essential-incubator-uk/?cf-view&cf-closed, last accessed 14 July 2024
19. Shenzhen Comen Medical Instruments Co., Ltd.: https://fr.comen.com/pages/BQ80.html
20. Babyleo Incubator: https://www.draeger.com/Content/Documents/Products/babyleo-tn500-pi-9102128-en-master.pdf, last accessed 14 July 2024

The Patient-Mounted Ambulatory Infusion Pump – At-Home Treatment and Patient Freedom Solutions

Anca Iulia Nicu(✉) and I. P. Ilișuan

Universitatea Tehnică din Cluj-Napoca, Cluj-Napoca, Romania
anca.nicu@ethm.utcluj.ro

Abstract. The paper describes the development of a portable, patient-mounted infusion pump prototype aimed at increasing patient mobility during long-term intravenous treatments. The design allows patients to move freely while receiving IV (IntraVenous) medication, addressing the limitations of traditional stationary infusion pumps. It also opens the possibility of at-home treatment by enabling long-distance configuration and disallowing patient control access.

The project utilized 3D modeling software and printing technology to create a functional prototype, featuring a stepper motor-driven peristaltic pump, a wireless control interface, and a wearable textile support system. The infusion pump is configured through a web-based interface, enabling users to adjust infusion parameters such as volume and duration.

The project successfully established the foundation for a practical wearable infusion device, with potential for future refinement and application in medical settings.

Keywords: infusion pump · ambulatory care · patient-mounted · telemedicine · wearable technology

1 General Overview

The modern definition of the world around us can be distilled into a single word: abundance. The trajectory of human society's development revolves around answering the question, "How can we have more without compromising quality?" The field of medical engineering is no exception to this.

The primary goal of medical engineering is to ease the workload of healthcare professionals while ensuring the highest quality of care for patients, all while achieving the best cost-quality ratio. Societal abundance has forced medicine to adapt: the number of patients is increasing due to population growth and the prevalence of unhealthy lifestyles. Unfortunately, the medical workforce cannot expand at the same rate.

The response of medical engineers has been to enhance the capacity of medical staff to manage and resolve daily challenges. Medical devices have become faster, smaller, and even capable of taking over some of the responsibilities of healthcare workers.

S. Vlad and N. M. Roman (Eds.): MEDITECH 2024, IFMBE Proceedings, 130, pp. 42–49, 2025.
https://doi.org/10.1007/978-3-031-95671-3_5

The ***infusion pump*** is a prime example of a device that has assumed one of the most time-consuming and attention-demanding tasks traditionally handled by nurses [1].

The infusion pump, typically mounted on a pole or placed on a work surface, delivers a steady, programmable flow of medication, nutrients, or blood to the patient via an IV line, without requiring frequent intervention or monitoring by nursing staff. However, these devices restrict patient mobility, as they are stationery and patients are often prohibited from handling them to avoid accidental changes to the settings or disruptions in operation [2].

As a result, patient mobility is significantly reduced, requiring the presence of medical staff to disconnect them from the device. This limits the patient's freedom to move at will or meet their personal needs, and the treatment may be interrupted if movement is necessary. Even without an infusion pump, the issue persists, as patients remain dependent on the IV pole that holds the infusion bag [3]. The independence of patients receiving infusion therapy could be achieved through the development of a wearable infusion pump, mounted on a support that accommodates all necessary accessories for its operation. Its autonomy does not need to be extended but should be sufficient to allow the patient to move freely for 1–2 hours, enabling them to perform various tasks away from the hospital bed.

2 Device Development

A prototype is proposed for a portable, patient-mounted infusion pump designed to be worn on the shoulder and arm, with additional anchoring on the chest. It features a wireless interface, operates via a stepper motor, and is powered by a Li-Ion battery.

The portable, patient-mounted infusion pump must meet the following criteria:

1. Provide all essential safety functions [4].
2. Perform the same basic functions as a traditional infusion pump [5].
3. Prevent unauthorized changes to settings.
4. Attach securely and comfortably to the patient.
5. Accommodate various infusion bag models without impeding their functionality.
6. Be as compact as possible while maintaining the performance standards of a traditional infusion pump.
7. Have a minimum battery life of 1 hour, allowing the patient to move freely for various tasks.

The infusion pump project was designed and simulated using SOLIDWORKS® (Fig. 1). Many aspects of the design were guided by existing infusion pumps, such as form, standard workflow and safety features, but their integration in a small and portable device belongs to the authors. The prototype was 3D printed and executed with the help of hobby integrated circuits, notably an ESP-32WROOM development board as the controller.

The design accounted for several safety and ergonomics principles by incorporating rounded edges and smooth contours. Material tolerance issues presented challenges in achieving precise dimensions, as the expansion and contraction of the 3D printed thermoplastic material introduced small geometric errors. These issues led to occasional mechanical conflicts between components that had to be fixed by sanding or filing.

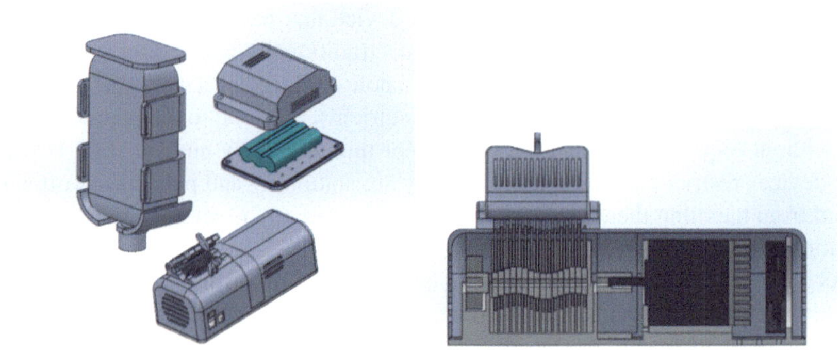

Fig. 1 The proposed infusion pump

Another challenge was the uneven weight distribution of the device. The motor, which made up a significant portion of the device's mass, was placed as low as possible vertically to ensure that the pump remained stable when mounted on the patient's body. This design choice helped minimize undesirable rotational momentum that might cause discomfort to the patient by dragging the whole assembly laterally.

The final prototype was divided into three distinct parts: the pump assembly (Fig. 2), the battery case and the flask receptacle. These were placed around the patient to ensure even weight distribution. Placing the accumulator case on the patient's back also functions to ensure the patient does not assume a horizontal or a low diagonal position, reducing the risk of air bubbles inside the perfusion tube.

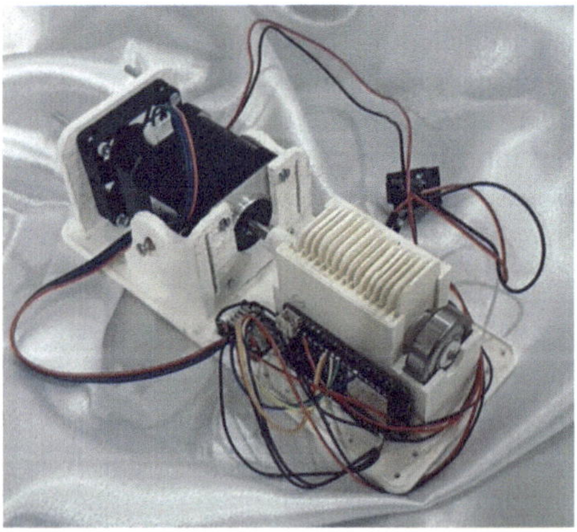

Fig. 2 The infusion pump without its upper casing

2.1 The Textile Saddle and I.V. Equipment Placement

The support system for the infusion pump was designed around a textile saddle (Fig. 3). This structure provided both stability and flexibility, allowing the device to be comfortably mounted on the patient, conforming to their body, while maintaining essential functionality. The flask containing the intravenous substance can be secured in its designated receptacle with straps that do not exert pressure on the flask's wall while also ensuring minimal movement while in use.

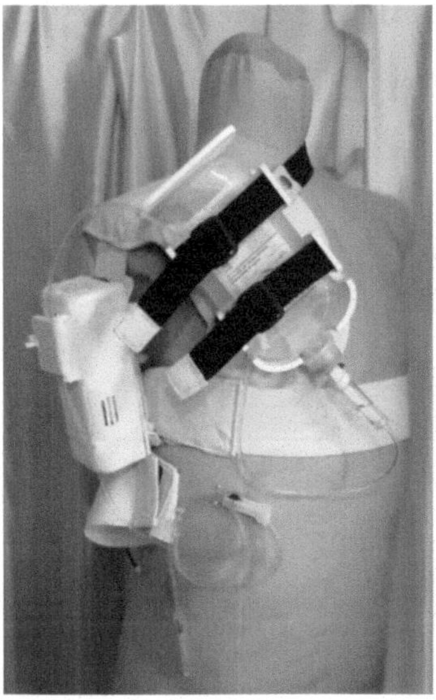

Fig. 3 A fully equipped infusion pump, mounted

The saddle is mounted on the patient on their shoulder and is secured with straps around the chest and arm, with an additional support point anchored to the neck on the opposite side. Elastic belts ensure that the saddle can accommodate varied patient dimensions.

2.2 Electromechanical and Electronic Details

Most of the electronic design process revolved around powering and running the stepper motor that spins the peristaltic pump. This is done through a stepper driver module that moves the axle according to the required flow. Two power lines exist within the circuit, a main 12 V supply and a step-down 5 V logic supply, both powered by a Li-Ion battery pack. An ESP-32WROOM controller provides the necessary logic signals.

A small cooling fan and the strategically placed air vents ensure a good air flow over the modules and the stepper's heatsink to reduce the risk of overheating. While the axle turns at very low frequencies, corresponding to equally small flow rates, the idle current still generates an important amount of heat.

The device was equipped with a start/stop button to allow the patient to quickly turn the device on or off as needed. A buzzer was included to provide audio feedback, notifying the patient of changes in the pump's status.

2.3 Programming the Controller

A logical control scheme for an infusion pump typically includes several key components to ensure precise fluid delivery, patient safety, and device reliability. Below is the proposed logical control flow (Fig. 4).

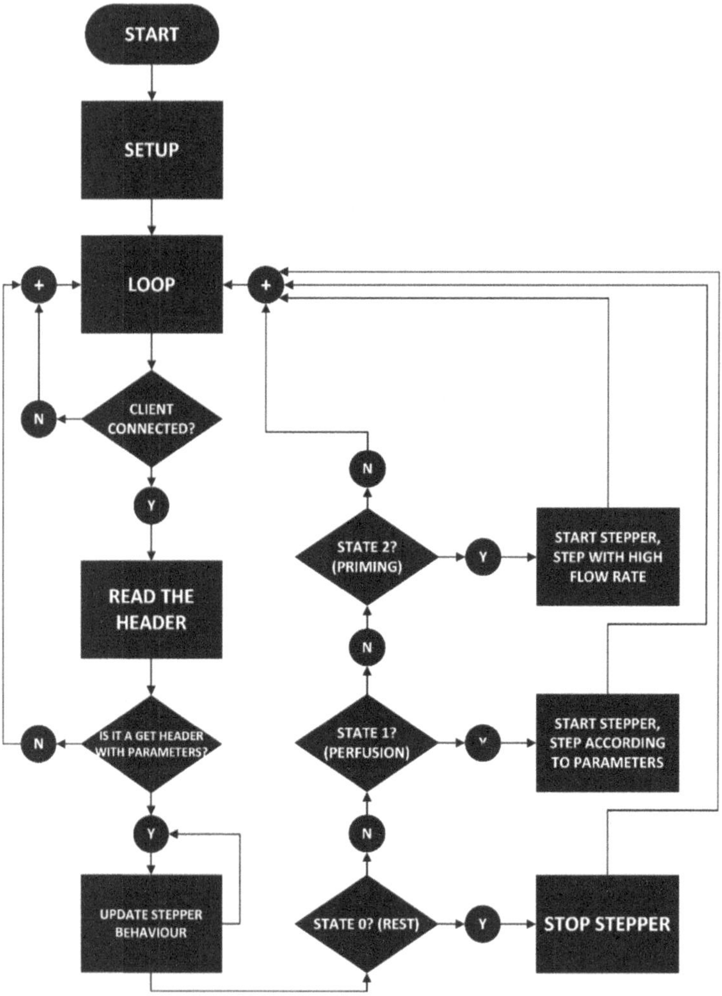

Fig. 4 The proposed logical control flow

The infusion pump's control system was implemented using Arduino IDE with libraries specific to the ESP32-WROOM's development platform. The program defines three possible states for the infusion pump: standby, priming, and infusion. The device's state and parameters can be modified through a wireless interface, using a simple website hosted on the microcontroller. The controller allowed for internet based wireless communication, meaning that the device used to configure the pump must be on the same wireless network and that the pump must be provided with an SSID and a password for it's own connection.

In standby, the stepper is powered off and no significant power consumption is generated. Priming and standby are similar in execution, powering up the stepper and generating a flow of liquid through the perfusion tube. Priming differs from normal usage by generating a large flow rate to rid the tube of any air bubbles and fill it with liquid. During the infusion mode, the device generates the specified flow rate.

Flow rate is determined by taking the user's input parameters, time and volume, and converting this information to steps per minute. An interrupt function linked to a timer pulses the steps accordingly.

As for any device, calibration must be done after assembly and periodically during its lifetime. This can be done by stepping a set number of steps to pump liquid with a known density. By weighing the dispensed amount with a precision scale and calculating the corresponding volume, the steps to volume ratio can be calculated easily.

2.4 Key Features

The portable infusion pump offers significant benefits for patients by enabling mobility and independence during treatment. Unlike traditional stationary pumps that confine patients to their beds, this wearable device allows them to move freely, perform their daily tasks and maintain a higher quality of life while receiving intravenous therapy. It reduces the need for constant medical supervision, giving patients more self-sufficiency. Additionally, the device can potentially facilitate at-home care, minimizing hospital stays and associated costs, while improving overall comfort and convenience during treatment.

This is not possible without making sure that the patient cannot change the pump's parameters, leading to possibly dangerous situations where drugs are administered at unsafe flow rates. The patient may only start and stop the device, everything else being configured by a qualified person through the wireless interface.

The saddle incorporates a big power bank, which gives it sufficient autonomy to accommodate long treatment durations. The flask receptacle also accepts standard flasks, meaning the device can be incorporated into already-existing systems, without the need for patented supplies. This makes it more accessible both financially and logistically for hospitals and patients alike.

As an infusion pump, this device inherits the downsides of all infusion pumps: the mechanical components will wear over time, periodic lubrication and cleaning should be performed and the necessary cooling airways make the device non-waterproof. More complex solutions, such as liquid cooling, synthetic membranes over the peristaltic pump active region and high-performance lubricants can favor the durability of the device. By implementing these, the device's critical parts can be isolated from the outside environment, removing the need for air circulation through the device. This improves the device's resilience in the home environment, which is less controlled than a hospital environment, but they will also greatly increase the cost of the device.

3 Possible Improvements and Alternative Uses

There are several improvements that could be made to enhance its performance and durability. Alternative materials, such as medical-grade plastics that are more durable and easier to sterilize would be needed for a commercial product. Due to the use of commercially available components, the infusion pump was larger than it needed to be. Custom-made parts, especially for the pump mechanism, would reduce the overall size of the device. Miniaturization would be key to improving comfort for the patient. An occlusion sensor would also improve the safety and efficacy of the product.

Being a miniaturized and portable medical system, this device could be repurposed for research in sport performance and conditioning. The device can be equipped with additional sensors and infuse stimulating or supporting nutritional liquid while an athlete is involved in their training routine. This opens the door to data-driven augmented training, pushing athletes to their limits and beyond.

The device could also be repurposed to aid patients that require lifelong parenteral nutrition. Currently, patients must allow a long period of rest in their schedule to be able to administer their nutritional solution and avoid unnecessary activity. A portable infusion pump would allow them to feed at any time, taking advantage of its portability.

A better configuring and communication platform would allow for efficient and safe at-home care, allowing doctors and nurses to configure devices from their hospital office. This moves some of the hospital's workload at the patient's home, freeing up resources and time for the medical staff. A cheap final product will also ensure that the product can be readily distributed and borrowed to patients or bought by those that wish to have their own device.

As an accessibility feature, communication should happen through a Bluetooth interface, as not all patients may have a Wi-Fi network at home. The device should also have a form of long-term storage, such as an SD card, to resume treatment between sessions or after communications are interrupted.

4 Conclusions

The development of the portable, patient-mounted infusion pump represents a significant advancement in medical device innovation, providing patients with otherwise unavailable mobility and independence during intravenous therapy.

The prototype successfully demonstrates the feasibility of a wearable solution, offering flexibility in both hospital and home care settings. While further refinements are needed to optimize material durability, size, and user comfort, the project lays a strong foundation for future improvements and potential commercialization, ultimately enhancing the patient's experience and the efficiency of medical care.

Conflict of Interest. The authors declare that they have no conflict of interest.

References

1. Roseborough, A, Sanatham, D.: An infusion of history: transfusion medicine from the 17th century to current emergency practices. UWOMJ [Internet]. https://ojs.lib.uwo.ca/index.php/uwomj/article/view/10516, last accessed 15 May 2024
2. Ambulatory infusion solutions: https://www.bd.com/en-eu/company/cme-infusion. last accessed 16 May 2024
3. https://nicetopsm.best/product_details/52668146.html, last accessed 15 May 2024
4. Sensors in infusion pumps: https://hitec.humaneticsgroup.com/perspectives/sensors-infusion-pumps, last accessed 17 May 2024
5. How to use an infusion pump: https://www.unitedinfusion.com/blogs/news/how-to-use-an-infusion-pump. last accessed 17 May 2024

Rehabilitation Sensory Device

Alina-Elena Ver(✉), Istvan Ver, and Silviu-Dan Mandru

Technical University of Cluj-Napoca, Cluj-Napoca, Romania
Alina.Ver@mdm.utcluj.ro

Abstract. The paper presents a device with sensors, adaptable for correcting vicious postures, preventing conditions due to them, or recovering conditions such as scoliosis and other types of spine-specific conditions. Also, the device with sensors can be used to recover coordination, re-educate the movements of selfcare gestures and those of other activities. The device is composed of tilt sensors, warning modules: light (led), sound (buzzer) and sensitive (vibration sensor) and optionally, development board, bluetooth module and an electrostimulation device. The novelty of this device is the fact that it combines the action of the warning system with the one of correction, allowing the voluntary correction of the posture based on the feedback provided by the device, or the involuntary correction of the posture, with the help of electrical stimulation. In the case of using the device in various objects, for example, cups, cutlery, hygiene items, etc., for carrying out exercises on the recovery of coordination, re-education of movements from self-care gestures or other recovery exercises from various ailments, electrical stimulation is not used. Also, the device allows the therapist to follow the patient's progress remotely, thus allowing him to adjust the recovery program depending on the pathology and the goals pursued.

Keywords: sensors · vicious postures · scoliosis · flat foot · recovery · movement coordination

1 Introduction

1.1 Vicious Posture

Posture is of major importance, whether we are talking about body aesthetics or movements that have a major impact on a person's self-image. As a rule, the vicious or incorrect posture, is the effect of a weakened musculature, due to the habit of adopting and maintaining a wrong posture, which leads to degenerative changes in the body, [1]. Posture is also the body's response to the effect of gravity, it is the way we maintain our standing, sitting or lying position. A single posture does not allow us to accomplish everything we want to do, so we must adopt many different postures to accomplish different tasks. "Good" posture allows us to move as we wish, subjecting the body to minimum effort and damage, achieving a balance between the muscles, the skeleton and the nervous system, [2, 3].

S. Vlad and N. M. Roman (Eds.): MEDITECH 2024, IFMBE Proceedings, 130, pp. 50–56, 2025.
https://doi.org/10.1007/978-3-031-95671-3_6

Maintaining the posture is not a conscious activity, but an automaticity; by being aware of one's own position and knowing the correct posture, conscious correction of posture can be achieved, [4].

The consequences of a bad posture are manifested especially at the level of the spine, but they are reflected in the functioning of the whole body. The most common manifestations of a vicious posture are, [5]:

- locomotor dysfunctions and joint problems - incorrect posture leads to abnormal wear of joint surfaces, muscle weakness, joint alignment disorders, chronic pain in the lower back, hips, knees and neck, [6, 7];
- disc herniations – incorrect posture, additional strain on the intervertebral discs, causing pain, decreased muscle strength, paresthesia; pressure on the intervertebral discs, may increase the risk of herniation, or worsen already herniated intervertebral discs, [8, 9];
- vision problems - due to incorrect posture, for example in the case of using the computer for a longer period of time (children, pupils, students, office staff), vision problems occur; thus, through the in-adequate position of the head and neck, additional pressure is exerted on the eye muscles, gaps appear between the image visualized through the eyes, its interpretation in the brain and the body's response, causing eye strain, visual fatigue and headaches; also poor vision, can determine the installation of a wrong position, [10, 11];
- carpal tunnel syndrome - occurs predominantly in people who adopt vicious, incorrect positions while working, or whose activity involves repetitive movements; manifestations range from pain in the hands to numbness of the upper limbs, [12, 13].

An example of a vicious posture is the head flexed position when using a mobile phone [14]. This incorrect posture influences the cervical spine. Due to the increasingly complex facilities and functions offered by mobile phones, the number of hours of use has also increased significantly, practically standing with the head in flexion, which led to the development of the "*text neck*" syndrome. The human head weighs between 4.5 – 5.5 kg, so in anatomical neutral position this is the weight supported by the cervical spine. As the head is flexed, reaching in front of the shoulders, for longer periods of time, the pressure on the cervical spine, respectively on the vertebral discs, will increase. Thus, a flexion of 15° creates a load of 12 kg, at 30° – 18kg at 45° – 22 kg and at 60° – 27 kg.

2 Constructive and Operation Features

2.1 The Structure of the Proposed Sensory System

The proposed sensory device can be used for the correction of vicious positions of the spine, correctable idiopathic scoliosis in children and adolescents, attached with kinesiological tapes on the spine) and can be inserted into various objects used in exercises for recovery of movement coordination, also. Thus the device can be used for the recovery of poststroke patients, or with neuromotor disorders, with degenerative and inflammatory rheumatic disorders, in the case of post-traumatic and post-surgical rehabilitation, both in a hospital environment and at home.

The connections between the constitutive elements of the device are shown in the block diagram, given in Fig. 1.

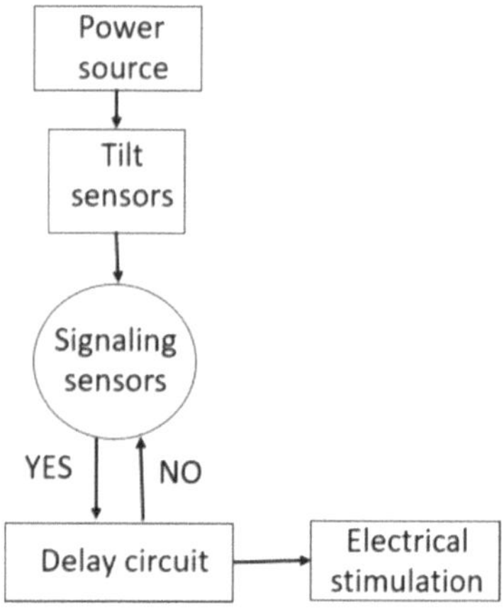

Fig. 1 The block diagram of the proposed device.

2.2 Using the Device for the Correction of Vicious Posture of the Spine

In the case of using the device for correcting vicious posture of the spine, the device with sensors has the following characteristics:

- includes the whole assembly in a case;
- uses its own power source (battery);
- to fix the device on the body, use kinesiological tape,
- to detect the deficient position, it uses a tilt sensor that can contain one or more ball or mercury switches, to detect a vicious attitude;
- uses a system to warn the patient if he adopts a vicious position, this can be made up of vibration sensors (sensitive warning), buzzer (sound warning), or led (light warning), in various combinations, chosen according to needs;
- if the patient does not perform the voluntary correction of the posture, within a predetermined and adjustable time interval, the involuntary correction of the vicious position is performed, with the help of electrical stimulation (functional electrical stimulation or other current type).

The objectives to be achieved through this device are the following:

- preventing, correcting and recovering vicious postures;
- scoliosis and kyphosis recovery, which uses physical therapy as a method of recovery;
- recovery of coordination after stroke, neurological conditions, etc.;
- increasing patients' autonomy.

2.2.1 Action Mode of the Sensory Device

The elements in the structure of the device are represented in Fig. 2.

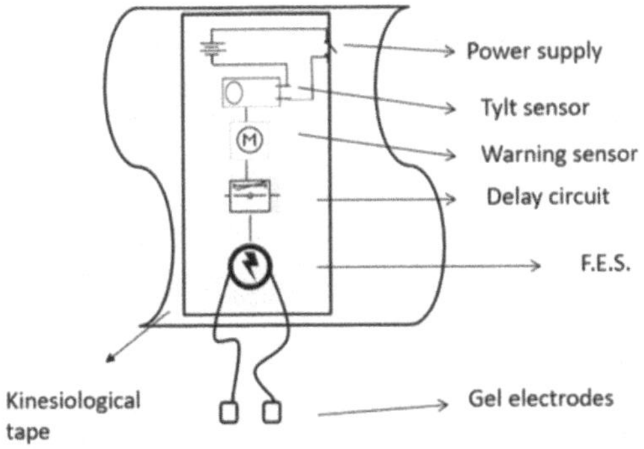

Fig. 2 The main constitutive elements

The functioning of the sensory device in the case of scoliosis is as follows: the kinesiological tape is applied to the concave side of the spine, to relax the shortened muscles, and when the inclination sensor senses that the patient is adopting a vicious posture, the warning system comes into action (in this case, only the vibration sensor can be used), which allows the patient to self-correct. If the patient does not self-correct or maintains his vicious position, after a predetermined time, the delay circuit switched on the electrical stimulator, for making the involuntary correction of the posture, the electrical stimulation [16] produces the contraction of the elongated muscles on the side of the convexity, toning it them. The delay circuit, also ensures the protection of the patient against the accidental triggering of the impulse.The advantages of the device with sensors, for vicious posture correction are, [15]:

- small dimensions and weight;
- easy to apply and remove;
- easy to hide under a clothing item or accessory;
- stimulation of deficient muscles for toning.

2.3 Using the Device for Movement Coordination Recovery

If using the tilt sensor in various objects for movement coordination recovery exercises, it will contain a minimum of 4 mercury or ball switches. In the variant for the recovery of coordination, the device will no longer contain the components used for electrical stimulation (Fig. 3).

The inclination of the mercury switch(es) influences the sensitivity of the device, so the closer the inclination is to the horizontal, the greater the sensitivity, which can be individually adjusted according to the condition, the stage of the condition and the goals pursued. For example, to relearn the movements required for feeding, the device can be placed in objects such as cups or spoons, made of materials like plastic, that reduce the risk of injury. The objectives are:

- recovery of movement coordination,
- prevention of the establishment of the syndrome of unilateral neglect of the hemibody affected after stroke,
- favoring neuroplasticity,

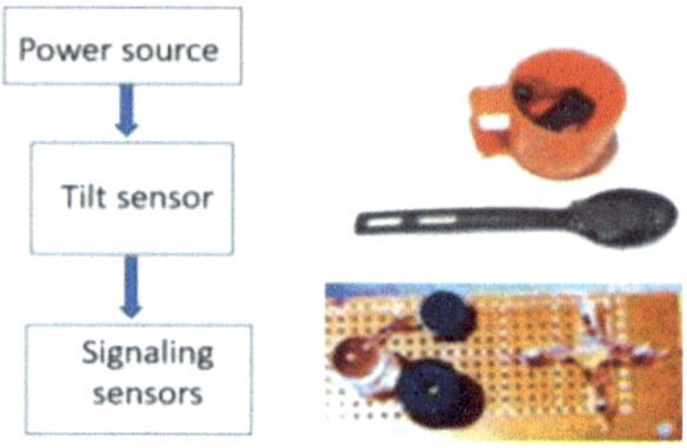

Fig. 3 Block diagram, component arrangement and objects examples

- acquiring self-care skills,
- increasing the patient's independence, ensuring independence in carrying out -daily activities or other activities,
- increasing self-esteem.

  Through repetitive mobilizations, and the visual feedback provided throughout the exercises, the movement performance is improved. Especially in patients with neurological conditions, movement coordination is a process that can be achieved after many repetitions. Coordinated activity is an unperceived, conscious automatism, but it can also be conscious. For example, typing, operating on the keyboard, etc. Thus, the patient using the affected limb for a longer time compared to an exercise program carried out only in the treatment base, a faster recovery is favored.The advantages in this case would be:
- regaining control over lost movements through sensory-motor feedback, based on learning and memorization, also supported by visual and tactile feedback,
- developing self-care skills,
- increasing patients' independence, self-esteem,
- minimizing the installation of unilateral neglect syndrome of the affected hemibody and complex regional pain,
- a faster recovery.

2.4 New Possibilities for Using the Device with Sensors

Compared to the first variant developed, during the research study new ways were sought to diversify and improve the applicability of the sensor device in recovery exercises with the help of various objects. Thus, as seen in the block diagram in Fig. 4, a development board and a bluetooth module were attached to the device with sensors, with which the therapist can remotely follow the patient's progress in real time on a smart device and thus being able to modify the therapeutic program, depending on the condition, the goals achieved and the patient's needs. The sensory device can be made with different degrees of sensitivity, so that after reaching a goal, the device can be changed from the object and changed for the next stage of rehabilitation program.

The use of objects containing the sensory device in recovery exercises alongside of portable/stationary exoskeleton robotic devices has been studied, also. In the example below (Fig. 5) is shown the use of a cup and a spoon with sensors alongside a prototype of a wearable exoskeleton device, developed in another research study. Thus, the use of

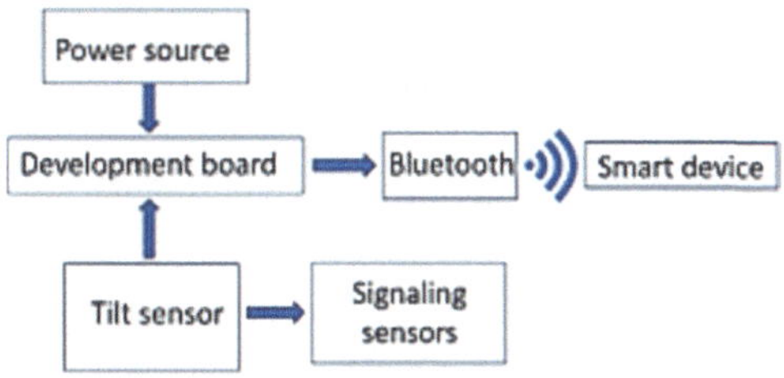

Fig. 4 The block diagram

the sensor device in patients with F2 force, which allows them to perform mobilizations in a zero-gravity plane, helps them to perform various coordination recovery exercises, scenario-based exercises aimed at developing self-care skills.

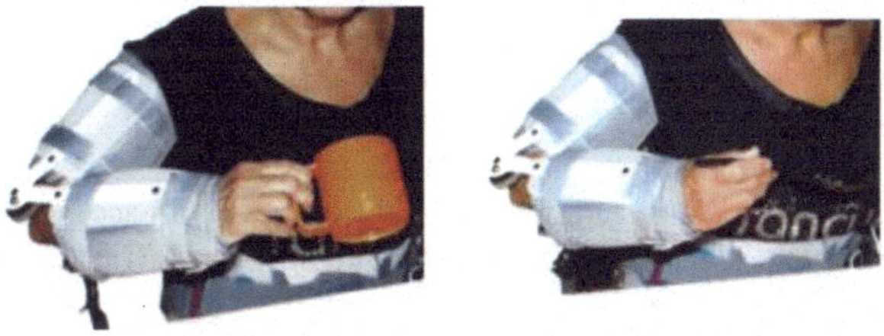

Fig. 5 Using instrumented objects with an exoskeleton rehabilitation system

The device can be used by patients with hearing impairments - in this case, the warning sensor used will be the light (led) and vibration (buzzer) one, and by patients with visual impairments - the warning sensor used, being the vibration and the sound (buzzer).

3 Conclusions

The sensory device, according to this paper, can be used with minimal training of the patient and relatives, including at home without the need for supervision by qualified medical personnel. Thus, the patient is given the opportunity to increase the exercise time, respectively the possibility of performing a large number of repetitive corrective movements, allows the restoration of motor engrams and the reduction of the time required for recovery.

Also taking into account the fact that according to the latest statistics presented by the Ministry of Health, Romania records the highest stroke mortality in Europe (21.64%), and at the national level being the second cause of death, after cardiovascular diseases [17] , also the increase life expectancy has led to the aging of the population, and the disability is installed from younger and younger ages, the device is useful because it contributes to reducing the pressure on the medical staff, because several patients can be treated at the same time, and the exercises can also be performed at home, adaptable depending on the condition, the stages of the condition, and the needs specific to each patient, offers a simple and effective recovery variance, achievable at costs that would allow its acquisition by many more patients.

Conflict of Interest Statement. The authors have no conflict of interest to declare.

References

1. https://kinetoiasi.ro/deficientele-de-postura/, last accessed 10 June 2024
2. https://www.gov.im/media/1355782/understanding-and-improving-your-posture.pdf, last accessed 10 June 2024
3. https://mstrust.org.uk/a-z/understanding-and-improving-your-posture/good-and-bad-posture, last accessed 10 June 2024
4. https://chiromedicahealthcenter.ro/postura-corecta-cum-o-mentii/, last accessed 10 June 2024
5. https://kinetodariana.ro/cum-ne-mentinem-corect-postura-spatelui/, last accessed 10 June 2024
6. Wernli, K., O'Sullivan, P., Smith, A., Campbell, A., Kent, P.: Movement, posture and low back pain. How do they relate? A replicated single-case design in 12 people with persistent, disabling low back pain. Eur. J. Pain. **24**, 1831–1849 (2020)
7. https://soptri.com/your-pain/posture-movement-dysfunction/, last accessed 10 June 2024
8. https://www.paincarecenter.com.sg/can-poor-posture-aggravate-herniated-disc/, last accessed 10 June 2024
9. Zielinska, N., et al.: Risk factors of intervertebral disc pathology-a point of view formerly and today-a review. J. Clin. Med. **10**(3), 409 (2021)
10. Rodrigues, S., Gotardi, G., Aguiar, S.: Effects of Vision on Postural Control in Neurologically Healthy Individuals. (2017)
11. Lee, J., Choi, H., Yum, K., Kim, J.: Classifying poor postures of the neck and spine in computer work by using image and skeleton analysis. Appl. Sci. **13**(19), 10935 (2023)
12. https://oali.com/blog/could-my-desk-job-cause-carpal-tunnel-syndrome/, last accessed 10 June Meta-analysis: association between wrist posture and carpal tunnel syndrome among workers,
13. You, D., et al.: Meta-analysis: association between wrist posture and carpal tunnel syndrome among workers. Saf. Health Work. **5**(1), 27–31 (2014)
14. https://www.durerispate.ro/blog/7-sfaturi-pentru-a-evita-durerile-de-spate-cauzate-de-folosirea-telefonului, last accessed 10 June 2024
15. Ver, I., Ungur, R.A., Ver, C., Ver, A.E: Device with sensors for the correction and recovery of vicious positions and coordination disorders in the spine and limbs, A00855/2018, Patent Decision no. 43/7 (2024)
16. Rădulescu, A.: Electrotherapy, 2nd edn. Editura Medicală, Bucharest (2017)
17. https://mfe.gov.ro/programul-sanatate-actualizare-la-ghidul-solicitantului-pentru-apelul- de-proiecte-fse-cresterea-accesului-si-eficacitatii-serviciilor-de-ingrijire-medicala-dedi- cate-pacientului-critic-pac/, last accessed 12 Sept 2024

The Study of a Low-Cost System Used to Evaluate the Body Mass Index

Traian Paduret[1] and Mariana-Rodica Milici[2(✉)]

[1] Suceava, Romania
[2] Stefan cel Mare University of Suceava, Suceava, Romania
dam@usm.ro

Abstract. This study presents anovel, low-cost system designed to evaluate Body Mass Index (BMI), a widely used indicator of health. The system leverages affordable, readily available components to create a user-friendly and accessible tool for BMI assessment. By combining (specific components, e.g., weight scale, height measurement device, and microcontroller), the system offers a practical and cost-effective alternative to traditional clinical methods. The system's accuracy was evaluated through rigorous testing against standard BMI calculation methods, demonstrating a high degree of correlation. Furthermore, the system's portability and ease of use make it suitable for various settings, including home use, community health centers, and remote areas with limited access to healthcare. Potential applications of this system include population health monitoring, early detection of obesity, and personalized health management. The development of such a low-cost system has the potential to significantly improve access to BMI assessment, thereby promoting healthier lifestyles and reducing the burden of obesity-related diseases. Future research directions include exploring additional features, such as real-time feedback and integration with mobile health applications, to enhance the system's functionality and user experience.

Keywords: body mass index · accurate · low-cost

1 Introduction

Adolphe Quetelet, a Belgian astronomer, mathematician, statistician and sociologist, introduced the basis of BMI (body mass index) between 1830 and 1850, while developing what he called "social physics" [1]. Quetelet himself never intended the index, then called the Quetelet Index, to be used as a means of medical evaluation. Instead, it was a component of his study of l'hommeyon or the common man. Quetelet considered the common man as a social ideal and developed the body mass index as a mean of discovering the socially ideal human person [2]. According to Lars Grue and Arvid Heiberg in the Scandinavian Journal of Disability Research, Quetelet's idealization of the common man would be elaborated by Francis Galton a decade later in the development of eugenics [3].

The modern term "body mass index" (BMI) for the ratio of human body weight to height squared was invented in a paper published in the July 1972 issue of the Journal

S. Vlad and N. M. Roman (Eds.): MEDITECH 2024, IFMBE Proceedings, 130, pp. 57–64, 2025.
https://doi.org/10.1007/978-3-031-95671-3_7

of Chronic Diseases by Ancel Keys and others. In this paper, Keys claimed that what he called BMI was "if not entirely satisfactory, at least as good as any other index of relative weight as an indicator of relative obesity" [4, 5].

In addition to measuring body mass and relative weight, Keys explained the relationship between obesity and metabolic energy. He showed that modest "overweight" can have advantages, especially in the elderly, determined that a low amount of body fat causes cardiovascular problems. Keys' study took place at a time when obesity was rare even though cultures were rich. Furthermore, little was known then about the relationship between body weight and cholesterol (HDL), nor was it appreciated that fat deposits, although requiring little energy to maintain, were metabolically active and secreted metabolically important substances, with feedback to the sensitivity of cells to insulin.

Keys noted that BMI is closely related to waist circumference and visceral adiposity, concepts that have proven difficult to separate. He concluded that BMI is very useful in industrial and traditional populations of men and women over the "adult" age range, let's say from 20 to 65 years [6]. He arrived at this opinion through meticulous and careful analysis, weighing his findings against the prejudgments of others. All things being considered, he recommended this method for estimating body fat because it is simple and, in most cases, accurate [8].

Over the years, since Keys's 1972 study, many have pondered the limitations of BMI, questioned its independence from height, thus the coefficients for the "new BMI" have been proposed. An Oxford mathematician, Nick Trefethen (like Quetelet, neither a physiologist nor an epidemiologist), recently proposed a reinterpretation of BMI in terms of tall or short people, corrected by a new ratio: weight multiplied by 1.3 (i.e. the square root of the average height or 1.69 m), product of the height raised to the power of 2.5. This new ratio much better reflects the relationship between mass and height [9].

Interest in an index that measures body fat has come with the increase in obesity seen in affluent Western societies. Keys explicitly considered BMI to be appropriate for population studies and inappropriate for individual assessment. However, due to its simplicity, it has come to be widely used for preliminary diagnostics [7].

The World Health Organization (WHO) has proposed the following BMI categories:

- Underweight—BMI below 18.5. People in this category are considered underweight and may be at increased risk of malnutrition and other health problems.
- Normal weight—BMI between 18.5 and 22.9. Considered the normal range for a healthy weight. People in this category have a reduced risk of weight-related health problems.
- Overweight (Pre-obesity)—BMI between 23 and 24.9. This category indicates overweight, with a slightly increased risk of developing weight-related health problems such as diabetes and cardiovascular disease.
- Obesity

Class I—BMI between 25 and 29.9. People in this category are considered obese and have an increased risk of developing chronic conditions, including diabetes and cardiovascular disease.

Class II—BMI between 30 and 34.9. This indicates moderate obesity with a significantly higher risk of health problems.

Class III—BMI of 35 and above. This is classified as severe obesity, associated with a very high risk of health complications.

Body mass index (BMI) ranges highlight the relationship between body weight and the likelihood of various diseases or even death [10]. Overweight and obese people, characterized by an increased BMI, are at increased risk for the following diseases [11]: coronary artery disease also known as ischemic heart disease, dyslipidemia, type 2 diabetes, hypertension known as high blood pressure increased or hypertension, osteoarthritis, sleep apnea, cerebral ischemia, endometrial cancer, colon cancer or colorectal cancer.

A study published by the Journal of the American Medical Association (JAMA) in 2005, entitled "Excess Deaths Associated with Underweight, Overweight, and Obesity" showed that overweight people had a death rate like that of normal weight people, as defined by BMI, while underweight and obese individuals had a higher death rate [12].

2 Description of the Experimental Stand

The device made to carry out this study is a device capable to evaluate the amount of body fat by means of the body mass index. The main quantities involved in the calculation of the body mass index are the weight and height of the subject being measured.

The system is presented in Fig. 1 and contains: domain limiter used to determine the height of the evaluated person, force sensors are used to determine the weight of the person being evaluated and are of the type of force cells with strain gauges used in common scales for people, touch sensors is used for not waking up in the display menu, and the coin sensors is used to charge people who want to use the system.

To determine the weight, a scale made with the help of four resistive force cells, connected in a bridge and which transmits data to the microcontroller by means of an analog-digital conversion and amplification module, was used. Its maximum admissible capacity is 200 kg. A vertical linear axis driven by a stepper motor was used to determine the height. Thanks to the use of the DRV8825 driver, it was possible to control the motor by micro-stepping, which determined its movement with a precision of 0.11 degrees, respectively 0.01 linear millimeters. The measurement range is between 120–219 cm (Fig. 2).

The human-device interface is very simple and intuitive, allowing the user to go through several parameterization menus so that the result is as accurate as possible. The display is made by means of a bright, color screen with a diagonal of 4 inches so that the experience throughout the duration of the determination is an unforgettable one. At the end of the determination, the medical report is displayed on the screen, showing: the person's gender, age, mass, height, body mass index accompanied by a message corresponding to its value (Underweight/Normal/Overweight/Obesity I/Obesity II/Obesity III) (Fig. 3).

3 Experimental Data

Eleven subjects participated in this study. All subjects were male, students of the Faculty of Electrical Engineering and Computer Science, aged 20–24 years. The first set of data was obtained using the reference device Tanita WB-380 H, and the second set of data was obtained using the device under study.

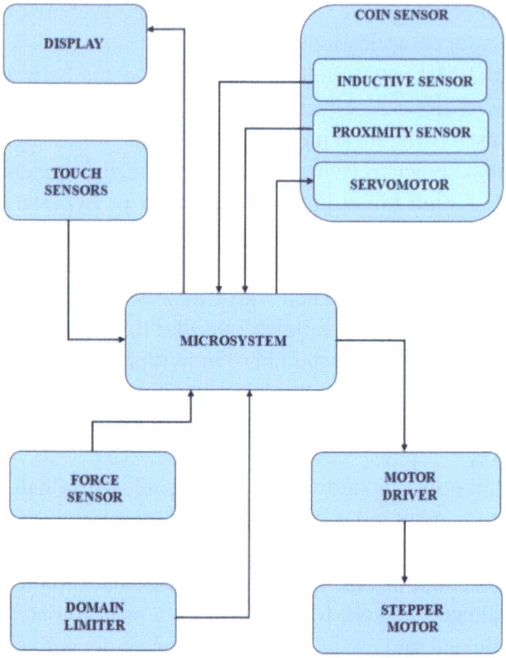

Fig. 1. Block diagram of the system designed, built and tested

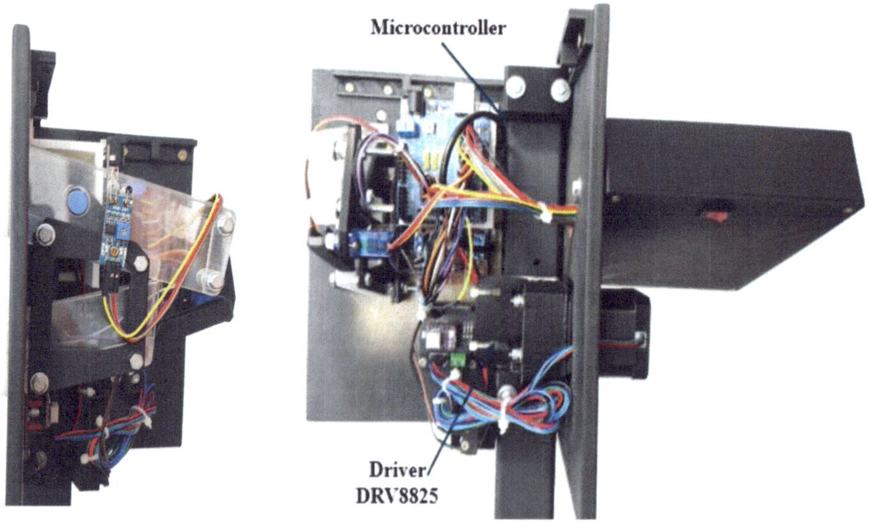

Fig. 2. Display, command, and control unit-rear view

Therefore, the obtained data were centralized in Table 1 and the body mass index was also calculated in the two cases.

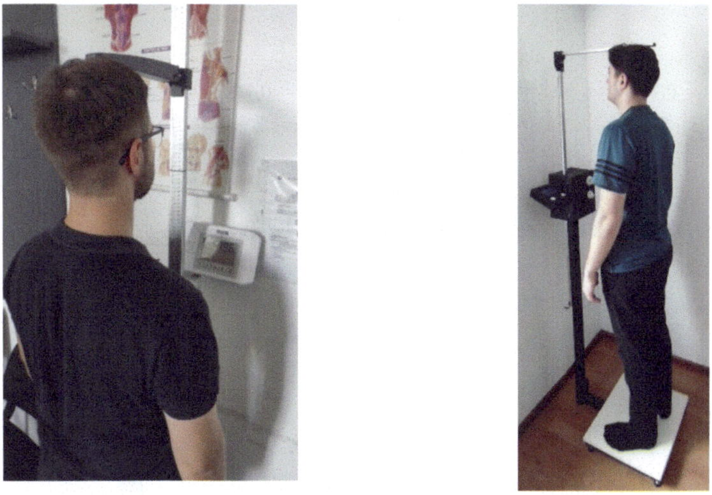

Fig. 3. The Tanita WB-380H reference device and the device made and tested

Table 1. Data centralization

Person	Standard device			Studied device		
	Weight [kg]	Height [cm]	BMI [kg/m^2]	Weight [kg]	Height [cm]	BMI [kg/m^2]
1.	84.2	175.8	26.71	84.3	175.5	26.86
2.	81.8	186.5	22.39	81.9	186.3	22.47
3.	87.2	179.5	26.26	87.4	179.1	26.47
4.	89.7	175.5	28.58	89.8	175.2	28.73
5.	82.3	175.5	26.22	82.5	175.8	26.17
6.	79.6	168.5	28.08	80	168.5	28.22
7.	68.1	174	22.17	68.1	173.7	22.39
8.	80.4	175.5	25.62	80.4	175.2	25.73
9.	69.6	185	19.44	70	185.1	19.53
10.	75.4	180	22.55	75.1	179.5	22.62
11.	64.4	168	22.89	64.7	167.7	23.09

To achieve a detailed analysis and an overview of the accuracy of the body mass index device, the errors related to the measurement of height and weight will be considered, as well as the influence of these errors on the final values of the body mass index (Figs. 4 and 5; Tables 2, 3, and 4).

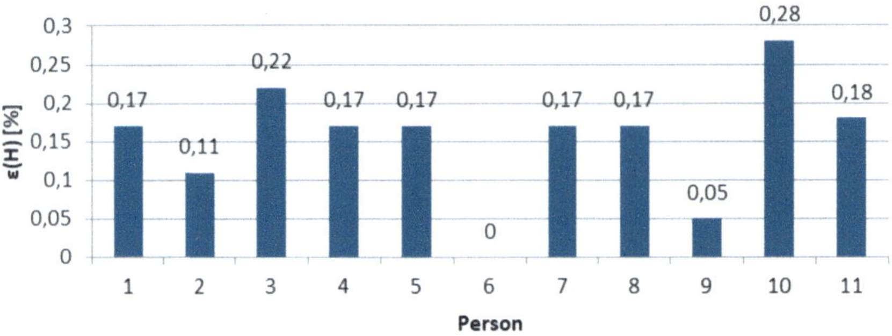

Fig. 4. Diagram of height measurement errors

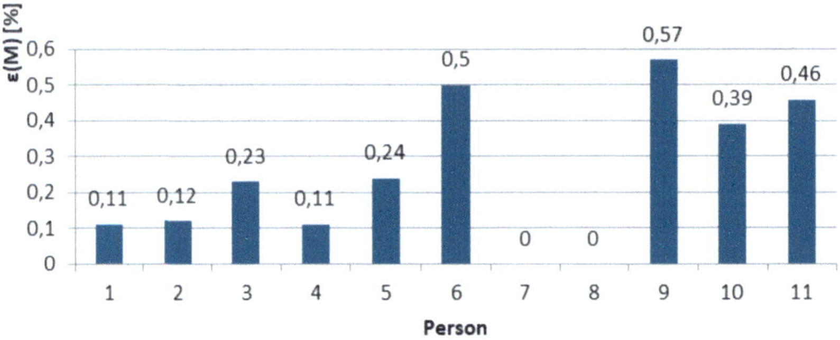

Fig. 5. Chart of relative weight errors

Table 2. The relative height errors

Person	1	2	3	4	5	6	7	8	9	10	11
ε(H) [%]	0.17	0.11	0.22	0.17	0.17	0	0.17	0.17	0.05	0.28	0.18

Table 3. Relative weight errors

Person	1	2	3	4	5	6	7	8	9	10	11
ε(M) [%]	0.11	0.12	0.23	0.11	0.24	0.5	0	0	0.57	0.39	0.46

For the calculation of the relative height errors, we considered the following calculation relationship:

$$\varepsilon(H) = \frac{|H(standard) - H(studied)|}{H(standard)} \cdot 100 \quad (1)$$

Table 4. Relative errors of body mass index

Person	1	2	3	4	5	6	7	8	9	10	11
ε(IMC) [%]	0.54	0.39	0.79	0.54	0.18	0.5	0.43	0.42	0.43	0.29	0.91

where:

ε(H) is the relative height error in [%],

H(standard) is the height measured with the standard device in [cm],

H(studied) is the height measured with the studied device in [cm].

To observe the influence of measurement errors (height and weight) on the final values of the body mass index, we determined the absolute error between the body mass index calculated with the values provided by the reference device and the one calculated with the values provided by the analyzed device (Fig. 6).

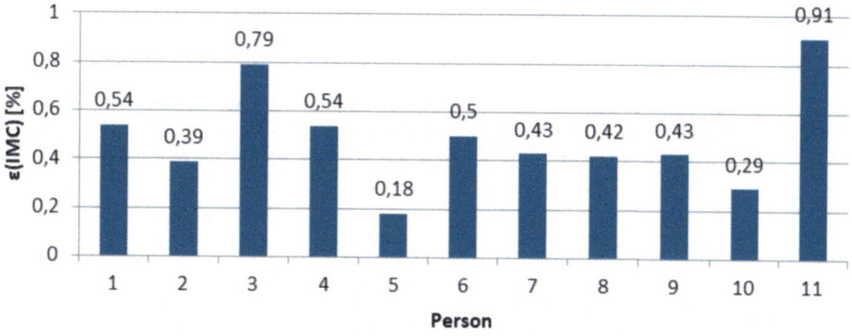

Fig. 6. Body mass index error plot

4 Conclusions

The designed, manufactured, and tested equipment is characterized by low price, reliability, satisfactory precision and small dimensions. It is a portable equipment that does not require connection to an electrical network thanks to the built-in battery. It can be transported and placed in areas of interest such as public spaces or institutional lobbies.

Analyzing both the tables with the measurement errors corresponding to height and mass, as well as the diagrams corresponding to them, it is observed that the results of the measurements obtained with the device under study are close to the values obtained with the help of the standard device and fall within the margin of error of the currently marketed devices.

In terms of height, we obtained relatively small differences compared to the reference device Tanita WB-380 H, the largest difference being 5 mm, which resulted in a relative error of 0.28%. Analyzing the table with the relative errors corresponding to the height we can see that the average error is about 0.15%.

In terms of mass, the differences are relatively small with a small exception in the case of persons 6 and 9 for which we obtained a maximum difference of 400 grams and some relative errors of 0.5% and 0.57% respectively. In this case, we can see that the average error is around 0.25%.

Despite the measurement errors of height and weight, the relative error, considering the body mass index obtained with the reference device and the body mass index obtained with the tested device, is surprisingly small, reaching a maximum value of 0.91%. As with the other measurements, the average error is around 0.49%, apparently insignificant considering the body mass index reference ranges. So, considering the errors obtained, we can state that the device for determining the studied body mass index falls within the standards in force regarding the accuracy of measurements, being characterized by accuracy class 1, a very good value if we compare it with the accuracy class of current equipment.

Conflict of Interest. The authors declare that they have no conflict of interest.

References

1. Grue, L., Heiberg, A.: Notes on the history of normality—reflections on the work of Quetelet and Galton. Scand. J. Disabil. Res. **8**(4), 232–246 (2006)
2. Singer-Vine, J.: Beyond BMI: Why Doctors Won't Stop Using an Outdated Measure for Obesity. Slate. Com (2009)
3. Keys, A., Fidanza, F., Karvonen, M.J., Kimura, N., Taylor, H.L.: Indices of relative weight and obesity. J. Chronic Dis. **25**(6–7), 329–343 (1972)
4. Trefethen, N.: New BMI (body mass index). Ox. ac. uk. Mathematical Institute, University of Oxford. Retrieved, 5 (2019)
5. National Institutes of Health. Assessing your weight and health risk
6. World Health Organization: The surveillance of risk factors report series (SuRF). SuRF Rep. **2**, 22 (2005)
7. Di Angelantonio, E., Bhupathiraju, S.N., Wormser, D., Gao, P., Kaptoge, S., De Gonzalez, A.B., et al.: Body-mass index and all-cause mortality: individual-participant-data meta-analysis of 239 prospective studies in four continents. Lancet. **388**(10046), 776–786 (2016)
8. Milici, M.R., Mihai, M.R., I., Milici L.D.: Aspects of engineering education in signal technology using virtual instrumentation, Elektronikair. Elektrotechnika. **94**(6), 113–116 (2009)
9. Milici, L.D., Placinta, V.M., Bujor, L., Milici, M.R.: System for highlighting the emotional states, used in assessing the teaching methods. In: 9th International Symposium on Advanced Topics in Electrical Engineering, pp. 965–968 (2015)
10. https://en.wikipedia.org/wiki/Body_mass_index, last accessed 1 June 2024
11. https://sonkascale.com/product/sonka-height-weight-bmi-scale-blood-pressure-machine/, last accessed 1 June 2024
12. https://romaniancoins.org/ro1ban2005.html, last accessed 1 June 2024

Low Cost Thermic System for Transfusable Substance Improving

Barbu Cristian Braun(✉), Corneliu Nicolae Drugă, Ionel Șerban, and Mitu Leonard

Transilvania University of Brasov, Brasov, Romania
braun@unitbv.ro

Abstract. The paper presents a stage of research regarding the design, development and testing of a flexible, low-cost system that allows improving the process of transfusion of vital substances. It is primarily about the process of heating and bringing them to an optimal temperature from the point of view of thermal biocompatibility for the beneficiary. From the implementation point of view, the heating solution proved to be an effective one both in terms of process duration and in reducing energy consumption. Regarding the functional testing, the method of recording and measuring the temperature values of transfusable substances using an infrared thermal imaging camera was used. Based on the results related to temperature measurement, finding values very close to the ideal ones specific to such processes, it was possible to conclude that the method is efficient, reliable, low-cost, and it can be implemented in the near future in hospitals and of blood transfusion centers, for all categories of subjects, regardless of the conditions that require a transplant.

Keywords: Transfusable Substances · Thermal Imaging Camera · Temperature · Blood · Biocompatibility

1 Vital Substances Transfusion—Solution and Trouble

Nowadays, the importance of transfusion is absolutely undeniable, in the context where daily thousands and thousands of people need vital substances to survive serious illnesses or accidents. It is well known internationally that a major problem, especially with regard to blood transfusion, is the insufficient number of donors [1, 2].

But another problem, perhaps equally important, which, until recently, was not given too much importance is that concerning the conditions under which a transfusion must take place. A particularly important aspect, apart from respecting the blood group, regarding the transfer is represented by sanitary hygiene conditions. But these aspects are far from completely solving the problem of transfusion of vital substances in general (blood, lymph, etc.). In addition to the issue of biocompatibility, a particularly important aspect concerns the environmental conditions regarding the transfusion (temperature, flow rate, pressure of the transfused substances). If until now the problems related to hygiene, blood group, transfusion conditions from the point of view of flow and pressure

S. Vlad and N. M. Roman (Eds.): MEDITECH 2024, IFMBE Proceedings, 130, pp. 65–77, 2025.
https://doi.org/10.1007/978-3-031-95671-3_8

have been largely solved, the problem of ensuring an ideal temperature for transfusion is the subject of much more recent research. For this reason, the problem of heating transfusable substances (especially blood) to a temperature as close as possible to that of the beneficiary person is currently increasingly addressed. Thus, recent research has led to the solution of this aspect as well, but the major problem that remains unsolved for the time being is related to the costs and the duration of the heating process, which, too often, is far too long in relation to the time it must have place the transfusion [3, 4].

2 Proposed Solution for Transfusion Condition and Safety Improvement

2.1 Critical Review of Existing State-of-the-Art Solutions for Heating Transfusable Substances

In order to find the better solution in terms of costs and efficiency, a state of art meaning the actual solutions was performed in the first stage of the research. For this purpose different heating transfusable systems were analyzed [5]. The first one refers to the FLUIDO COMPACT device system for warming blood and fluids from low to moderate flow rates, using an infrared technology as a heating method [6, 7]. Another systems is HOTLINE that provides blood and warm fluids needed by patients, with the heating technique based on a unique coaxial water recirculation method, namely through a triple lumen tube [8, 9]. M WARMER is another type of blood and fluid warmer, specialized for emergency transfusions. The system has a portable fluid warmer that warms blood from cold (4°C) to body temperature in seconds and has the capacity to warm up to 1.5 liters of fluid. BH-600A blood warmer, infusion solutions and nutrition, being suitable for any type of infusion support. To minimize temperature loss during general anesthesia, intravenous fluids are warmed to body temperature before infusion [11]. BARKEY PLASMATHERM is an electronic system used primarily for heating and thawing bags or bottles of fluids to be administered to patients by infusion or transfusion [12, 13]. 3M™ RANGER™ SYSTEM is built to keep warm blood, blood products and other infusible fluids which delivers them with a flow rate of up to 500 ml/min. This device uses dry heating technology, which means faster and more accurate temperature control, minimizing the risk of fluid overheating [14, 15].

In the Tables 1 and 2, the most relevant technical characteristics of the 7 best-performing transfusable substance heating systems on the market are summarized [5].

Starting from the information summarized in Tables 1 and 2 regarding the technical-functional characteristics of existing solutions on the market regarding the solution to the problem of heating transfusable substances, was done a detailed critical analysis. For this purpose, a small software application was programmed, in the LabVIEW environment, for the objective and efficient establishment of the optimal solution. The purpose of this stage of the study is to develop a somewhat similar, low-cost and more accessible solution for transfusion center staff [5, 16].

The software application for critical analysis involved the definition of some text boxes (marked in red) as input variables in the window-panel, the one for the user interface, which represent all the criteria for which the critical analysis is done. In blue,

Table 1. Main technical specifications for the 7 best-performing systems_Part1.

Device	Supply [V]	Maximum power consumed [W]	Transfusible line length [mm]	Maximum heating temperature [°C]	Temperature precision [°C]
FLUIDO COMPACT	220–230	160	700	39	0.1
HOTLINE	220–230	115	1000	41.5 ± 0.5	0.2
S-LINE BARKEY	220–230	75	1500	39 ± 2	0.2
M WARMER	Portable battery (12 V)	15	500	39 ± 3	1
BH-600A	220–230	45	800	42 ± 0.5	0.5
BARKEY PLASMATHERM	220–230	1600	–	$37 \div 45$	0.1
3 M RANGER	220–230	900	1200	41	0.1

Table 2. Main technical specifications for the 7 best-performing systems_Part2.

Device	Solution heating time [min]	The number of bags that can be heated	Heating method	Price [Euro]
FLUIDO COMPACT	1–2	1 bag of 450 ml	Infrared	2000
HOTLINE	4	1 bag of 450 ml	Coaxial water recirculation	1500
S-LINE BARKEY	2	1 bag of $250 \div 300$ ml	Dry heating (metallic plate)	2000
M WARMER	Up to 2	1 bag of 150 ml	Dry heating (metallic plate)	600
BH-600A	5	1 bag of $250 \div 300$ ml	Dry heating (metallic plate)	450
BARKEY PLASMATHERM	$7 \div 9$	4 bag of 500 ml or 8 bag of 250 ml	Dry heating (2 thermal pillows with recycled water)	10,000
3 M RANGER	2	1 bag of 450 ml	Dry heating (metallic plate)	3000

all the output variables, of numerical type, have been defined to display the scores associated with each analysis criterion, separately. Separately, a numeric output variable was provided to display the total score for each analyzed solution. For example, the BARKEY PLASMATHERM model was chosen and its specific characteristics were entered in the text boxes marked in red, and in the blue boxes the customized score was automatically generated, finally displaying the critical analysis grade. The example is illustrated in Fig. 1 [5, 12, 16].

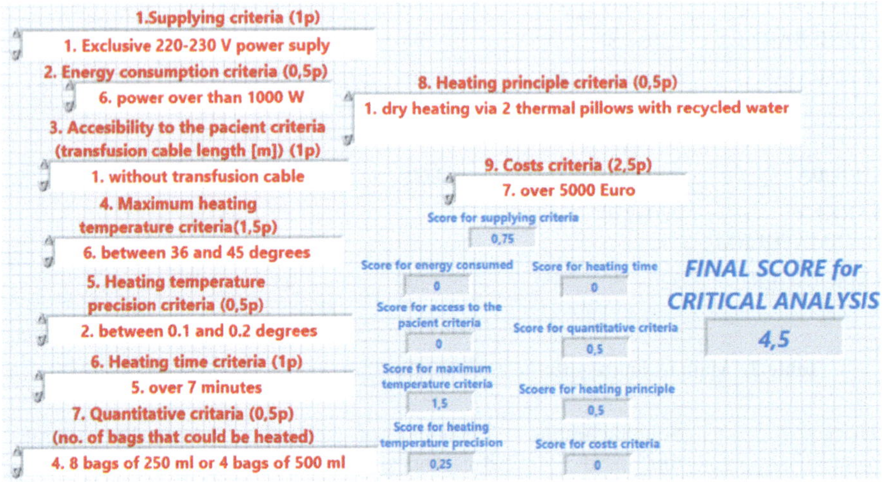

Fig. 1. Assisted critical analysis example for BARKEY PLASMATHERM heating system.

Based on the information entered in the table that reviews the technical-functional characteristics for each heating system, each entry-type entity, as an analysis criterion was set to include all possible feature variations. For this purpose, for each type input variable dialog box has created a predefined list of options that facilitate easier access to the necessary information. Each of these analysis criteria, as a selective input variable, was related to a multi-case programming structure (Switch), including, for each option, a separate case within that structure.

An example of the relationship of the criteria presented above in the graphical programming window can be seen in the Fig. 2. Similarly, partial scoring algorithms were programmed for all criteria analysis.

Once completed the application it, was used, through successive runs, for the critical analysis through all the criteria, for each individual solution, as shown in the heating systems description table. The final scores associated with each heating system were thus centralized in Table 3 [5].

As a result, none of the devices reviewed scored higher than 7.90 points out of 10, which means we can't pick a "winning" device as none met all the criteria reviewed.

Thus, only the best features of each device can be chosen, the optimal solution being that of designing, developing and testing a device taking into account the most suitable criteria or improving one of the mentioned devices. For this purpose, it is known, with

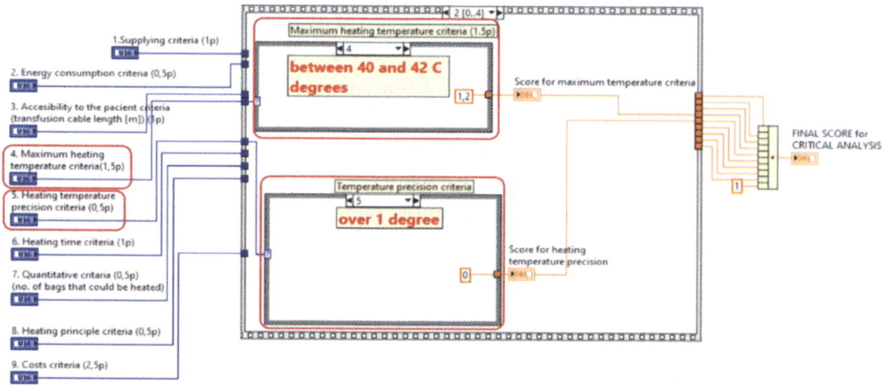

Fig. 2. Exemplifying the programming of multiple cases of partial scoring for the variables associated with analysis criteria no. 4 and 5.

Table 3. Awarding the final score obtained following the critical analysis for each variant of the heating system, by using the developed software application

No.	Heating system	Final obtained score
1	FLUIDO COMPACT	6.45 from 10.00
2	HOTLINE	7.50 from 10.00
3	S-LINE BARKEY	6.75 from 10.00
4	M WARMER	7.05 from 10.00
5	BH-600A	7.90 from 10.00
6	BARKEY PLASMATHERM	4.50 from 10.00
7	3 M RANGER	6.55 from 10.00

the help of critical analysis, what exactly should be adjusted (for example, too high cost of the device, better temperature accuracy, etc.).

In order to start the conception of the installation and subsequently its design and realization, the choice of the most suitable method of ensuring the optimal temperature of the substance on the route to the beneficiary must first be taken into account.

The company Smiths Medical, through a comparative study presents the differences between their HOTLINE device and the other heaters by conventional fluids. Unlike the HOTLINE® Blood and Fluid Warmer, conventional fluid warming systems suffer from an unwanted cooling process between the heater and the patient connection, especially at lower routine flow rates. In these systems, active heating ceases after the blood or fluid has passed through the conventional heater, allowing the fluid to cool in the intravenous line before being delivered to the patient [5, 8, 9].

In the study conducted by the University of Florida, the Department of Anesthesia, the heating-based Ranger device was compared dry and the previously mentioned HOTLINE device, evaluating the flows and heating efficiency. This study is based on several tests and experiments on these devices. One of these consisted of using saline at room temperature (21 °C) and packed red blood cells at 10 °C using a pressure of 90 mmHg (gravity equivalent) and 300 mmHg. The outlet flow from each device was connected to an in-line thermistor and simultaneous measurements of outlet temperature and flow volume per minute were recorded. Additional data points were obtained with a peristaltic pump that provided flow rates of 1–6 l/h through each device. Calculations were made on the effect of these flow rates and outlet temperatures on the average body temperature of a 70 kg patient [5, 8, 9].

To demonstrate the clinical effect of the observed differences in flow rate and temperature output with each blood/fluid warmer, the mean body temperature drop for a 70 kg patient receiving RBCs at 10 °C or saline at 21 °C was calculated with Hotline, Ranger, or no warmer at 1–6 L/h and at maximum flow rates, as would occur in a major fluid transfusion, with the following formula:

$$\Delta \mathrm{MBT} = \left(\left(\mathrm{T_F{-}{-}T_{pt}}\right) \mathrm{S_F}^{*}\ \mathrm{V_b}\right) / \left(\mathrm{S_{pt}}^{*}\ \mathrm{W_t}\right) \tag{1}$$

where: ΔMBT indicates the change in average body temperature, T_F indicates the temperature of the infused fluid, T_{pt} indicates the patient's core temperature (37 °C), S_F indicates the specific heat of the infused fluid for blood, V_b indicates the volume of infused fluid (in liters), S_{pt} indicates the specific heat of human tissue and W_t indicates the patient's weight (in kilograms) [5].

From the studies reviewed previously, we were presented with the disadvantages and advantages of the most popular methods of warming blood and fluids for transfusion, but no method was satisfactory. Devices based on dry heating through a metal plate are among the most widely used, but fluids have been found to suffers from cooling between the heater and the patient connection, especially at lower routine flow rates, because active heating ceases after the blood or fluid has passed through the heater, which will cause the fluid to cool through the intravenous line before it is delivered to the patient [5].

Thus, another type of device was studied based on the coaxial water recirculation method (through a triple lumen tube), which solves the problem of cooling the liquid on the way to the patient, but another problem appears with this device, namely that in rapid transfusions with higher flow rates, the outlet temperature of the fluid decreases, because the recirculation solution would not have enough time to warm up at increased and faster flow rates.

The German company BARKEY has developed a new device for warming blood and thawing plasma using a dry heating method using two thermal pads with recirculated water that seems to be the solution to the problems of the other existing devices. For optimal use, up to four bags of blood or plasma have been experimented with, placed between two water-filled cushions, where the water is pumped into the cushions continuously [12, 13].

In Table 4 the main advantages and disadvantages of these methods of ensuring the optimal temperature to the patient are briefly presented for a better interpretation of the final choice [5].

Table 4. Main advantages and disadvantages for the different heating solutions

Heating method	Advantages	Disadvantages
Dry heating using a metallic plate	– Only 2 minutes for heating; – Low cost, easy using	– Direct contact with the sanguine bags; – Fluid cooling along the way to the patient
Co-axial water recycling (by a triple lumen tube)	– Maintains the constant temperature on the way to the patient; – There is no direct contact with the sanguine bags	– High risk of contamination of the transfused substance; – Low exit temperature for high debits
Dry heating using two thermal pillows with recycled water	– There is no contamination risk of the transfused substance; – No direct contact with the fluid bags; – Mixing anti-coagulation fluids	– High costs; – Longer heating time

2.2 Design and Development of the Low-Cost System for Heating Transfusable Substances

In conclusion, following all these comparative studies, the solution was chosen to design a device based on the dry heating method using two thermal pillows with recirculated water, with the aim of making the product at a lower cost compared to the current purchase price of a device that uses the same heating method [5].

For system designing, in a first stage, the problem of heat transfer calculation, the volume of fluid subject to heating and the sizing calculation of the tubing was raised. For the heat transfer, first of all the thermal conduction was taken into account. Thermal conduction is based on Fourier's law, whereby the direction of heat transfer is from higher temperatures to lower temperatures, and which can be expressed by the following equation:

$$q_x = -K\,(dT/dx) \tag{2}$$

where: The heat flux [W/m^2] is the rate of heat transfer in the X direction per unit area perpendicular to the direction of transfer and which is proportional to the temperature gradient, *dT/dx*, in this direction; The parameter *K* is the thermal conductivity [W/m·K]. The minus sign is a consequence of the fact that heat transfer occurs in the direction of decreasing temperature [5].

From the point of view of calculating the volume of the recirculated water type heating agent, the following hypothesis was taken into account: for the use of the two thermal pillows, a volume of water sufficient to fill the 2 containers must be used, but not too much because it will require a higher water heating power from the electrical resistance part. According to the capacities of the 2 manufactured thermal pillows, namely 2000 ml each, a tank with a volumetric capacity of 5000 ml was chosen, because filling the bags will require a volume of 4000 ml and plus 1000 ml that will have to remain in the tank, a volume of 5000 ml that would be suitable for the heating fluid volume.

For the sizing of the piping, it was taken into account that the transmission of water from the tank to the thermal pads will require a predetermined route by means of connecting hoses. These hoses will have the diameters required by the inlet channels and output of the thermal cushions, respectively of the pump which are 6 mm. Therefore, hoses with Ø 5 × 7 mm (5 mm inside by 7 mm outside) will be used, which will be forced to be clamped. The diagram of the heating route of the recirculation solution is represented in Fig. 3 in which the order of transmission of the heated water from the tank and back to it is highlighted in a continuous recirculation process [5].

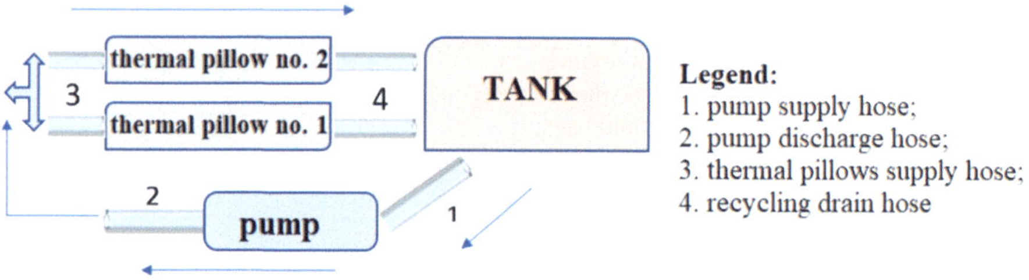

Fig. 3. Scheme of the recirculation solution heating path

To calculate the flow rate, the fluid pressure was taken into account, through the Hagen–Poiseuille equation that describes the pressure drop due to the viscosity of the fluid. To calculate flow from pressure, the formula is expressed as:

$$Q = \left[\pi R^4 (p_1 - p_2)\right]/8\,\mu L \tag{3}$$

where: $(p1 - p2) = \Delta p$ is the pressure difference between the ends of the pipe (pressure drop), μ is the dynamic viscosity of the fluid, L and R are the length and radius of the pipe segment [5, 17].

To calculate the flow rate, the fluid pressure was taken into account, through the Hagen–Poiseuille equation that describes the pressure drop due to the viscosity of the fluid. To calculate flow from pressure, the formula is expressed as:

$$Q = \left[\pi R^4 (p_1 - p_2)\right]/8\,\mu L \tag{3}$$

where: $(p1 - p2) = \Delta p$ is the pressure difference between the ends of the pipe (pressure drop), μ is the dynamic viscosity of the fluid, L and R are the length and radius of the pipe segment [5, 17].

Regarding the constructive principle, the device was designed and made of the following components: the water tank, the outer casing, the electronic command and control block, the thermal pillow storage tray.

The electronic control scheme was initially designed in the form of a block diagram highlighting the most important routes of the component elements. The block diagram includes the most important blocks and systems of the installation, namely: the command and control block, the detection block that transmits the information taken from the 2 temperature sensors to the microcontroller, the actuation system presents the servomotor that operates based on the information received from microcontroller, the user interface includes the LCD screen, buttons and buzzer where the information is visualized, and finally the power system based on a 12 V power supply that will power the command and control block and the tank that supplies water to the pump (Fig. 4) [5].

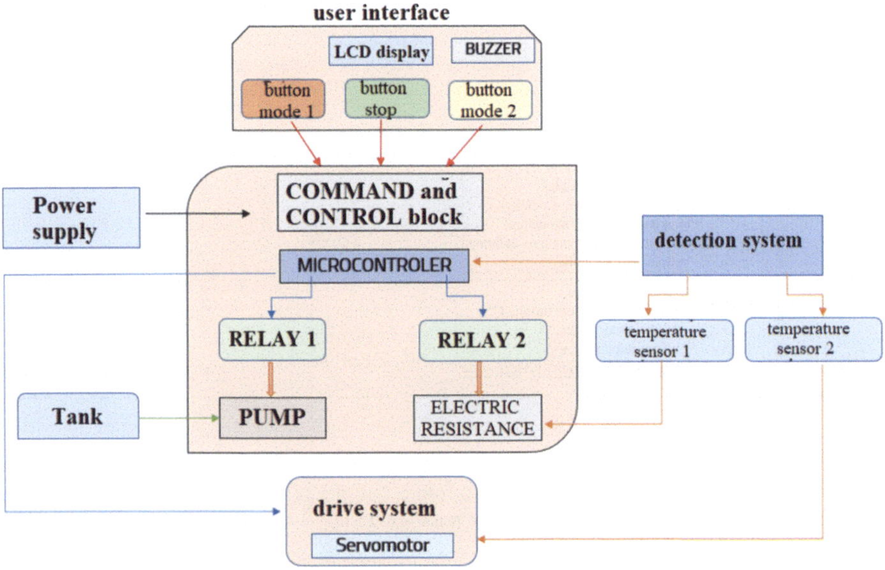

Fig. 4. Block scheme of the designed system.

After the design of the system, the necessary components were prototyped and procured, then their assembly followed, in Fig. 5 the final version is shown, following the practical implementation [5].

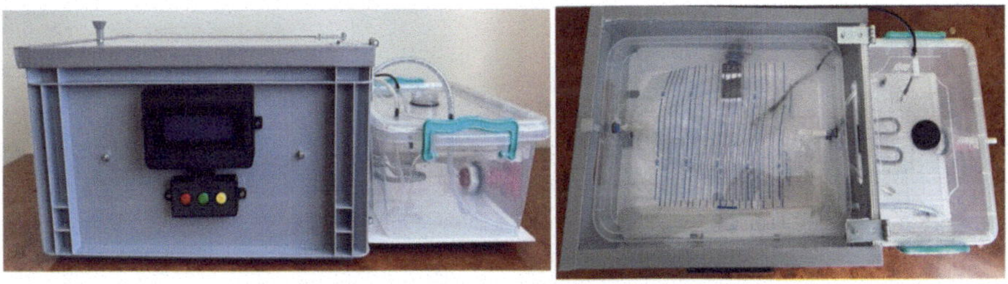

Fig. 5. Final practical realization of the system.

Once the system was realized and tested from a functional point of view, a software application was programmed, by interfacing the hardware development board used to name the Arduino UNO. In the first stage of creating the program code, predefined libraries were introduced for certain components in the project, such as the LCD screen, the I2C module, the DALLAS 18B20 temperature sensor and the servo motor. The next step was to define the pins for making the connections with the previously mentioned components, along with buttons, relays, water pump and buzzer. At this stage the addresses for the two temperature sensors were declared and for the LCD screen as well as the upper and lower limits for the temperature sensors [5]. A flowchart for programing algorithm is shown in Fig. 6.

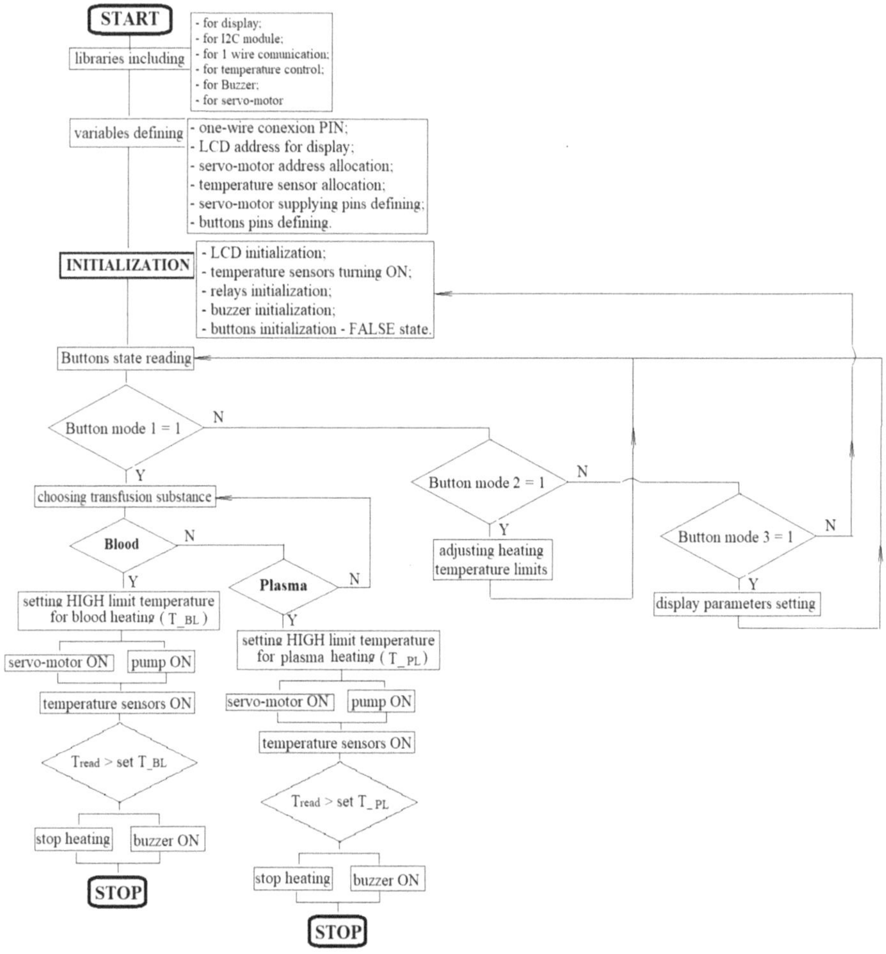

Fig. 6. The first stages of the source code to include libraries and pins defining.

3 Results and Discussion on the Proposed Heating Transfusion System

In this research, to ensure the sufficiency and accuracy of the developed heating system, a comparative study between the values given by temperature sensors and real temperature before and after heating has been done. For this purpose, a FLIR Systems thermal imaging camera was used for more accurate system testing. It generates a thermographic image of a scanned object based on the radiation it emits. The thermographic chamber is shown in Fig. 7, next to the containers subjected to the test [5, 18].

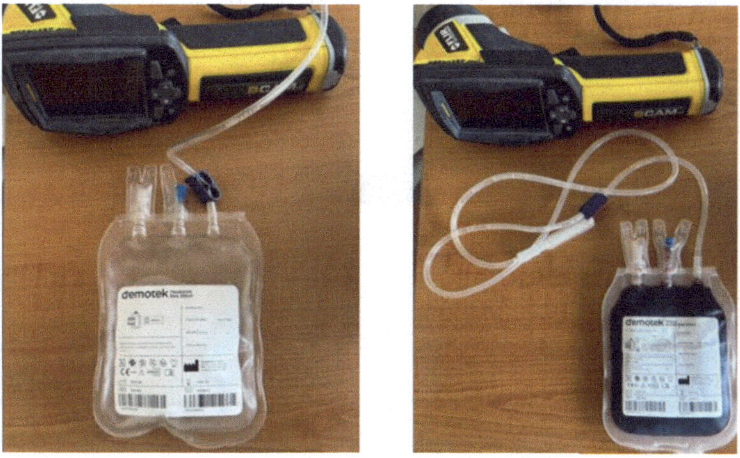

Fig. 7. The thermal imaging camera and the containers to be tested.

The initial temperatures of the containers were 9.4 °C for distilled water and 15.8 °C for blood. Distilled water was used to simulate plasma, and therefore it was desired to have a lower temperature than blood, since plasma is stored at lower temperatures than blood. These containers were inserted into the machine and approximately 10–15 minutes were waited until the operating programs were completed. Meanwhile, the temperature of the water tank was also measured and it was found that initially the water in the tank had a temperature of 22.6 °C, and at the end of the heating process it reached 48 °C, because in the plasma heating mode a higher water temperature is desired to shorten the waiting time, the plasma being frozen in storage.

Until the final measurement of the temperatures of the containers, the measurement of the thermal pads was also carried out, to see what temperature they have as well. Thus, with the thermal imaging camera, the temperature of the thermal pads was measured and it reached a temperature of 40 °C in the case of the blood heating module and 44 °C in the case of the plasma heating process.

The final point was to measure the temperatures of the 2 containers at the end of the heating process, and during the process, the temperature reached 26.7 °C. After an interval of about 5 minutes it reached the final temperature of 35.6 °C [5, 18].

It can be seen that the temperature gradation scale in the second image of Fig. 8 reaches up to 35.6 °C in the red areas of the bag, which means that an average temperature of the whole bag (there being only measured locally) would be 35 °C, temperature close to 37 °C, desired value reached. Due to the heating functioning and also thermographic measuring it was proved than the differences between measured heating temperature and thermographic measured temperatures did not exceed 1.5 ÷ 2 degrees. In conclusion, following the tests carried out, the device fell within the required requirements, achieving good functionality, meaning only small improvements [5].

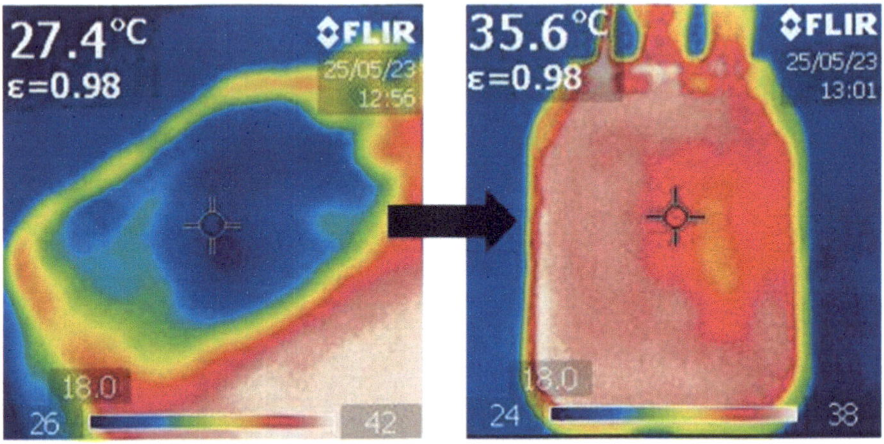

Fig. 8. The temperatures of the containers measured with the thermal imaging camera: the initial and the final temperature (after heating).

4 Conclusion on the Developed Heating Transfusion System

From the point of view of the operation of the designed system, it will have to follow certain steps towards proper use and will work according to certain predetermined parameters and characteristics.

To certify the efficient and safe operation of the designed and built system, it was tested for all its operating modes. First of all, the testing was done from the point of view of the transfusion regimes, for both substances, both blood and plasma. For this purpose, all the necessary work stages were completed, respecting the imposed technological conditions.

It was found that the heating system is functional and reliable, and can be implemented on a large scale, in clinics and hospitals, especially in case of emergency interventions [5].

Acknowledgments. The research described in the paper was largely due to the contribution of Medical Engineering graduate students, 2023-2024 promotion, within the Diploma Project, the practical activity being carried out in the laboratories of Transilvania University.

Conflict of Interest. The authors declare they have no conflict of interest.

References

1. Geneva, S.: The Clinical Use of Blood Handbook (2002). https://apps.who.int/iris/bitstream/handle/
2. W. H. Organization: Educational Modules on Clinical Use of Blood (2021). http://www.transfusion.ru/2022/03-04-1.pdf
3. P. E. Marik: Transfusion of Blood and Blood Products, PubMed Central (2014).
4. D. M. Harmening: Modern Blood Banking & Transfusion Practices 6th edition (2012). https://www.pdfdrive.com/modern-blood-banking-transfusion-practices-sixth-edition e60355202.html
5. Mareş, I.: Proiectarea unei instalații pentru pregătirea pentru administrare către pacient a sângelui, produselor din sânge şi a soluțiilor perfuzabile, Diploma Project, coordinator: Prof. dr. ing. Luciana CRISTEA, Şef lucr. dr. ing. Barbu BRAUN (2023)
6. T. T. S. Company: Fluido® Compact (2022). https://www.tsc-group.com/patient-temperature management/products/fluido-compact/
7. T. 37 Company: Fluido® Compact Fluid and Blood warming, Manual de utilizare (2022). https://www.tsc-group.com/app/uploads/sites/3/2022/05/INT-R683-EN-3-10-17-Fluido Compact-User-Manual-English.pdf
8. U. Medical: HOTLINE Blood and Fluid Warmer - LEVEL1 Brochure.pdf (2021). https://www.uslmedical.co.nz/temperature-management/blood-and-fluid-warmers/hotline blood-and-fluid-warmer-level-1.html#!tab-files
9. I. Medical: HOTLINE® Blood and Fluid Warmer (2022). https://www.smiths medical.com/products/temperature-management/blood-and-fluid-warming-systems/routine flow-fluid-warming/hotline-blood-and-fluid-warmer
10. MEQU: Warm blood saves lives (2021). https://mequ.dk/
11. Medfarm: Încalzitor sange si solutii perfuzabile BH-600A (2022). https://medfarm.ro/aparatura-medicala/produs/detalii-produs-incalzitor-sange-si-solutii perfuzabile-bh-600a-2105.html
12. MedWrench: Barkey Plasmatherm User Manual.pdf. https://www.medwrench.com/documents/view/13013/barkey-plasmatherm-user-manual-pdf
13. Barkey: Barkey plasmatherm I plasmatherm V.https://barkey us.com/us/product/barkey-plasmatherm%2D%2Dplasmatherm-v/24
14. Hugger™: 3M™ Ranger™ Blood and Fluid Warming System (2022). https://www.bairhugger.com/3M/en_CA/bair-hugger-ca/experience-manager-content/3m ranger-blood-and-fluid-warming-system/
15. Repanovici, A., Cristea, L.: Bloc operator, Notițe de curs, Universitatea Transilvania din Braşov (2023)
16. Cottet, F., Ciobanu, O.: Bazele programării în LabVIEW, Editura MATRIX ROM Bucureşti, ISBN 973-9390-56-0 (2008).
17. Ministerul Sănătății Publice: Ghidul National de utilizare terapeutica rationala a sangelui si componentelor sanguine umane (2007). https://www.hosptm.ro/files/ghiduri/Ghidul%20National%20de%20utilizare%20terapeutica%20rationala%20a%20sangelui%20si%20a%20componentelor%20sanguine%20umane.pdf
18. FLIR Systems: User manual FLIR Ex series (2019). https://docs.rs-online.com/bd2d/A700000006725722.pdf

Measuring Angle with an Arduino Based Device for a Foot Drop Orthosis Adjustment

Ionel Şerban(✉), Corneliu Nicolae Drugă, Barbu Cristian Braun, Alexandru Tulică, and Ileana Constanta Roşca

Faculty of Product Design and Environment/Product Design, Mechatronics and Environment Department, Transilvania University, Brasov, Romania
ionel.serban@unitbv.ro

Abstract. This paper aims to describe a solution that might be used to measure, and assess, the optimal angle, in the ankle joint, of a foot drop orthosis. This is done using a low-cost device composed of an Arduino UNO board; MPU6050 accelerometer and gyroscope module; 16x2 LCD; servo motor SG90; a protractor and a portable battery. The solution described might be done as simple DIY work with accessible materials and some technical knowledge. The device described in this paper is not an innovation, it could be identified on other open-access sources in various configurations. The paper describes what this device is made of and how it is used to adjust the optimal angle in the ankle joint for those who use foot drop orthosis. The critical study indicates there exist many devices and methods that can be used to assess the range of motion in joints. The paper also presents a bibliographic study made to understand the relevance of some keywords/terms, specific to this study, in a recognized database.

Keywords: ankle angle · Arduino · foot drop

1 Introduction

Foot drop, also called drop foot, defines difficulty in lifting the front part of the foot, as it might drag on the ground when you walk [1].

Several different factors might cause foot drop, which relate to the weakness or paralysis of the foot muscles, like, for example: nerve injuries, brain or spinal cord disorders, muscle or nerve disorders.

Some of the symptoms of a foot drop are foot drags on the ground while walking, unable or difficulty lifting the foot, thigh rise higher than normal (to avoid dragging, high stepping gait), pain or numbness in the foot when moving or lifting [2].

Searching Science Direct (in July 2024) using terms indicated in Table 1, it found a great interest in this area of research regarding "Foot drop measure angle" and "Foot drop adjust angle" terms. Although there are few results regarding terms: "Arduino foot drop" it is considered a good opportunity to use an Arduino-based device, low cost, that might be used in the angle evaluation for foot drop when adjusting the orthosis/brace.

S. Vlad and N. M. Roman (Eds.): MEDITECH 2024, IFMBE Proceedings, 130, pp. 78–83, 2025.
https://doi.org/10.1007/978-3-031-95671-3_9

Table 1. Critical study, in direct science database, regarding the number of studies, nowadays, considering some specific terms

Terms used	No. of results	No. of review articles	No. of research articles
Measuring angle with an Arduino-based device for a foot drop orthosis adjustment	4	–	2
Foot drop orthosis adjustment	845	114	360
Foot drop measure angle	18,513	1457	9587
Arduino foot drop	241	30	155
Foot drop adjust angle	12,632	1002	5819

After filtering the results, for the terms "Foot drop measure angle" considering the research done in the last 10 years, it was found several (7896 articles), which represent approx. 43% of the total. There are many studies on other databases that treat this subject, but it was more suitable for the authors to use Science Direct [3].

Regarding the first research paper recommended in Science Direct, when searching the terms "Foot drop orthosis adjustment", it states that an ankle-foot orthosis (AFO) with plantar flexion resistance improves the first rocker function during gait and concludes that medium plantar flexion resistance grade can enhance balance parameters like the average velocity of the center of pressure (COP), path length, cadence, improve maximal peak ankle dorsiflexion, and average velocity during gait [4].

Nowadays, many solutions and methods might be used to measure and predict the angle in a joint. From these, we specify goniometer [5, 6]; accelerometer or gyroscope with Arduino or other microcontroller board [7]; video kinematics with specific software (Visual 3D, Kinovea, OpenSim, Tracker, etc.) [8]; wearable conductive fiber sensors [9]; using machine learning and applied biomechanical inverse dynamics [10]; smartphone applications [11].

Considering the above, this paper aims to describe a solution, based on an accelerometer and an Arduino microcontroller board, to measure the angle, and possible range of motion (ROM), of the ankle joint to adjust the angle in the ankle joint for a foot drop orthosis.

2 Realization of the Proposed Device Used to Measure the Angle in the Ankle Joint

The purpose of the orthosis, for foot drop, is to prevent the development of this pathology (the foot hitting the ground and dragging the toes on the ground while walking) by supporting the foot and keeping it at an optimal angle. An orthosis will be tested using a device based on an accelerometer and gyroscope module that will show the angle of the foot during gait to obtain the optimal angle.

The system (see Fig. 1) has the following components: Arduino UNO board; MPU6050 accelerometer and gyroscope module; 16x2 LCD; 10 K potentiometer; servo motor SG90; a protractor; portable battery. Fritzing is a good tool for obtaining wiring diagrams.

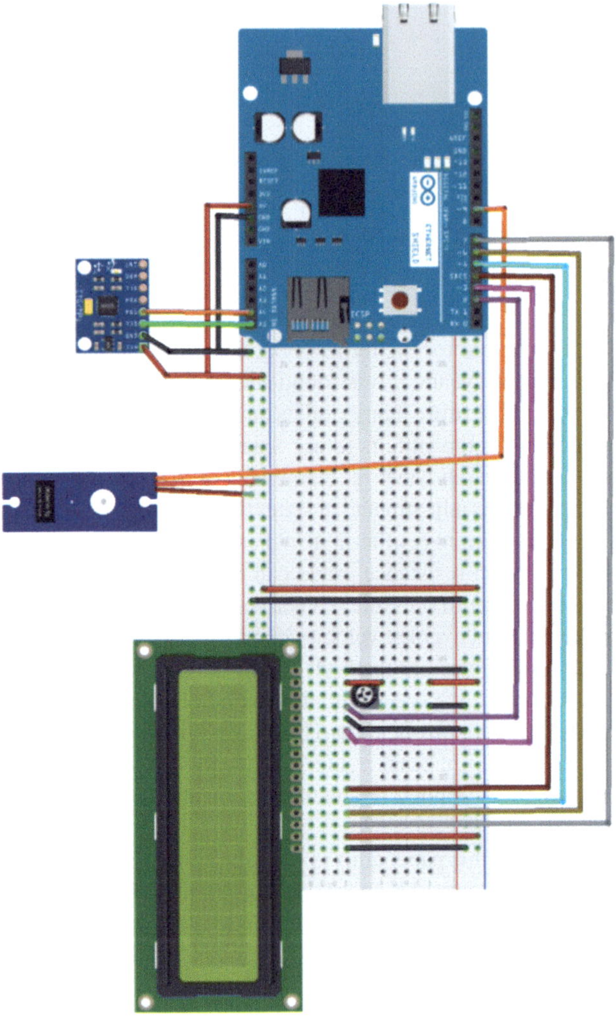

Fig. 1. Wiring diagram of the measuring device.

Arduino, according to the producer, is an open-source electronic platform based on easy-to-use hardware and software. This platform is often used by people new to electronics because, unlike most programmable circuit boards, Arduino uses a USB cable to upload code to the board, so it does not require separate hardware components for this. Both the hardware and software side are easy to learn even for beginners, which has resulted in a large community of users contributing to a variety of Arduino-based projects. The hardware part consists of a microcontroller (connections with the elements, of our proposed system, are indicated in Table 2), and the software part is used to write and load the codes on the physical board. Arduino offers flexibility, it can interact with LEDs, buttons, motors, and speakers, but also with the TV or smartphones. The Arduino software can be found on the official website and is free, and the hardware boards can be purchased at a low price.

Table 2. Connections between Arduino Uno pins and the other elements (MPU6050, SG90, LCD) pins

MPU6050 module pins	SG90 servo motor pins	LCD display pins	Arduino UNO pins
VCC	VCC	VDD, A	+5 V
GND	GND	VSS, RW, K	GND
SCL			A5
SDA			A4
	PWM		9
		V0	The center pin of the potentiometer
		RS	2
		E	3
		D4	4
		D5	5
		D6	6
		D7	7

Currently, this device is provided with a wheeled stand and a video camera (see Fig. 2 a and b). The measured angle value is displayed both on the LCD (digital) and on the protractor (analog) which is connected to the SG90 servo motor. The device can be used to adjust the optimal position of the orthosis, static, as well as to evaluate angles during gait dynamic. The appropriate angle, optimal when prescribed by clinicians, for an AFO is between 5° and 10° [12]. This is the angle that was assessed with the above-described device. The angle is very important to be measured, and assessed, as it impacts the locomotion/gait cycle. Setting up an optimal angle will optimize the gait cycle. The angle, in this case of orthosis, can be adjusted by tightening, releasing the quick-release band and velcro (see Fig. 2b).

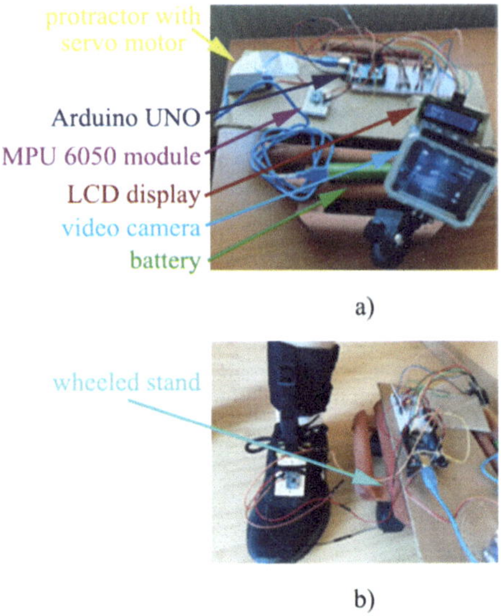

Fig. 2. The final device with breadboard: (a) on the wheeled stand; (b) connected to the shoe

3 Conclusions

According to the critical study, there is much interest in this area. Nowadays solutions change following advances in technics. This paper describes a device that can be used to measure/adjust the optimal angle for a foot drop orthosis. This solution can be improved but it is low cost and might be done as DIY work with accessible materials and some technical knowledge. There might be considered a further study regarding static and dynamic (during locomotion/gait) angles that appear in the ankle joint in various situations with and without different types of orthoses. A biomechanical analysis might be done to complete the study. By adding a wireless or Bluetooth module, the process of measuring angles during locomotion/gait can be improved.

Acknowledgments. We would like to thank a former student, Eng. Mureşan Andreea Alexandra, for its involvement in the development of this paper.

Conflict of Interest. The authors declare they have no conflict of interest.

References

1. https://www.mayoclinic.org/diseases-conditions/foot-drop/symptoms-causes/syc-20372628, last accessed 7 Aug 2024
2. Shahar, F.S., Sultan, M.T.H., Lee, S.H., Jawaid, M., Shah, A.U.M., Safri, S.N.A., Sivasankaran, P.N.: A review on the orthotics and prosthetics and the potential of kenaf composites as alternative materials for ankle-foot orthosis. J. Mech. Behav. Biomed. Mater. **99**, 169–185 (2019)

3. https://www.sciencedirect.com/, last accessed 31 July 2024
4. Pradhan, D., Mohanty, R.K.: Impact of plantar flexion resistive moment of dynamic ankle foot orthosis on measures of center of pressure and clinical gait outcomes in individuals with post-stroke hemiparesis. Gait Posture. **113**, 58–66 (2024)
5. Product sheet: https://www.biopac.com/wp-content/uploads/Goniometers-Torsiometers.pdf, last accessed 7 Aug 2024
6. Gandbhir, V.N., Cunha, B.: Goniometer. https://www.ncbi.nlm.nih.gov/books/NBK558985/, last accessed 7 Aug 2024
7. El-Gohary, M., Pearson, S., McNames, J.: Joint angle tracking with inertial sensors. In: 30th Annual International Conference of the IEEE Engineering in Medicine and Biology Society, pp. 1068–1071, Vancouver (2008)
8. Chida, K., Inami, T., Yamaguchi, S., Nishioka, T., Yoshida, Y., Kohtake, N.: Assessing the validity of two-dimensional video analysis for measuring lower limb joint angles during fencing lunge. Front. Sports Active Living. **6**, 1–9 (2024)
9. Gibbs, P.T., Asada, H.H.: Wearable conductive fiber sensors for multi-Axis human joint angle measurements. J. Neuroeng. Rehabil. **2**, 1–18 (2005)
10. Choffin, Z., Jeong, N., Callihan, M., Sazonov, E., Jeong, S.: Lower body joint angle prediction using machine learning and applied biomechanical inverse dynamics. Sensors (Basel). **23(1):228**, 1–15 (2022)
11. Wang, K.Y., Hussaini, S.H., Teasdall, R.D., Gwam, C.U., Scott, A.T.: Smartphone applications for assessing ankle range of motion in clinical practice. Foot Ankle Orthop. **4**(3), 1–9 (2019)
12. Chang, S., Lee, Y., Kim, K.: Effects of the ankle angle of an ankle foot orthosis on foot pressure during the gait in healthy adults. J. Phys. Ther. Sci. **27**(4), 1033–1035 (2015)

A Photoplethysmography—Based Affordable Demonstrator for Monitoring Muscle Oxygenation Using Infrared Light

Ana Cristina Feneşan(✉), Alexandru Ianoşi-Andreeva-Dimitrova, and Silviu-Dan Mandru

Department of Mechatronics and Machine Dynamics, Technical University of Cluj-Napoca, Cluj-Napoca, Romania
Ana.Fenesan@campus.utcluj.ro, {Alexandru.Ianosi, Dan.Mandru}@mdm.utcluj.ro

Abstract. This paper examines the application of near-infrared spectroscopy (NIRS) in monitoring skeletal muscle function and metabolism. NIRS is a non-invasive, portable technology that measures muscle oxygenation and hemodynamics by detecting the absorption of near-infrared light. This research evaluates the efficacy of NIRS in determining muscle oxygenation levels, metabolic activity, and blood perfusion during various physical activities. Despite its advantages, such as ease of use and its non-invasive nature, NIRS presents certain limitations, including limited light penetration and individual variability in hemodynamic responses.

Keywords: muscle oxygenation · photoplethysmography · hemodynamic response

1 Introduction

Photoplethysmography is a technique for measuring the absorption or reflection of light by blood vessels in living tissue. Photoplethysmography technology is increasingly accessible, affordable, and easy to integrate into portable devices. Recent advancements have led to the incorporation of pulse oximeter signal collection into smartphones and wearable devices [1].

Photoplethysmography is an optical technique that uses a light source and a photodetector to function. The light emitted by the source illuminates the tissue, and the photodetector captures small variations in the intensity of the reflected or transmitted light [2]. These variations are associated with changes in the blood perfusion volume in the illuminated tissue [3]. The fundamental principle of photoplethysmography is based on the sensitivity of different optical wavelengths to blood and other tissue components. Thus, the technique allows for non-invasive monitoring of blood flow and pulse, being frequently used in portable devices [4].

In this paper, we outline the methodology employed in the creation of a demonstrator designed to monitor muscle contractions utilizing near-infrared light within the near-infrared spectrum. At this moment, the focus was on the biceps brachii muscle, as

S. Vlad and N. M. Roman (Eds.): MEDITECH 2024, IFMBE Proceedings, 130, pp. 84–89, 2025.
https://doi.org/10.1007/978-3-031-95671-3_10

this presents interest for staging future development of a NIRS-based human-machine interface for people with upper-limb disabilities.

2 State of the Art

2.1 Muscular Gas Exchange Mechanism

The gas exchange at the blood level is a vital process, primarily occurring in the pulmonary alveoli and at the tissue level. In the lungs, inhaled air reaches the alveoli, where oxygen (O_2) from the air crosses the thin alveolar walls and enters the pulmonary capillaries through diffusion, a process based on the differences in gas concentration on either side of the alveolo-capillary membrane. Hemoglobin in red blood cells binds to oxygen, forming oxyhemoglobin (HbO_2), which gives the blood a bright red color [5].

As oxygenated blood circulates through the body, it reaches the capillaries in the tissues, where the partial pressure of oxygen is lower than in the blood. Oxygen dissociates from hemoglobin and diffuses into the tissue cells, where it is used for energy production through cellular metabolism. Carbon dioxide (CO_2), produced as a result of this metabolism, diffuses from the cells into the blood capillaries. In the blood, CO_2 is transported in three ways: dissolved in plasma, bound to hemoglobin (forming carbaminohemoglobin), and as bicarbonate ion ($HCO_3{}^-$). When CO_2 binds to hemoglobin, this form is called carboxyhemoglobin and has a darker, burgundy color. Hemoglobin without oxygen, known as deoxyhemoglobin, has a dark red color [6].

The blood loaded with carbon dioxide returns to the lungs, where CO_2 is released from hemoglobin and diffuses into the alveoli to be exhaled. Simultaneously, a new cycle of oxygenation occurs, ensuring a constant supply of oxygen to the body and the efficient removal of carbon dioxide [5].

Photoplethysmography (PPG) is a method used to measure blood parameters, based on detecting volumetric variations of blood within tissues. This technique involves using a light source, typically LEDs emitting infrared or red light, and a photodetector that records changes in the intensity of light reflected or transmitted through tissues. By analyzing these variations, important parameters such as heart rate and blood oxygen saturation can be determined. The principle relies on the differences in light absorption by oxyhemoglobin and deoxyhemoglobin at specific wavelengths, allowing blood oxygen saturation assessment. Photoplethysmography can be integrated into wearable devices, providing a continuous and non-invasive means of monitoring these vital parameters [7].

2.2 Related Research

Photoplethysmography is used to measure blood volume and oxygenation levels in the muscles of patients suffering from cardiac conditions [8].

However, this method is also employed in monitoring physiological parameters during physical exercise, not just for clinical purposes. Photoplethysmography provides valuable insights into real-time blood volume and oxygenation levels in muscles, facilitating a comprehensive understanding of cardiovascular and muscular function during various physical activities. By utilizing this technology in both clinical and non-clinical settings, researchers and healthcare professionals can gather critical data that enhances the assessment of an individual's health status and physical performance [9, 10].

There are PPG-based, cuff-less blood pressure monitoring devices that can be widely adopted in clinical practice and everyday health monitoring. The review made by El-Hajj and colab. Also addresses the key challenges faced in this field, such as the variability in PPG signal quality due to motion artifacts, differences in skin tone, and physiological condition [11].

Barstow (2019) provides a comprehensive overview of near infrared spectroscopy (NIRS) technology and its applications in skeletal muscle research. Barstow explains the operational principles of NIRS, detailing how it measures muscle oxygenation and hemodynamics through the absorption of near-infrared light, and highlights the advantages of this non-invasive, portable method. The cited author discusses the applications of NIRS in assessing muscle oxygenation, energy metabolism, and blood perfusion, emphasizing its utility in monitoring athletic performance and rehabilitation; and also the limitations of the technology, such as limited light penetration and individual variability, and suggests directions for future research to enhance the accuracy and applicability of NIRS [12].

3 Materials and Methods

The demonstrator consists of a Velcro strip V (refer to Fig. 1) designed to surround the arm A and position the emitter module E and receiver module R at a suitable distance. The emitter module has three infrared LEDs L, each with a radiant intensity of 21 mW/sr, with a peak wavelength at 940 nm. These LEDs continuously illuminate the muscular tissue M, which selectively absorbs the light based on the hemodynamic equilibrium inside the tissue. The receiver R senses these changes with a LL503 photodiode PD, which also has its peak sensitivity at 940 nm; the dark current is maximum 10 nA, and the rise/fall time is 45 ns, values that allows for a precise measurement.

The emitter-receiver pair is placed strategically on opposite side of the biceps brachii. The method facilitated precise observation of the changes in blood flow in the muscle sanguine capillaries and oxygenation levels associated with muscle activity, providing insights into the physiological processes occurring during muscle contraction. At this stage, the signal was sampled with NI cDAQ 9172, using the NI 9219 analog-to-digital converter; the acquired signal was sampled with a frequency of 30 Hz and a precision of 100 nV. The amplifier gains was set to 1, as no amplification was required. For this purpose, the NI MAX software was used, with which all the acquisition parameters were set up; the exported data file was, at this stage, processed in a custom Microsoft Excel file. The processing consisted of smoothing the signal with a moving-average window with an empirically determined width of 5 samples. In order to highlight the muscle contractions a simple function was implemented which compares the current gradient of the curve with the previous one.

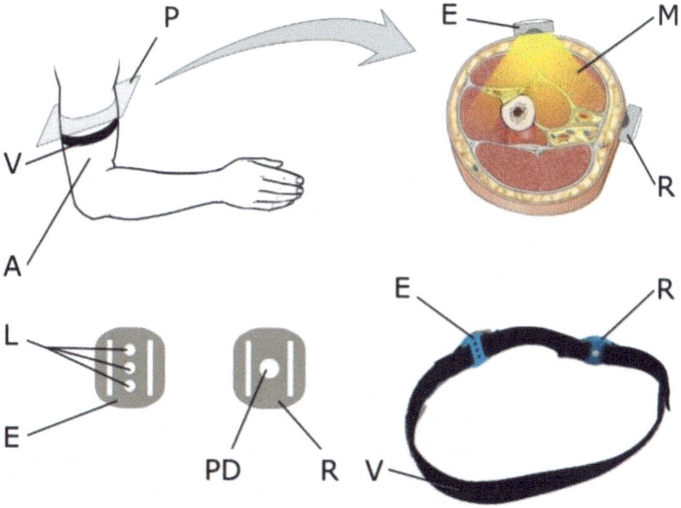

Fig. 1. The demonstrator and its placement on the arm.

4 Results

Upon completing the assembly phase of the demonstrator, we proceeded to the testing phase. This phase was conducted in several stages. In the first stage, we recorded hemodynamic activity during muscle contraction of the arm without any additional weight. The subject was instructed to perform isometric muscular contractions consistently and with similar intensity. In the second stage, the subject was instructed to hold a 5 kg weight, maintaining the same mode of contraction.

In detail, the first stage involved monitoring the hemodynamic responses of the biceps brachii during voluntary muscle contractions performed in a controlled manner, ensuring uniformity in contraction intensity and duration across all trials. This baseline data allowed us to assess the intrinsic hemodynamic variations associated with unweighted muscle activity. In the subsequent stage, by introducing a 5 kg external load held in hand allowed us to evaluate the impact of increased resistance on hemodynamic parameters. The contractions were performed with the same consistency and intensity as in the first stage, enabling a comparative analysis between the unweighted and weighted conditions. As seen in Fig. 2, in the first stage (upper part), in absence of an external load, the oxygen consumption is much smoother, as indicated by the less abrupt slope of the recorded signal. Comparatively, an external load leads to a rapid change in the oxyhemoglobin levels. Another observation is that under external load, the signal is more robust than under no-load conditions, which poses additional challenges if one wants to use it as an alternative human-machine interface.

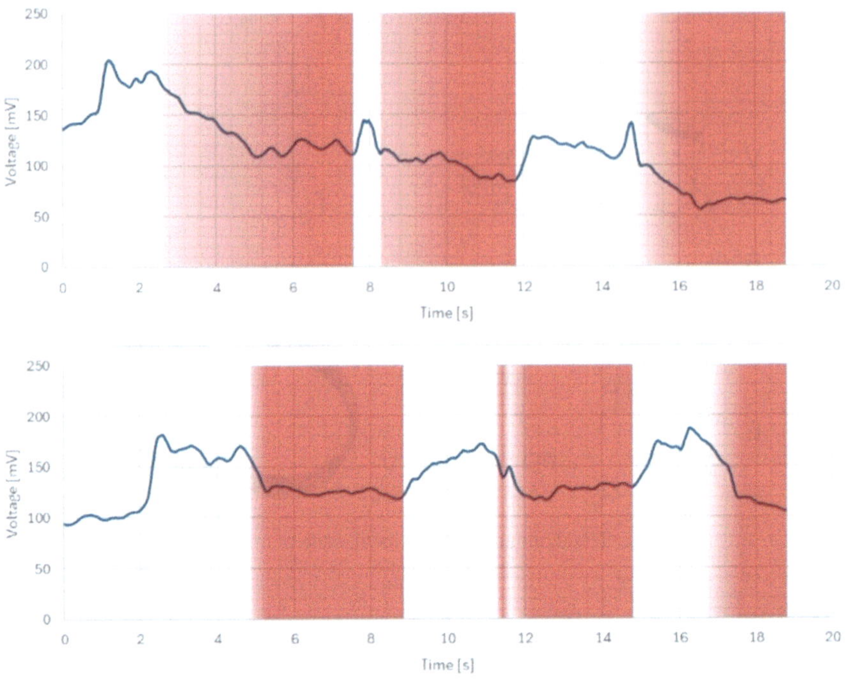

Fig. 2. The first and second stages of the results. At the top are the results without the 5 kg weight, while at the bottom are the results after the addition of the 5 kg weight.

5 Conclusions

Our results, obtained through the analysis of muscle contraction using infrared light, demonstrate the efficacy of the photoplethysmography technique that underscores the efficacy of using infrared light in penetrating muscle tissue, thereby offering a non-invasive means to study muscle function and hemodynamics in real-time as well as demonstrating a good avenue for future research of a NIRS-based human-machine interface.

A surprising find was the robustness of the acquired signal, as well as the undemanding way of obtaining a strong signal without additional amplification, granted, though, we used a quite capable acquisition device. Nevertheless, the raw amplitude of the signal points to the fact that future research for a whole self-contained NIRS-based human-machine interface is well within reach.

The device's affordability and ease of use make it a valuable tool for both clinical and research applications, offering potential advancements in the monitoring of muscular health and performance. Further studies could enhance the accuracy and expand the applications of this technology, contributing to the fields of sports science and rehabilitation medicine.

Acknowledgements. This study was financially supported by IOSUD-UTCN.

Conflict of Interest. The authors declare no conflict of interest.

References

1. Elgendi, M., Fletcher, R., Liang, Y., Howard, N., Lovell, N.H., Abbott, D., et al.: The use of photoplethysmography for assessing hypertension. NPJ Dig. Med. **2**(1), 60 (2019)
2. Sun, Y., Thakor, N.: Photoplethysmography revisited: from contact to noncontact, from point to imaging. IEEE Trans. Biomed. Eng. **63**(3), 463–477 (2015)
3. Fine, J., Branan, K.L., Rodriguez, A.J., Boonya-Ananta, T., Ajmal, Ramella-Roman, J.C., et al.: Sources of inaccuracy in photoplethysmography for continuous cardiovascular monitoring. Biosensors. **11**(4), 126 (2021)
4. Nitzan, M., Nitzan, I., Arieli, Y.: The various oximetric techniques used for the evaluation of blood oxygenation. Sensors. **20**(17), 4844 (2020)
5. Petersson, J., Glenny, R.W.: Gas exchange and ventilation–perfusion relationships in the lung. Eur. Respir. J. **44**(4), 1023–1041 (2014)
6. Kaile, K., Fernandez, C., Godavarty, A.: Development of a smartphone-based optical device to measure hemoglobin concentration changes for remote monitoring of wounds. Biosensors. **11**(6), 165 (2021)
7. Mukkamala, R., Hahn, J.O., Chandrasekhar, A.: Photoplethysmography in noninvasive blood pressure monitoring. In: Photoplethysmography, pp. 359–400. Academic Press (2022)
8. Hamaoka, T.: Muscle oxygenation monitoring using near-infrared spectroscopy. J. Phys. Fit. Sports Med. **2**(2), 203–207 (2013)
9. Ferrari, M., Muthalib, M., Quaresima, V.: The use of near-infrared spectroscopy in understanding skeletal muscle physiology: recent developments. Philos. Trans. R. Soc. A Math. Phys. Eng. Sci. **369**(1955), 4577–4590 (2011)
10. Rissanen, A.P.E., Tikkanen, H.O., Koponen, A.S., Aho, J.M., Hägglund, H., Lindholm, H., Peltonen, J.E.: Alveolar gas exchange and tissue oxygenation during incremental treadmill exercise, and their associations with blood O2 carrying capacity. Front. Physiol. **3**, 26080 (2012)
11. El-Hajj, C., Kyriacou, P.A.: A review of machine learning techniques in photoplethysmography for the non-invasive cuff-less measurement of blood pressure. Biomed. Sig. Process. Control. **58**, 101870 (2020)
12. Barstow, T.J.: Understanding near infrared spectroscopy and its application to skeletal muscle research. J. Appl. Physiol. **126**(5), 1360–1376 (2019)

Virtual Reality Soccer Game for Lower Limb Rehabilitation

Silviu George Ciorap[1], Monica Blănaru[2], George Iftimie[3], Eugen Crețu[2], and Radu Ciorap[2(✉)]

[1] Faculty of Automatic Control and Computer Engineering, Gheorghe Asachi Technical University of Iasi, Iasi, Romania
[2] Faculty of Medical Bioengineering, Grigore T. Popa University of Medicine and Pharmacy, Iasi, Romania
radu.ciorap@umfiasi.ro
[3] Faculty of Physical Education and Sports, Alexandru Ioan Cuza University of Iasi, Iasi, Romania

Abstract. Optimizing procedures for neuro-muscular rehabilitation involves a multidimensional approach that combines theoretical and clinical aspects from a broad spectrum of fields, favoring the development of innovative therapeutic strategies. A modern and effective way to combine entertainment with therapeutic exercises is Virtual Reality (VR) applications, which play an essential role in facilitating the locomotor rehabilitation process in an engaging and motivating way for patients. This paper proposes a combined solution of virtual reality and a serious game that can be useful in neuro-muscular rehabilitation therapeutic procedures at a patient's home. It was designed as a soccer game where the patient has to hit a ball and strike a target, and exercise helps the patient recover the mobility of the lower limb. The software application was developed using the Unity 3D engine and subsequently deployed on a smartphone integrated into a virtual reality headset. Lower limb kinematics are captured by an MPU9250 inertial measurement unit (IMU) interfaced with an ESP-WROOM-32 development board. The ESP32 module, operating in WiFi mode, establishes a wireless communication channel with the smartphone application, enabling real-time data exchange and immersive VR interactions.

Keywords: Virtual Reality · Wearable Sensor · Home-based Rehabilitation

1 Introduction

Rehabilitation often involves manual exercises or therapist-assisted techniques, which are labor-intensive for both the patient and the therapist. Sessions can be time-consuming, leading to patient fatigue and limited therapist capacity to attend to multiple patients [1]. Standardized exercise programs may not fully adapt to an individual's unique needs, such as specific biomechanical or neurophysiological conditions.

S. Vlad and N. M. Roman (Eds.): MEDITECH 2024, IFMBE Proceedings, 130, pp. 90–96, 2025.
https://doi.org/10.1007/978-3-031-95671-3_11

Virtual Reality (VR) in rehabilitation presents numerous advantages that contribute to improved therapeutic outcomes and enhanced patient experiences. VR facilitates targeted and personalized rehabilitation by enabling customized exercises to address individual patient needs and specific impairments (such as lower-limb functionality and gait training). VR environments' interactive and gamified nature fosters increased patient engagement and motivation, promoting active therapy participation. Additionally, providing real-time feedback during VR tasks cultivates a sense of achievement, encouraging sustained commitment to rehabilitation [2]. Furthermore, VR offers a safe and controlled environment where patients can practice complex tasks without the risk of injury, ensuring greater exercise consistency and standardization than traditional therapist-led sessions [3, 4]. Many studies indicate that VR improves upper and lower limb function, balance, gait speed, and daily function outcomes [5, 6]. In [7], 26 systematic reviews were carried out that included 758 RCTs and over 22,000 participants. This review indicated that VR rehabilitation programs lasted for 5–8 weeks, with sessions lasting at least 45 minutes and for more than 15 hours. Rajashekar et al. describe [8] the contribution of robotics and virtual reality to motor recovery, gait, balance, and engagement in stroke patients. VR enhances rehabilitation due to increased intensity of therapy, engagement, and neuroplasticity; immersive VR seems to have a greater benefit compared to non-immersive VR. Some of them are also using Nintendo Wii for the development of low-cost virtual reality platforms, providing personalized exercises and games that can be calibrated based on the needs of the individual patient [9].

When combined with gaming technologies, VR demonstrated high levels of patient engagement and a feeling of connectedness to the VR system, with participants describing heightened motivation and increased movement precision [10, 11].

The presented rehabilitation system will improve ambulatory capacity in stroke patients, aiming at a new virtual rehabilitation system. Using passive sensors on lower limbs and specific software, the user can interact with a virtual avatar inside a virtual environment. Active training protocols enable individualized gait and balance improvement, enhancing neuroplasticity and motor function recovery on an individual level through specially designed virtual activities. Thereby, the patient also maintains more motivation throughout rehabilitation due to the fact that the treatment process is immersing and fully interactive.

2 Material and Methods

The term "serious games" was introduced to take advantage of video games beyond mere entertainment and to use them for educational, engaging, and therapeutic purposes. In the case of rehabilitation of the lower limbs, such games are developed in order to enhance physical activity by guiding patients to carry out specific exercises in a virtual but controlled environment [12]. The entire VR rehabilitation system must accurately measure the patient's movements while ensuring intuitive and natural interaction with the virtual environment.

This rehabilitation application aims to improve mobility and coordination in patients with lower-limb disabilities. A virtual player kicks a ball toward a goal in the created soccer game in order to score points. A sensor records the patient's leg motions (extension, flexion, abduction, and adduction), which are then wirelessly sent to the computer that is playing the game. This allows the patient to engage with the virtual world in real-time.

The VR rehabilitation system components include VR glasses, a complex motion sensor that combines a 3-axis gyroscope, 3-axis accelerometer, 3-axis magnetometer, and a Digital Motion Processor, ESP32 microcontroller, that ensures an optimal and accurate user experience (see Fig. 1).

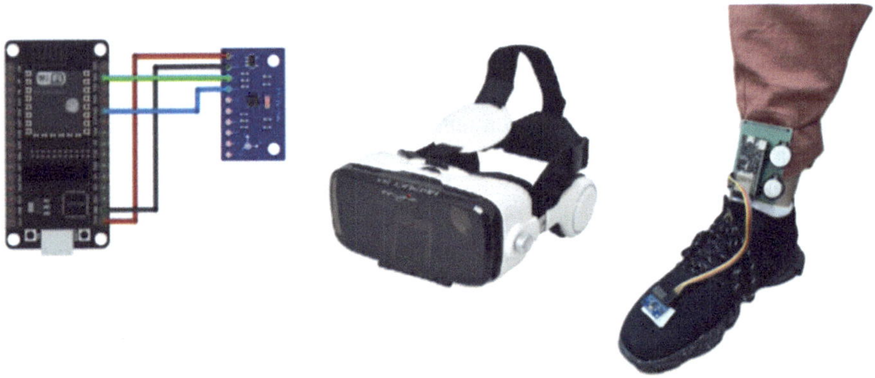

Fig. 1. Hardware components of the VR rehabilitation system

The ESP-WROOM-32, a strong and adaptable microcontroller with a dual-core processor, built-in WiFi and Bluetooth, and numerous peripheral interfaces, is at the heart of this configuration. Serving as the central hub, it retrieves sensor data from the MPU9250 and passes it on for real-time use within the software application.

The MPU9250 sensor, connected via an I2C interface, supplies the measurements for leg movements. In this system, the sensor initialization and the real-time acquisition and processing of acceleration, rotation, and magnetic field data are handled by code developed using the Arduino IDE. Once processed, the data are transmitted over WiFi to the software application using the WebSocket protocol, ensuring a reliable and bidirectional communication channel. This configuration allows patient movement data to be immediately forwarded to the application and, in turn, enables rapid feedback from the virtual environment to be sent back to the hardware. As a result, patient movements are seamlessly synchronized with virtual objects, providing a smooth and efficient rehabilitation experience.

The Myria MY9802 VR headset is employed in this system to transform a smartphone screen into an immersive virtual reality portal, enhancing rehabilitation therapy's effectiveness. With a wide field of view of approximately 100 degrees, the headset delivers depth and immersion essential for creating a realistic virtual environment. The headset has a soft protective sponge that conforms to the user's facial contours, minimizing pressure on the nose and forehead. This design enables extended sessions without causing discomfort, making it well-suited for therapeutic applications.

The software framework for this project integrates multiple components to enable the collection, processing, and visualization of motion data within an interactive and immersive virtual environment. Unity 3D was used to develop the virtual environment and implement the game logic, while Visual Studio was employed for writing and integrating C# scripts that support the application's functionality.

The Arduino IDE programs the sensor and ESP32 board, ensuring enough data acquisition and transmission. Data exchange between sensors and VR applications will be done through the WebSocket protocol, a fast, reliable means of communication needed in real-time interaction.

Within this framework, a Unity 3D-powered rehabilitation game for individuals with lower limb motor difficulties is controlled by Visual Studio. The game will offer a fun therapeutic exercise by using a soccer-based scenario with a virtual player, ball, goal, and other user interface features. To improve immersion and provide a calming ambiance, this virtual setting will feature buildings, greenery, and a mountain landscape. Because of established 3D textures and alterations in object placement, this creates a realistic, appealing atmosphere that will help patients feel less stressed and provide a pleasant, immersive experience. A cross-platform programming tool called Unity 3D is widely used to make interactive 3D apps and video games. Because of its strong features that guarantee the delivery of complex user-environment interactions and realistic visual experiences, it was selected for this project. Its straightforward yet intuitive graphical user interface enhances the effectiveness of the creation and update procedures a developer goes through when creating 3D item and environmental models. However, the inclusion of the C# programming language demonstrates the engine's versatility and allows for complex implementations of the game's logic and sensor integrations, as well as real-time object manipulation capabilities if desired.

3 Results

Data from the motion sensor on the patient's leg is used to control the virtual player's motions. More specifically, flexion and extension regulate the rotations along the X and Y axes, while adduction and abduction movements of the leg dictate the motion of the 3D avatar along the X-axis. In this regard, the code underlying the processing of sensor data into corresponding in-game actions will make sure that accurate synchronization between the physical movements of the patient and the responses of the avatar is maintained.

At game initialization, the script verifies the proper assignment of all necessary references to prevent execution errors that might disrupt the user experience. In cases where references are missing, error messages are generated to enable prompt troubleshooting. The script's primary function updates the score by adding the appropriate points whenever the patient successfully scores a goal. This action immediately reflects on the graphical interface, providing the patient with continuous, real-time feedback essential for an engaging and effective rehabilitation process. The scoreboard is designed to integrate seamlessly into the game environment, employing clear and high-contrast fonts to ensure quick and accurate score readability, regardless of the visual conditions within the virtual setting. This visual clarity is critical for maintaining the patient's focus and engagement during rehabilitation exercises. From a technical perspective, the scoreboard's functionality is managed by a dedicated script that ensures the score's accurate real-time display (see Fig. 2).

Fig. 2. Screenshot of the videogame-developed rehabilitation

The code is structured to enable avatar movement and rotation only after the game has commenced, ensuring that uncontrolled avatar actions are prevented before the start of the rehabilitation session. This design choice enhances the safety and control of the user experience. Each leg movement (flexion, extension, adduction, and abduction) is accurately mirrored by the avatar's corresponding motions and rotations within the virtual environment. The patient's physical movements and the avatar's responses are precisely correlated thanks to this methodical, well-structured methodology, which also provides quick and clear visual feedback—a essential component of successful rehabilitation.

Using the goal as a target in the rehabilitation game has a precise therapeutic purpose. Kicking at the goal involves movements that activate and strengthen the muscle groups essential for the recovery of lower limb mobility, thus helping to improve motor functions. The goal is strategically placed at the end of the virtual playing field, oriented to facilitate essential therapeutic movements such as adduction, abduction, flexion, and extension of the lower limbs. This positioning helps to achieve the exercises needed to recover mobility while providing a clear and motivating goal. These exercises are essential for patients who need to regain strength and coordination in the lower limbs.

A timer has been implemented to evaluate and monitor the patient's progress, set to 90 seconds, during which the points obtained during each training session are recorded. A script that controls the running state and the time display is used to implement the timer. The timer is reset and started as the session begins, counting the amount of time that has passed. Patients are always aware of how much time is left because the time is updated in real-time and shown on the screen. This continuous update is essential to keep patients informed and focused on the activity. After the 90 seconds have elapsed, patient information and current and previous training session points are displayed (see Fig. 3).

Validated data then stores the name, age, and gender of a patient and the foot on which the sensor is on for future use. Then, the game starts, and the input panel is hidden, freeing the visual space for the rehabilitation activity. The timer begins the count down of the 90 seconds and accumulated score is displayed on screen in real time. This starts the rehabilitation session, where everything in the game is adjusted depending on the patient's needs and limitations. The game has a STOP button that can momentarily halt the game in situations where the patient feels painful or any medical emergency during the game. When this occurs, the timer stops operating and all the activities developed for the game.

Fig. 3. Print screen from game initialization where the timer and STOP button are highlighted

4 Conclusions

The conclusions of the current paper show that linking the conventional and advanced methods into an integrated, multidisciplinary approach would offer better performance in both the effectiveness and personalization of treatment for patients who suffer from motor disabilities. This work highlights the role played by the integration of technology in enhancing motor rehabilitation, showing great prospects to seriously change the faces of the recovery process in an application-based approach. These findings also opened up prospects for enhancing patients' quality of life due to faster, nicer, and more efficient rehabilitation paths. Beyond physical rehabilitation, it is an emotional approach through which patients can surmount the challenges that every rehabilitation process implies and strengthen all the conditions necessary for perfect social and professional reinsertion. The study underlines the need for a combined, technology-supported rehabilitation paradigm that will revolutionize treatment outcomes.

Conflict of Interest. The authors declare that they have no conflict of interest.

References

1. Zhang, X.Q., Rong, X.Y., Luo, H.W.: Optimizing lower limb rehabilitation: the intersection of machine learning and rehabilitative robotics. Front. Rehabil. Sci. **5** (2024)
2. Janatová, M., Petioky, J., Hoidekrová, K., Vesely, T., Hána, K., Smrcka, P., Stepánek, L., Lippert-Grünerová, M., Jerábek, J.: System for game-like therapy in balance issues using audiovisual feedback and force platform. Electronics. **11**(8), 1179 (2022)
3. Lu, H.Y., Wang, X.Y., Hu, C.P., Lau, C.C.Y., Tong, R.K.Y.: Home-based guidance training system with interactive visual feedback using kinect on stroke survivors with moderate to severe motor impairment. J. Neuroeng. Rehabil. **21**(1), 189 (2024)
4. Nagashima, Y., Ito, D., Ogura, R., Tominaga, T., Ono, Y.: Development of Virtual Reality-based Gait Training System Simulating Personal Home Environment. In: 2021 43rd Annual International Conference of the Ieee Engineering in Medicine & Biology Society (Embc), pp. 5764–5767 (2021)

5. Ciorap, R., Andriţoi, D., Casuţă, A., Ciorap, M., Munteanu, D.: Game-based virtual reality solution for post-stroke balance rehabilitation. IOP Conf. Ser. Mater. Sci. Eng. **1254**(1), 012037 (2022)
6. Fusco, A., Giovannini, S., Castelli, L., Coraci, D., Gatto, D.M., Reale, G., Pastorino, R., Padua, L.: Virtual reality and lower limb rehabilitation: effects on motor and cognitive outcome - a crossover pilot study. J. Clin. Med. **11**(9) (2022)
7. Zhang, B.H., Wong, K.P., Qin, J.: Effects of virtual reality on the limb motor function, balance, gait, and daily function of patients with stroke: systematic review. Medicina-Lithuania. **59**(4) (2023)
8. Rajashekar, D., Boyer, A., Larkin-Kaiser, K.A., Dukelow, S.P.: Technological advances in stroke rehabilitation robotics and virtual reality. Phys. Med. Rehabil. Clin. N. Am. **35**(2), 383–398 (2024)
9. Tsekleves, E., Paraskevopoulos, I.T., Warland, A., Kilbride, C.: Development and preliminary evaluation of a novel low cost VR-based upper limb stroke rehabilitation platform using Wii technology. Disabil. Rehabil. Assist. Technol. **11**(5), 413–422 (2016)
10. Iosa, M., Verrelli, C.M., Gentile, A.E., Ruggieri, M., Polizzi, A.: Gaming technology for pediatric neurorehabilitation: a systematic review. Front. Pediatr. **10** (2022)
11. Tosto-Mancuso, J., Tabacof, L., Herrera, J.E., Breyman, E., Dewil, S., Cortes, M., Correa-esnard, L., Kellner, C.P., Dangayach, N., Putrino, D.: Gamified neurorehabilitation strategies for post-stroke motor recovery: challenges and advantages. Curr. Neurol. Neurosci. Rep. **22**, 183–195 (2022)
12. Singh, Y., Rodrigues, V., Prado, A., Agrawal, S.K., Vashista, V.: Lower-limb strategy assessment during a virtual reality based dual-motor-task. In: 2020 8th Ieee Ras/Embs International Conference for Biomedical Robotics and Biomechatronics (Biorob), pp. 252–257 (2020)

Biomedical Signal and Image Processing; Biomedical Imaging

Search for Common Physiological Responses to Videos Evoking Unpleasant Emotions with Different Arousal Levels

Masaji Yamashita(✉) and Yoshitaka Itoh

Hokkaido University of Science, Sapporo, Japan
yamashita@hus.ac.jp

Abstract. In order to establish an objective method for evaluating emotions, we sought to identify the physiological parameters that change in response to various unpleasant emotions. Experiments and analysis were conducted using video stimuli that evoked the unpleasant emotions of "fear" (high arousal), "disgust" (medium arousal), and "sadness" (low arousal), as well as the pleasant emotion "exhilaration" (high arousal). The results of our questionnaire confirmed that the emotions evoked by each stimulus were valid. Because the physiological response of the autonomic nervous system differs depending on the arousal level of each emotion evoked, analysis of general physiological responses did not reveal any physiological values common to unpleasant emotions with different arousal levels. "Fear" and "exhilaration," which have high arousal levels, showed similar physiological responses, while "disgust" and "sadness," which have different arousal levels, responded with different physiological quantities. Furthermore, there was no commonality in the physiological responses to unpleasant stimuli. On the other hand, analysis using blood pressure waves showed that a physiological index linking two parameters that tend to change in common to the three unpleasant emotions increased in common to the three unpleasant emotions.

Keywords: Displeasure · Physiological response · Video · Blood pressure wave

1 Introduction

The improvement of environmental conditions and the management of workplace health requires the use of evidence-based measure that assess individuals' emotions objectively. Methods with which to objectively assess emotions using physiological parameters have long been studied [1, 2]. In this context, pleasure and displeasure are often considered as the basic elements of emotions [3]. Autonomic responses are often considered when assessing emotion, which differ during high-arousal states compared to responses during low-arousal states; similarly, responses during displeasure in high-arousal states differ from those during displeasure in low-arousal states. Therefore, it is extremely difficult to track physiological parameters that respond commonly to multiple displeasure.

Parameters such as alpha wave 1/f fluctuations and asymmetry in electroencephalograms (EEGs) [4] have been proposed as parameters for discriminating pleasure from

S. Vlad and N. M. Roman (Eds.): MEDITECH 2024, IFMBE Proceedings, 130, pp. 99–106, 2025.
https://doi.org/10.1007/978-3-031-95671-3_12

displeasure. However, these parameters have not been confirmed to respond to the emotions of pleasure or displeasure in high- and low-arousal states. Therefore, conclusive and satisfying physiological parameters for discriminating between pleasure and displeasure are currently lacking.

In this study, combined parameters of blood pressure pulse wave components that discriminate certain pleasures, even at different arousal levels, against displeasure in previous studies were identified [5, 6]. To evaluate emotions more broadly, there is a need to identify physiological responses not only to pleasant stimuli but also to multiple unpleasant stimuli with different arousal levels. To this end, we measured the physiological responses of subjects to video stimuli that evoked unpleasant and pleasant emotions and characterized their physiological responses via blood pressure wave analysis.

2 Methods

This study was approved by the ethics committee of Hokkaido University of Science. A simple survey of the video stimuli was conducted prior to the physiological experiment.

2.1 Preliminary Survey

A preliminary survey was conducted to select the unpleasant video stimuli. Since videos evoking "disgust" and "exhilaration" were reported in previous studies [5], videos evoking "fear" or "sadness" were investigated here. A total of ten young adults aged 18–22 years old were enrolled in the study. The participants were tasked with watching five fearful or sad videos each alternatively, after which they were presented with two questionnaires to assess the emotions elicited by the videos.

One of the questionnaires, i.e. "three-item questionnaire," was comprised of three items (pleasure–displeasure, tension–relaxation, and excitement–calmness), consisting 11-point scales that were rated +5 to −5. Regarding the pleasure–displeasure scoring process, a positive score was indicative of a pleasurable experience, while a negative score indicates a non-pleasurable experience. The other questionnaire, i.e. "displeasure questionnaire," was comprised four items (fearful, gross, angry, and disgusting) and consisted of a 11-point scale (0 to 10).

Finally, taking into consideration the scores attributed to each of the videos, those videos identified as having the highest scores for either "fear" or "sadness" and a low score for the other emotion were selected.

2.2 Experiment

Subjects and Conditions. In total, 15 healthy male volunteers aged 18–22 years were enrolled as participants in the study, all of whom provided written informed consent. Experiments were performed in a room that was temperature-controlled (24.5–25.5°C) and humidity-controlled (30–40%). In addition, the subjects were asked to refrain from smoking, eating, or drinking for 2 hours before the experiments.

Stimuli and Procedures. Stimuli consisted of videos projected onto a screen installed with stereo sound. The experimental protocol was as follows: After a 10-min habituation period, the participants were tasked with watching a familiar "campus" video for 90 s in the pre-rest period, an emotional stimulus video for 100 s in the stimulus period, and a similar "campus" video for an additional 40 s in the post-rest period. The emotional stimuli being targeted were either "fear," "disgust," "sadness," or "exhilaration," as well as a control. One scene of the horror movie "Ju-On" was used for "fear," an "animal carcass" video was used for "disgust," a video of a "dog mourning a dead dog" was used for "sadness," a "sports highlight" video was used for "exhilaration," and a similar but different "campus" video was used as the control. To compare the physiological responses to the unpleasant and pleasant videos, an "exhilaration" video was included. The participants were tasked with watching the videos and assessing the emotions they elicited in each session.

Measurements. Blood pressure wave was measured using arterial tonometry on the left wrist, while the respiratory wave activity was measured using a thermistor sensor placed in front of the subject's nostrils. Photoplethysmography data were obtained on the right index finger, and electrocardiography (ECG) data were obtained using standard lead II. Finally, cardiac stroke volume was measured by impedance plethysmography (Physioflow®; Manatec Biomedical, Paris, France) on the chest.

All data, except for stroke volume, were recorded with 200-Hz sampling and subsequently analyzed using several computer programs developed in the lab.

2.3 Analysis

Bonferroni and Steel-Dwass' multiple comparison tests were adopted for the statistical analysis.

Sections. In total, nine sections were assigned for the analysis. First, the pre-rest period was divided into two sections: 70 s [rest 0] and 20 s [rest 1]. Next, the stimulus period was divided into five sections, each comprising 20 s [stim. 1] to [stim. 5]. The post-rest period was divided into two sections, each comprising 20 s [rest 2] and [rest 3]. The mean of each data set was determined for each section. To adjust for individual variations, each mean was normalized using the [rest 0] period data.

Analysis of Physiological Data. Time series data for systolic blood pressure (SBP), diastolic blood pressure (DBP), and pulse pressure (PP) were obtained from the blood pressure wave. The respiration depth (RD) and respiration rate (RR) were obtained from the respiratory wave data. The pulse wave amplitude (PWA) was obtained via photoplethysmography. The heart rate (HR) and heart rate variability (HRV), respiratory sinus arrhythmia component (RSA; 0.15–0.4 Hz), and Mayer wave component (MW; 0.04–0.15 Hz) were obtained from the ECG data. Furthermore, each RSA and Mayer components of SBP or DBP were obtained either (SBP-RSA, SBP-Mayer, DBP-RSA, DBP-Mayer). Finally, stroke volume (SV), cardiac output (CO), and cardiac index (CI) were obtained using Pysioflow®.

Additional Respiration Wave Analysis. For an additional analysis of the respiratory wave data, the inspiratory time (Ti), expiratory time (Te), inspiratory speed (dTi), expiratory speed (dTe), and respiratory volume (RV) were obtained from the respiratory wave data.

Additional Blood Pressure Wave Analysis. Based on a previous study on pleasure [5, 6], a blood pressure pulse wave analysis method was conducted. In this study, ejection time (et), and pulse width (w), and dip level (h) were examined (Fig. 1). Moreover, in response to their results, combined parameters et*h as CP1 and et*h*w as CP2 were examined.

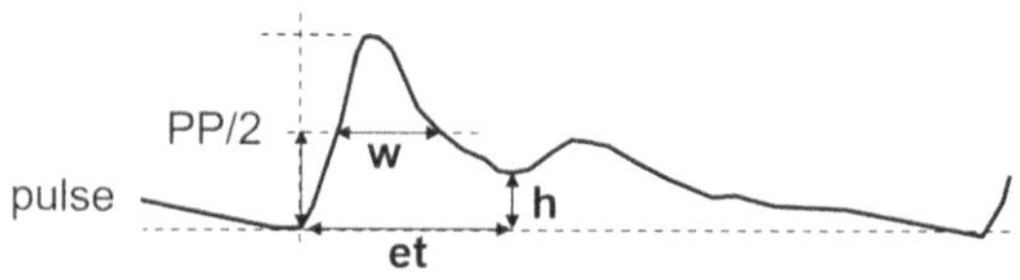

Fig. 1. Parameters for blood pressure wave analysis. et: ejection time w: pulse width defined as half level of pulse pressure h: dip level

3 Results

3.1 Self-ratings

Due to the nature of the questionnaire, which evaluated each emotion separately in comparison to the rest of the emotions, the scores indicating individual emotions were significantly higher than those indicating the other emotions ($p < 0.001$).

In addition, compared to the control, the unpleasant emotions "fear," "disgust," and "sadness" showed significantly higher unpleasant scores ($p < 0.001$), while the pleasant emotion "exhilaration" showed significantly higher pleasant scores ($p < 0.001$). Furthermore, a significant difference was observed between the pleasant emotion "exhilaration" and the unpleasant emotions "fear," "disgust," and "sadness" ($p < 0.001$).

3.2 Physiological Responses

The results of the physiological signals observed during the experiment are shown in Table 1, comparing each emotion based on the normalized stimulation interval average values. Notably, due to a malfunction in the continuous blood pressure monitor sensor, blood pressure-related data could not be recorded for the last two subjects, resulting in a total number of 13 subjects for the blood pressure-related analysis. Similarly, there was an artifact of sensor detachment in respiratory wave measurements, which resulted in the exclusion of one subject. The direction of change in many responses was inconsistent, and no significant differences were observed. However, SBP was significantly higher in response to "fear" compared to "disgust" ($p < 0.01$), and in "exhilaration" compared to "disgust" ($p < 0.01$) and "sadness" ($p < 0.05$). Similarly, PP was significantly higher in response to "fear" compared to the control ($p < 0.01$), "disgust" ($p < 0.01$), and "sadness" ($p < 0.05$), and in "exhilaration" compared to the control ($p < 0.01$), "disgust" ($p < 0.01$), and "sadness" ($p < 0.05$). Thus, similar physiological responses were observed for the unpleasant emotion "fear" and the pleasant emotion "exhilaration." HR was lower in response to "sadness" compared to the control ($p < 0.05$). RR was increased for "exhilaration" and showed significant differences compared to the control ($p < 0.01$) and "sadness" ($p < 0.05$), while RD was significantly decreased for "disgust" compared to the control ($p < 0.05$).

Table 1. Comparison of physiological responses to the video for each emotional stimulus.

Parameter	Emotion				
	c: Control	f: Fear	d: Disgust	s: Sadness	e: Exhilaration
SBP	0.993 ± 0.039	1.049 ± 0.092 ** vs d	0.970 ± 0.028	1.010 ± 0.053	1.061 ± 0.060 ** vs d, * vs s
DBP	0.983 ± 0.053	1.011 ± 0.102	0.956 ± 0.087	1.000 ± 0.069	1.028 ± 0.081
PP	1.002 ± 0.053	1.092 ± 0.092 ** vs c,d, * vs s	0.981 ± 0.045	1.019 ± 0.057	1.096 ± 0.070 ** vs c,d, * vs s
SBP-Mayer	1.255 ± 1.375	1.384 ± 1.078	1.066 ± 0.774	1.180 ± 0.774	1.012 ± 1.047
SBP-RSA	1.085 ± 0.501	0.855 ± 0.441	0.765 ± 0.321	0.699 ± 0.238	0.760 ± 0.360
DBP-Mayer	1.274 ± 0.969	1.768 ± 1.524	1.116 ± 0.931	1.288 ± 0.610	1.284 ± 1.245
DBP-RSA	1.032 ± 0.380	1.245 ± 1.319	0.748 ± 0.371	0.836 ± 0.816	0.904 ± 0.475
HR	1.045 ± 0.054	1.021 ± 0.086	0.999 ± 0.0436	0.0974 ± 0.043 * vs c	1.003 ± 0.053
Mayer	1.624 ± 1.396	1.459 ± 2.685	1.278 ± 1.105	0.830 ± 0.543	1.061 ± 0.876
RSA	0.977 ± 0.436	0.935 ± 0.588	0.814 ± 0.334	0.716 ± 0.334	0.944 ± 0.461
RR	0.956 ± 0.186	1.093 ± 0.172	1.084 ± 0.164	1.033 ± 0.147	1.255 ± 0.304 ** vs c, * vs s
RD	1.164 ± 0.326	0.979 ± 0.176	0.885 ± 0.178 * vs c	1.032 ± 0.252	0.900 ± 0.260
PWA	1.010 ± 0.202	0.838 ± 0.198	1.092 ± 0.134	0.990 ± 0.185	0.879 ± 0.181
TPR	0.962 ± 0.061	1.046 ± 0.118	0.975 ± 0.063	1.018 ± 0.063	1.046 ± 0.077
CO	1.034 ± 0.066	1.005 ± 0.076	0.992 ± 0.045	0.981 ± 0.044	1.000 ± 0.035
SV	0.991 ± 0.028	0.993 ± 0.038	0.999 ± 0.034	1.010 ± 0.032	1.002 ± 0.034
CI	1.034 ± 0.064	1.009 ± 0.075	0.991 ± 0.042	0.981 ± 0.045	0.999 ± 0.035

These results indicate that none of the physiological responses changed in tandem between the unpleasant emotions "fear," "disgust," and "sadness."

3.3 Features of Respiration Wave

In terms of the results of additional analysis of respiratory waves, those for inspiratory time Ti, which showed a common response in response to the unpleasant emotions of "fear," "disgust," and "sadness," are shown in Fig. 2. Although there was no significant difference, the direction of the response was the same for these unpleasant emotions. However, for "exhilaration," which was considered a pleasant emotion, there was a strong reaction in the same direction, showing a significant difference from the control ($p < 0.05$). RR also showed a similar tendency to Ti; however, for other physiological quantities, either no differences were observed in the response or there was no consistency between the unpleasant emotions.

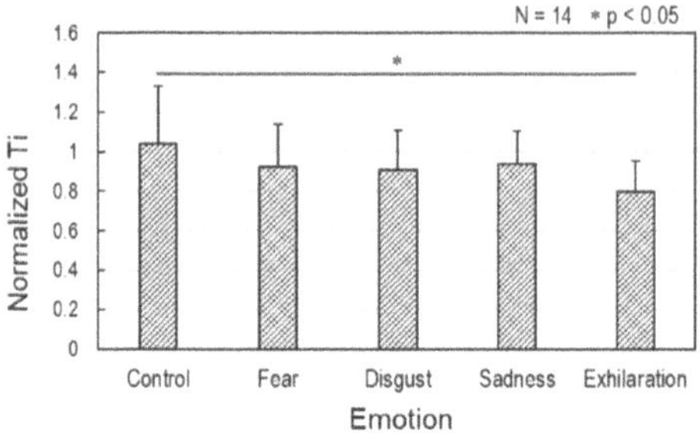

Fig. 2. Comparison of normalized Ti for each emotion.

Therefore, additional respiratory wave analysis did not reveal any physiological parameters common to unpleasant emotions.

3.4 Features of Pulse Wave

The analysis results of the ejection time et, pulse width w, and dip level h for each emotional stimulus are shown in Fig. 3. The value for et was significantly elevated in response to "disgust" ($p < 0.05$) and "sadness" ($p < 0.01$) compared to the control, while w was significantly elevated in response to "disgust" ($p < 0.01$) and "sadness" ($p < 0.01$) compared to "exhilaration." Furthermore, h was elevated in response to "fear" ($p < 0.05$) and "sadness" ($p < 0.05$) compared to the control.

The physiological responses to each unpleasant emotion tended to show an upward trend compared to the control and "exhilaration." Figure 4(a) and (b) shows the analysis results of combined parameters CP1 and CP2, which were created to find physiological parameters that showed a common response to the unpleasant emotion stimuli of "fear," "disgust," and "sadness." CP1 showed significant increases in response to "fear" ($p < 0.05$), "disgust" ($p < 0.05$), and "sadness" ($p < 0.01$) compared to the control. CP2 showed increases in response to "disgust" compared to the control ($p < 0.05$) or "exhilaration" ($p < 0.05$), and also increased "sadness" compared to the control ($p < 0.01$) or "exhilaration" ($p < 0.05$).

4 Discussion

In order to identify physiological parameters that change in response to various emotions and achieve an objective evaluation of emotions, we analyzed responses to video stimuli evoking the unpleasant emotions of "fear" (high arousal), "disgust" (medium arousal), and "sadness" (low arousal), as well as the pleasant emotion "exhilaration"

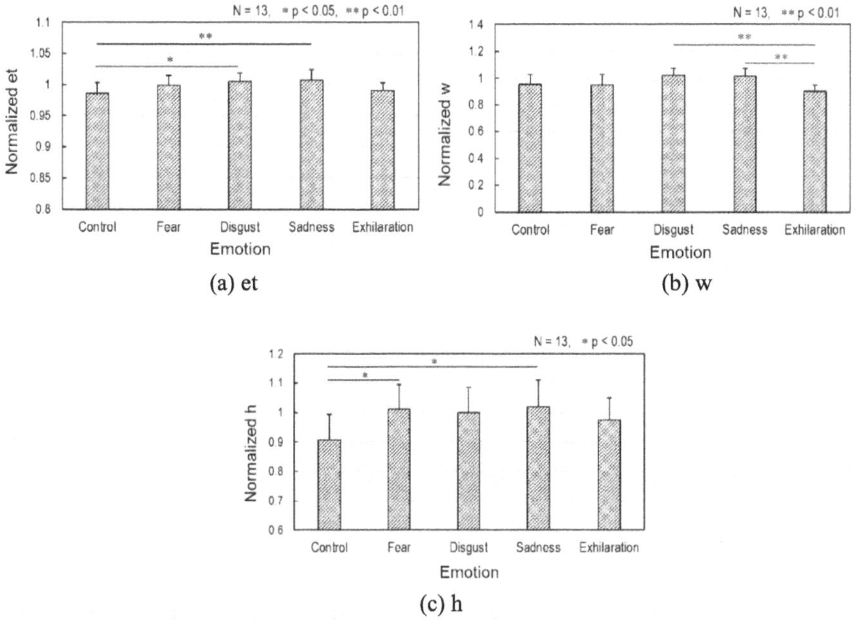

Fig. 3. Comparison of normalized et, w, and h for each emotion.

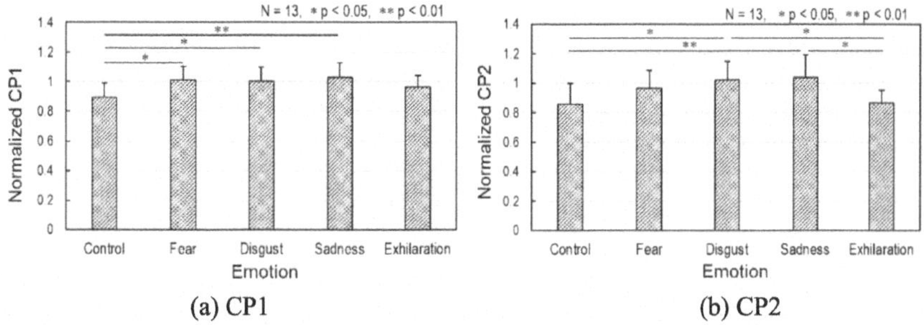

Fig. 4. Comparison of normalized CP1 and CP2 for each emotion.

via questionnaires. The results of the questionnaire evaluation of each emotion-evoking stimulus showed that each of the targeted emotions was evoked, had few other emotional elements, and was appropriate as a stimulus.

Since the response of the autonomic nervous system varies depending on the different arousal levels of each emotion evoked, we were unable to find any physiological quantities common to unpleasant emotions of different arousal levels. Despite this, similar physiological responses were observed for "fear" and "exhilaration," both of which are associated with high levels of arousal. However, the same was no observed for "disgust," which has a medium arousal level, and "sadness," which has a low arousal level, which showed different physiological quantities. This suggests that no commonalities exists in terms of physiological responses to unpleasant stimuli.

By contrast, analysis using blood pressure waves showed that CP1, which combines the parameters et and h, and which tended to change in tandem with unpleasant emotions, increased in the same direction for multiple unpleasant emotions.

We previously identified physiological parameters that change in common with pleasant emotions using blood pressure wave analysis [5, 6]. Therefore, arterial wave analysis may be used to identify differences between pleasant and unpleasant physiological responses. Although these results cannot be generalized due to the small number of subjects in this study, these results provide a basis for future research in the field.

Acknowledgements. This work was supported by JSPS KAKENHI (grant Number JP20K12743).

Conflicts of Interest. The authors declare no conflict of interest.

References

1. Kreibig, S.D.: Autonomic nervous system activity in emotion: a review. Physiol. Psychol. **84**, 394–421 (2010). https://doi.org/10.1016/j.biopsycho.2010.03.010
2. Yamashita, M.: Experimental verification of psychophysiological index of excitement. Int. J. Affect. Eng. **14**(1), 65–71 (2015)
3. Russell, J.A.: A circumplex model of affect. J. Pers. Soc. Psychol. **39**, 1161–1178 (1980). https://doi.org/10.1037/h0077714
4. Davidson, R.J., Fox, N.A.: Asymmetrical brain activity discriminates between positive and negative affective stimuli in human infants. Science. **218**(4578), 1235–1237 (1982). https://doi.org/10.1126/science.7146906
5. Yamashita, M., Aikawa, T., Kitama, M., Yokoyama, T.: Analysis of physiological response to pleasure elicited by video. IFMBE Proc. **68**, 207–211 (2019). https://doi.org/10.1007/978-981-10-9038-7_38
6. Yamashita, M., Itoh, Y.: Pilot study on the physiological responses to pleasant video stimuli in different intensities. In: KEER2022, pp. 551–557 (2022). https://doi.org/10.5821/conference-9788419184849.57

AI-Based 3D Segmentation of Pulmonary Vasculature of CTEPH Patients

Angela Lungu[1(✉)], Andrew J. Swift[2], Alina S. Danciu[1], Michael Sharkey[2], Rod D. Hose[2], and Maciej Malawski[3]

[1] Technical University of Cluj-Napoca, Cluj-Napoca, Romania
angela.lungu@ethm.utcluj.ro
[2] University of Sheffield, Sheffield, UK
[3] Sano Centre for Computational Medicine, Krakow, Poland

Abstract. Chronic thromboembolic pulmonary hypertension (CTEPH) presents challenges for pulmonary artery segmentation due to vascular remodeling, stenosis, and obstructions. This study evaluates a 7-layer dilated convolutional neural network (CNN) with Tversky loss, applied to computed tomography angiography (CTA) images that were preprocessed with image enhancement techniques. The model achieved a Dice score of 0.792 on non-CTEPH data but scored 0.693 on CTEPH data, reflecting the challenges of manual segmentation, where smaller branches are often missed. While the results align with other research, advanced 3D CNN models have shown higher accuracy. Future work should refine ground truth data and explore 3D models to better capture CTEPH-specific complexities.

Keywords: 3D segmentation · AI · CTEPH

1 Introduction

Pulmonary hypertension is a clinical condition characterised by an increase in the mean pulmonary pressure >20 mmHg, measured at right heart catheterisation. Imaging modalities such as magnetic resonance (MRI) or computed tomography (CT) play a crucial role during the diagnosis, assessment, and patient follow-up of PH patients [1]. Over the past two decades, tools based on computer modelling and artificial intelligence [2] have emerged, supporting the clinical decision-making process.

Among these tools, blood flow hemodynamics simulations could provide a deeper understanding of pressure and flow distribution, and they could be used to predict the changes that may occur after surgical intervention [3], as it is the case in chronic thromboembolic hypertension (CTEPH). Often, the development of personalised models of the pulmonary vasculature relies on quick and sensible image segmentation and geometry reconstruction. Various research teams [4–8] have proposed AI-based tools for detailed pulmonary arterial segmentation and reconstruction, showing high agreement with manually segmented ground truth. However, these methods are not focused on CTEPH patients, who present significant arterial remodeling, stenosis, webs, and bands [9], complicating the task.

S. Vlad and N. M. Roman (Eds.): MEDITECH 2024, IFMBE Proceedings, 130, pp. 107–114, 2025.
https://doi.org/10.1007/978-3-031-95671-3_13

This paper evaluates whether a 2D convolutional neural network, supported by image processing techniques, can offer accurate pulmonary arterial tree segmentation in CTEPH.

2 Materials and Methods

2.1 Datasets

The proposed segmentation method was developed using two image datasets. The first, referred to as SheffCTEPH has CTA images from 14 CTEPH patients, collected at the Sheffield Pulmonary Disease Unit. All patients provided written consent for data collection.

The images have matrix dimensions of 512 × 512 pixels, with stack heights ranging from 905 to 1256 slices per patient. The mean in-plane resolution is 0.72 mm/pixel, and the average slice thickness is 0.625 mm.

The second dataset consists of CTA volumetric images and corresponding masks from 100 subjects, available through the PARSE challenge [7]. The stack height of these images varies between 228 to 376 slices. Since no information on the clinical conditions of these subjects is available, this dataset is referred to as noCTEPH.

2.2 Generation of the Ground Truth Masks

The images were segmented using the MIMICS software package (www.materialise.com). 3D geometry rendering, visualisation and comparison of cases was performed with 3D Slicer (www.slicer.org/). All the 2D image processing as well as AI model implementation and evaluation were performed in MATLAB R2023b (www.mathworks.com), using various built-in toolboxes, including the Image Processing Toolbox and Deep Learning Toolbox.

It is important to note that the masks for the SheffCTEPH dataset were generated from 3D reconstructed geometries, which were cleaned and smoothed for haemodynamic model-based analysis. Due to this process, the level of detail at the branch level was reduced for some CTEPH patients.

2.3 Image Preprocessing

All the images went through several preprocessing steps including contrast enhancement, pixel values scaling and normalisation, image cropping and resizing. The image contrast enhancement was achieved after clipping the pixel values between −800 and 500 Hounsfield units (HU), followed by min-max pixel scaling and normalisation in the [0, 1] interval.

Cropping all the volumes with a rectangle that includes the structures delimited by the outside lungs border was further applied. The creation of this region of interest (ROI) included: (i) binarisation of the volume central slice with an optimal threshold value chosen based on the histogram using the Otsu method [10], (ii) cropping the grayscale image with a rectangle that frames the object with the largest perimeter in the binary image, (iii) binarising the newly cropped grayscale image and create its negative to outline the lungs, (iv) retaining the two lungs from the complement image, (v) cropping the lung mask with a rectangle that fully encloses them, and (vi) cropping all images in the volume according to the new coordinates. For each data volume, the cropping coordinates were determined based on the image corresponding to the middle slice. Finally, all the images were resized at 512 × 512 and individually saved as .png files. Figure 1 shows the 2D image preprocessing steps illustrated on a slice.

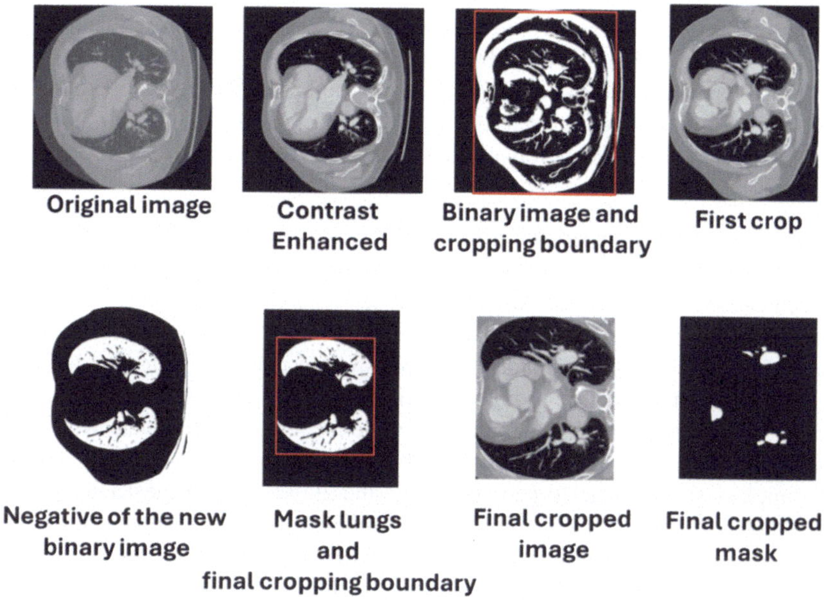

Fig. 1. Images preprocessing steps

2.4 AI Based Images Segmentation

For the task of image segmentation a dilated convolutional neural network (DCNN) [11] architecture was used. The dilated convolutions allow the network to capture features at multiple scales, which is essential for detecting the fine details of small arteries while also understanding the broader context required for segmenting larger arteries, both being present in the CT images of the pulmonary arteries. By expanding the receptive field without reducing spatial resolution, the network can accurately differentiate small, intricate vessels alongside larger structures, making it well-suited for this type of medical imaging task.

The network's first layer processes grayscale images with a 512 × 512 matrix size on a single channel, followed by six convolutional blocks. Each block includes a convolutional layer with 64 filters (3 × 3), batch normalization to stabilize training, and ReLU activation for non-linearity. To capture features at different scales, dilation factors (1, 2,

4, 8, 16, 32) were applied, increasing with network depth. The final 1x1 convolutional layer reduces the number of channels to match the number of classes, preparing the output for the softmax layer to generate pixel-wise class probabilities.

The training was carried out using the Adam optimizer, with a learning rate of 1e-4, for 100 epochs, in mini-batches of 4 images. The Tversky loss function [12], with $\alpha = 0.3$ and $\beta = 0.7$, chosen empirically after trying several configurations, was used to penalize false negative predictions (white pixels defined as background) in an attempt to compensate for the class imbalance.

The dataset was split into 70% for training, 10% for validation, and 20% for testing, ensuring that no .png images from the same subject appeared in more than one set to avoid data leakage and improve model generalization.

A final image post-processing step, based on simple morphological operations such as opening (using a structural disk element with a radius of 1) and hole filling, was applied to the test set masks before evaluating the final results.

3 Results and Discussions

The results were evaluated on the entire test set, as well as separately on each type of images to better underline the challenges in the CTEPH cohort. Figure 2 shows a comparison between the ground truth mask and the mask estimated by the trained neural network for images from both categories (CTEPH and noCTEPH). Generally, the network successfully identifies the large arteries, but the agreement with smaller blood vessels is weaker, especially for the SheffCTEPH data.

Examples of reconstructed geometry from the 2D segmented masks are shown in Figs. 3 and 4. It can readily be noticed that while the AI-based segmentation model provides results closer to those of the human expert for the noCTEPH set, it predicts more arterial vessels than the ground truth in CTEPH cases. The CTEPH examples in Fig. 4 highlight both the tortuosity of the vascular tree in these patients and the arterial path obstruction in the proximal vessels.

The qualitative results were supported by the quantitative data presented in Table 1, expressed as the mean Dice and Jaccard indices [13] as well as accuracy, based on binary pixel classification across all slices.

As it can be noticed, the highest artery Dice score of 0.792 was achieved for the noCTEPH set, while the lowest score of 0.693 was recorded for the CTEPH set. The

Table 1. Evaluation metrics computed on the test set

		JACCARD	DICE	ACCURACY
NoCTEPH	artery	0.656	0.792	0.897
	background	0.995	0.997	0.996
SheffCTEPH	artery	0.530	0.693	0.835
	background	0.992	0.996	0.994
NoCTEPH+SheffCTEPH	artery	0.557	0.715	0.850
	background	0.993	0.996	0.994

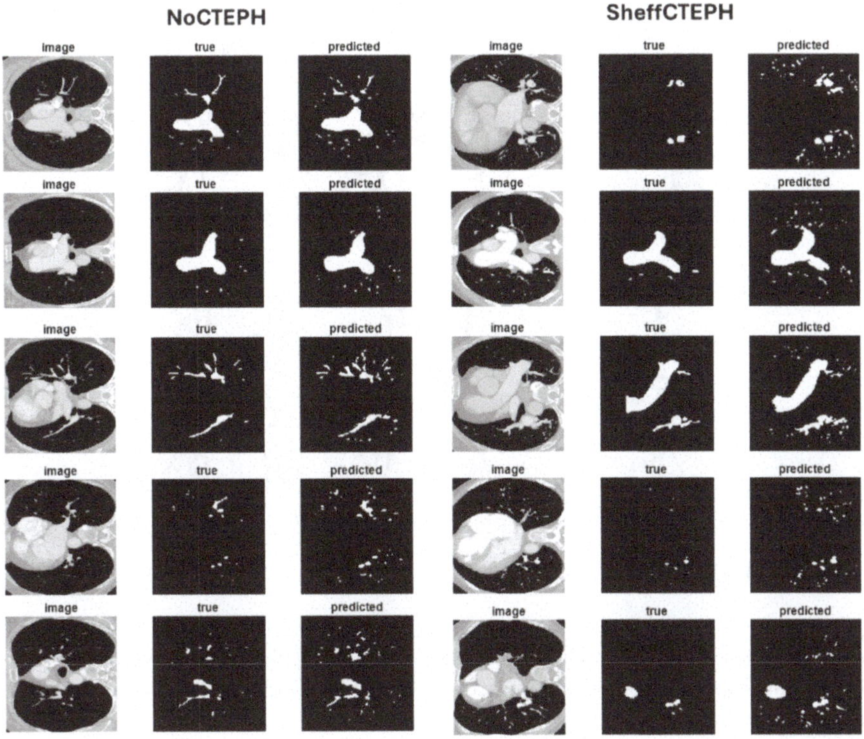

Fig. 2. Examples of 2D images from the test sets with corresponding ground truth and predicted masks

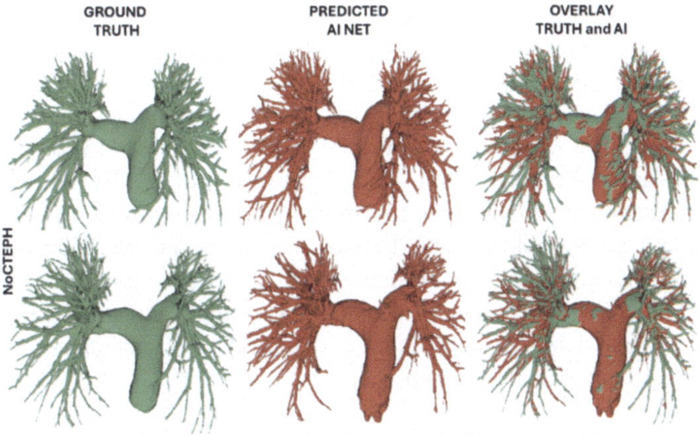

Fig. 3. Geometry reconstructions of the masks from the noCTEPH test set

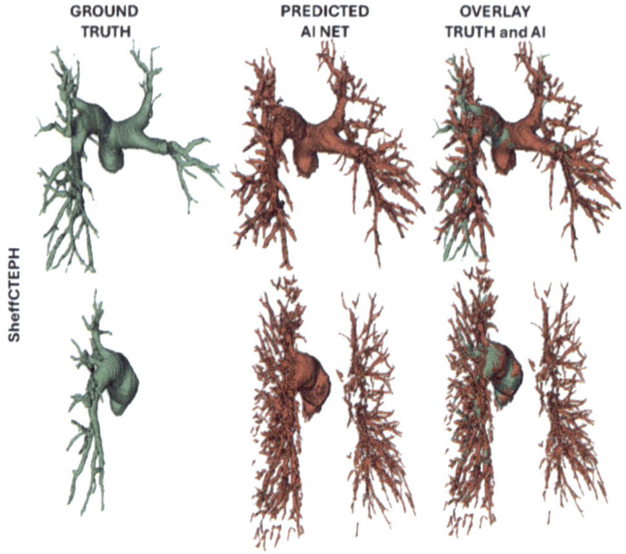

Fig. 4. Geometry reconstructions of the masks from the SheffCTEPH test set

mixed test set scored 0.715. The lower score for the CTEPH cases is influenced by the ground truth segmentation, which missed unconnected branches that were not identified during manual segmentation.

The results for the entire set are in the range of those reported by the 25 teams participating in the Parse Grand challenge [7], where the highest score was 0.7969. On the same Parse dataset, L. Lou. et al. [4] reported with their 2D-CNN a mean dice score of 0.7816. Superior results were showed for the more complex, 3D CNN, as it is the case of M. Zulfiqar et al. [6] or J. Han et al. [5], reporting mean dice scores of 0.8775 and 0.8659, respectively. It is important to note that these results were obtained from generic lung CTA images, with no mention of CTEPH cases. A few research groups reported results on CTEPH data, such as H. Suzuki et al. [14] who focused on main pulmonary artery segmentation, showing a dice score of 0.968, or Kim et al. [15] who presented a deep learning-based pulmonary arterial tree segmentations in CTEPH but did not provide evaluation scores. In contrast, our research utilizes specialized image enhancement techniques and a CNN modified with Tversky loss to specifically address the unique challenges of segmenting pulmonary arteries in CTEPH patients—a domain that has been relatively underexplored in existing studies. This approach not only provides critical insights but also achieves significant improvements in segmentation accuracy for this particularly complex patient group.

4 Conclusions

This study explored the use of a 7-layer dilated convolutional network with Tversky loss for segmenting pulmonary arteries in 2D CTA images. Image enhancement techniques were applied during preprocessing to support the segmentation of arteries in both CTEPH and non-CTEPH cases. The model achieved a Dice score of 0.792 in non-CTEPH cases but struggled in CTEPH cases, with a lower score of 0.693, likely due to missing details in the ground truth caused by complex vascular alterations typical of CTEPH. While the performance aligns with similar 2D CNN-based studies, more advanced 3D models and refined segmentation processes could improve accuracy and clinical relevance in CTEPH cases.

Acknowledgments. This work was supported through the research project PN-III-P1-1.1-PD-2021-0601. The authors would like to thank Sheffield Pulmonary Vascular Disease Unit, UK for providing the CTA image data of CTEPH patients used in this study.

This project has received funding from the European Union's Horizon 2020 research and innovation programme under grant agreement No 857533. The publication was created within the project of the Minister of Science and Higher Education "Support for the activity of Centers of Excellence established in Poland under Horizon 2020" on the basis of the contract number MEiN/2023/DIR/3796 and is supported by Sano project carried out within the International Research Agendas programme of the Foundation for Polish Science, co-financed by the European Union under the European Regional Development Fund.

Conflict of Interest Statement. The authors have no conflict of interest to declare.

References

1. Humbert, M., et al.: 2022 ESC/ERS guidelines for the diagnosis and treatment of pulmonary hypertension. Eur. Heart J. **43**(38), 3618–3731 (2022)
2. Lungu, A., et al.: Diagnosis of pulmonary hypertension from magnetic resonance imaging–based computational models and decision tree analysis. Pulmon. Circul. **6**(2), 181–190 (2016)
3. Colebank, M.J., et al.: A multiscale model of vascular function in chronic thromboembolic pulmonary hypertension. Am. J. Physiol. Heart Circ. Physiol. **321**(2), H318–h338 (2021)
4. Lou, L., et al.: A detail-oriented super-2D network for pulmonary artery segmentation. Biomed. Sig. Process. Control. **93**, 106183 (2024)
5. Han, J., et al.: 3D pulmonary vessel segmentation based on improved residual attention u-net. Med. Novel Technol. Dev. **20**, 100268 (2023)
6. Zulfiqar, M., et al.: DRU-net: pulmonary artery segmentation via dense residual U-network with hybrid loss function. Sensors. **23**(12), 5427 (2023)
7. Luo, G., et al.: Efficient automatic segmentation for multi-level pulmonary arteries: The PARSE challenge.. arXiv preprint arXiv:2304.03708 (2023)
8. Wu, Y., et al.: Transformer-based 3D U-net for pulmonary vessel segmentation and artery-vein separation from CT images. Med. Biol. Eng. Comput. **61**(10), 2649–2663 (2023)
9. Shahin, Y., et al.: Quantitative CT evaluation of small pulmonary vessels has functional and prognostic value in pulmonary hypertension. Radiology. **305**(2), 431–440 (2022)
10. Xu, X., et al.: Characteristic analysis of Otsu threshold and its applications. Pattern Recogn. Lett. **32**(7), 956–961 (2011)
11. Yu, F., Koltun, V.: Multi-scale context aggregation by dilated convolutions. arXiv:1511.07122. (2015). https://doi.org/10.48550/arXiv.1511.07122
12. Jadon, S., IEEE: A survey of loss functions for semantic segmentation. In: 2020 IEEE Conference on Computational Intelligence in Bioinformatics and Computational Biology (CIBCB) (2020)

13. Taha, A.A., Hanbury, A.: Metrics for evaluating 3D medical image segmentation: analysis, selection, and tool. BMC Med. Imaging. **15**, 29 (2015)
14. Suzuki, H., et al.: Aorta and main pulmonary artery segmentation using stacked U-net and localization on non-contrast-enhanced computed tomography images. Med. Phys. **51**(2), 1232–1243 (2024)
15. Kim, J., et al.: Abstract 17842: 3D visualization and quantitative assessment of the pulmonary arteries on CT using deep learning segmentation. Circulation. **148**(Suppl_1), A17842–A17842 (2023)

Blood Vessel Segmentation and 3D Reconstruction of Neovascularization in Age-Related Macular Degeneration from OCTA Scans

Pop Adrian[1](✉), Damian Ioana[2,3], Groza Adrian[1], and Simona Delia Nicoară[2,3]

[1] Computer Science Department, Technical University of Cluj-Napoca, Cluj-Napoca, Romania
adrian.pop@cs.utcluj.ro

[2] Department of Ophthalmology, "Iuliu Hațieganu" University of Medicine and Pharmacy, Cluj-Napoca, Romania

[3] Emergency County Hospital, Cluj-Napoca, Romania

Abstract. We propose here a novel method for vessel segmentation and 3D reconstruction of nAMD from OCTA images. The method has four steps: (1) creating vessel mask; (2) isolating the Macular Neovascular Membranes from noise; (3) finding the vessel depth in volume; (4) reconstructing the 3D volume. Blood vessels were identified by detecting and marking spikes from four different directions—row, column, and two diagonals—treated as a wave form, combined with cross-correlation among five neighbouring points. When the parameters are manually optimised for the available dataset, the method can achieve up to 100% accuracy. Furthermore, using a kernel extracted from the main image, the depth of a blood vessel was determined with an accuracy of 95.6%. By extracting a 3D volume centred around the point of interest, based on previously calculated coordinates, and retaining only high-intensity points, a 3D reconstruction of the neovascularization area in AMD pathology was achieved. The method has been tested on a dataset provided by the Emergency County Hospital, Cluj-Napoca, Romania.

Keywords: Age-related Macular Degeneration · Optical Coherence Tomography Angiography · Image Processing · Vessel Segmentation

1 Introduction

Age-related macular degeneration (AMD) is an acquired degeneration of the macula that induces significant visual impairment. It is considered the most common cause of severe loss of eyesight among people 50 and older. AMD impairs the ability to see fine details since it affects central vision. In the early stages, there are no signs or symptoms, so people may not suspect they even have it. In the more advanced stages, patients may lose their ability to see faces, drive and read smaller print.

There are two primary types of AMD: the dry form (non-exudative) which is the most common one (85–90%), where the visual acuity decrease is slow and gradual; and

S. Vlad and N. M. Roman (Eds.): MEDITECH 2024, IFMBE Proceedings, 130, pp. 115–128, 2025.
https://doi.org/10.1007/978-3-031-95671-3_14

the neovascular form, which is less common (10–15%) but may lead to profound vision loss. The latter is characterised by the growth of a macular neovascularization (MNV) at the interface between choroid and retina.

In terms of clinical diagnosis, the hallmark findings in neovascular AMD (nAMD) are subretinal fluid, hemorrhages or retinal pigment epithelium detachment. Other diagnostic procedures include fundus fluorescein angiography (FFA), which highlights the leakage secondary to the presence of MNV, optical coherence tomography (OCT) where the intraretinal or subretinal fluid can be seen, and optical coherence tomography angiography (OCTA) which is the only examination capable of direct visualisation of the choroidal membrane.

Currently, ophthalmologists can only perform a qualitative assessment of the membrane, describing the presence of: (1) tiny branching vessels (thin, tangled capillaries) vs. mature vessels (linear filamentous capillaries); (2) an anastomotic arcade (peripheral connection, lacy wheel appearance); (3) inner loops (inner anastomoses); (4) perilesional hypointense area around NV, local choriocapillaris alteration (5) feeder vessel.

2 Related Work

There is increased interest in blood vessel segmentation using deep learning methods, mainly focuses on the segmentation of healthy retinal vessels. Chang et al. [13] proposed an method for fine retinal vessel segmentation by combining Nest U-net and patch-learning. Beaudelaire at al [8]. which proposed a two step method, extracting a feature vector with eight measures characterising each pixel and process the vector through an cascade feed-forward neural network composed of an input layer, an output layer and four hidden layers. Datasets where provided from the DRIVE [11] and STAER [6] databases.

Choi et al. [4] have also used OCTA to analyse 3D images of neovascular age-related macular degeneration. The reconstructed 3D images were used to evaluate three markers: (i) the volume, (ii) average volume per slice, and (iii) z-axis of the volumetric structure. There are 32 patients with type 1 and 2 on nAMD, treated with prorenata (PRN), who were followed-up form 14 months. The method signaled that the nAMD volume has decreased significantly from baseline to 12 weeks. Technically, Angio tool has been used to quantify the MNV area and the count of open-ended capillaries. To render 3D MNV images, enface images of 4 μm thickness have been collected in the direction of inner retina. Fiji tool was used to convert to an eight-bit image. The Voxel counter tool was used to measure volumetric characteristics of reconstructed images. Finally, a 3D Viewer plugin was used to visualise the images.

For a quantitative approach, several biomarkers were previously tested, such as mCNV area (mm^2), vessel area (mm^2), vessel density, vessel length (mm), vessel diameter (μm), fractal dimension, vessel junction, junction density (n/mm) or vessel tortuosity. However, since the MNV is a 3D structure, their calculation proved to be insufficient to describe the membrane and its behaviour after intravitreal anti-VEGF injections. After observing the evolution of the membranes and their response to treatment, we will be closer to understanding why some patients respond better and faster to therapy than others. Therefore, a volumetric reconstruction and volume calculation seem imperative to deal with the burden induced by this disease [2, 3, 14].

3 Proposed Method and Experiments

3.1 Data Sources

Data was fully provided by Department of Ophthalmology from "Iuliu Hațieganu" University of Medicine and Pharmacy, Cluj-Napoca, Romania. Images were taken with Heidelberg Spectralis HRA + OCT system, an expandable diagnostic imaging platform which combines scanning laser fundus imaging with high-resolution OCT, and exported from Heidelberg Engineering Inc. Heyex2 [6] Software from a retrospective study population included thirty six patients with a total of 118 scans who were diagnosed with nAMD. When the images were exported from Heyex2 Software [6] the "Projection artifact removal" option was checked, the contrast was set to "Auto" and a full thickness section with MNV was manually selected. The dataset contains images captured by two different methods: *high-speed* and *high-resolution*. The *high-resolution* version is used as input for the proposed method because of the resolution ($11.34 \times 11.34 \times 3.87$ compared to $5.54 \times 5.54 \times 3.87$). The resulting data set exported for one case contains one main image and two sets of 90(±45) image sections with a $4\ \mu m$ or $25\ \mu m$ thickness, an image sample is illustrated in Fig. 1.

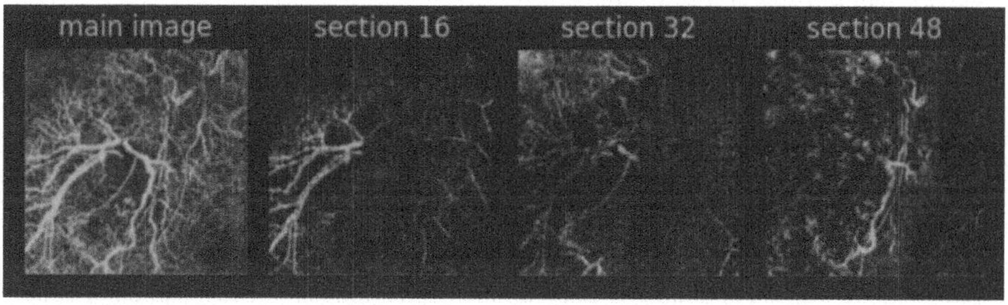

Fig. 1. Main image and depth sections 16, 32 and 48 from 84 sections exported from Heyex2 Software with MNV section at a thickness of 25μm, dataset *ie*

3.2 Technical Challenges

The first challenge in segmenting the neovascularized area is the elimination of noise. In the first image from Fig. 1, where the neovascularization area is highlighted, we can distinguish the area of interest with the naked eye. However, from the machine perspective: (i) the area is in grayscale encoding; (ii) the blood vessels are not continuously; (iii) for some regions, the pixel intensity and density is higher for noise rather than for blood vessels.

The second challenge is to identify blood vessels from a group of stacked volume sections (Fig. 1), where the neovascularization area is captured in depth. Here, the difficulty to distinguish the area of interest for at least two causes: (i) the blood vessels are not continuously at all, the section is a slice from a volume and blood vessels can go up and down or they have low intensity zones; (ii) the pixel intensity and density is on average the same for both noise and blood vessels.

The proposed method consists in a Python implementation with four steps: (1) creating vessel mask; (2) isolating the Macular Neovascular Membranes from noise; (3) finding the vessel depth in volume; (4) reconstructing the 3D volume. The workflow has the step in Algorithm 1.

Create the Vessel Mask Skeleton

First, the main and inner section images are preprocessed, since the Heyex 2 [6] software exports the image with whole captured area and a default border. Hence, the OCTA scans need to be cropped to get the interest area, and then a median filter is applied to get a small dilatation of the main image from Fig. 1.

Then, a preliminary mapping mask is created and applied to the set of image sections to find the blood vessel depth in the neovascularization volume. For this, we have tested four variants of blood vessel segmentation: (1) applying an intensity filter; (2) applying a neighbours filter; (3) applying the snake-expand strategy; (4) processing the pixels intensity as a wave.

1. *Applying an intensity filter*. For each point get the average intensity from a three different masks sizes (3×3, 9×9, 25×25) and check if the inner one is the most intense. This method shows average results because it can not differentiate closed vessels and the vessel is not complete.
2. *Applying the neighbours filter*. This method started from the ideas of Babin et al. [1], in which CT angiography images of brain blood vessels are segmented using a combination of pixel neighbourhood structure and intensities of the object intended for segmentation. Differently from CT angiography images, our OCTA dataset contains more noise and choriocapillaris. That is a thin and dense meshwork of large capillaries located in the innermost part of the choroid that serves as the major source of nutrition for the retinal pigment epithelium (RPE) and outer retinal layers [10]. Therefore, for our task, we mix this method by keeping the projections all around and count a path with at least 7 neighbours keeping the segmentation function $s = (mdr + avr + max)/3$, where minimum (*min*), maximum (*max*), average (*avr*) and mid-range (*mdr*), defined as the average of the minimum and maximum value). The resulted segmentation is illustrated in Fig. 2.
3. *Applying the Snake-expand strategy*. As a Snake game approach, starting from a single, high intensity and high probability to be a vessel pixel, then expand the path to all other connected pixels, where, in a best case scenario, all blood vessels should be connected. This strategy failed to differentiate both the small vessels and the noise.
4. *Processing the pixels intensity as a wave*. The most sensitive and accurate method was taking each row, column, two diagonals (from top left and top right) and process the pixels intensity as a wave. The results are illustrated in Fig. 3. The graph in Fig. 3(b) shows five prominent curves. The first spike is between ± 45 and ± 55 (Y axis), the second one from ± 80 till ± 90 (Y axis), the third and the fourth from ± 105 till ± 125 (Y axis), while the fifth one from ± 140 till the end of the Y axis. The Finding peaks algorithm was applied to both filtered and non-filtered waves, where filtered waves method was used for further processing. The Finding peaks algorithm takes a list of values as an input, in this case each row, column and two diagonals lightly smooth filtered and 5 parameters: (i) minimum distance between peaks; (ii) how prominent

the spike is related to his neighbours; (iii) window length to compare spikes; (iv) relative height at which the peak width is measured; (v) width, peak width range. The downside of the current implementation is that, for the best results, those parameters needs to be lightly adjusted for each scan. For further calculations, we keep the spike (middle point of the vessel), and spike width. The advantages of the peaks-based method are: (i) high sensitivity and accuracy in finding and marking the midpoint of the blood vessel; (ii) almost all vessels were marked and identified, (iii) provides additional information about a blood vessel, such as the midpoint and approximate width.

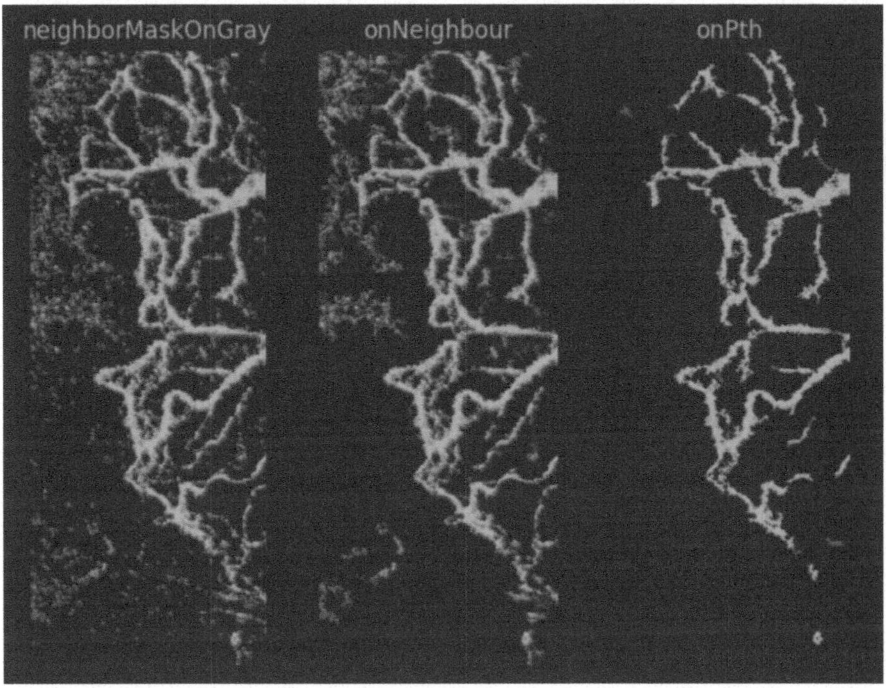

Fig. 2. Combinations of filtering methods. First image: *Neighbours Intensity filter*, second image: second iteration for *Neighbours Intensity filter*; third image: *Neighbour Intensity filter* applied after *Intensity filter*, on dataset *s2*

Identifying the Shape of Blood Vessels

Having all prominent peaks and their corresponding peak width in a matrix was the starting point to find the connected peaks. Three methods. Have been investigated: (1) Computing the Average cross-correlation between each set of 5 consecutive spike; (2) Searching for connected components; (3) Recursively analysing the neighbours.

Method$_1$. Computing the Average cross-correlation between each set of 5 consecutive waves. The most accurate method was to compute the average cross-correlation between each set of 5 consecutive spikes, and if the value is greater than a threshold, we consider

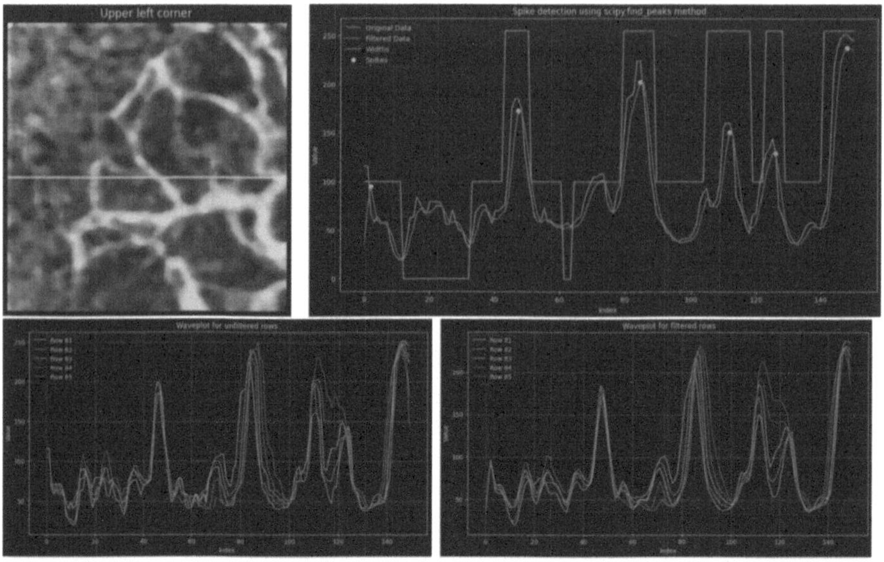

Fig. 3. (a) *Upper left corner* of the dilated image and test row marked with a white continuous line in the middle; (b) *Spike detection* method applied to test row; (c) *Wave plot* for test row and 4 neighbour rows; (d) *Wave plot filtered* for test row and 4 neighbours rows

that there is a blood vessel. As illustrated in Fig. 3(d), if we shrink the wave to fit in the minimum window spike of all 5 consecutive waves, the cross correlation for test row is presented in Table 1. Even if the points are not continuous, we can mark a high probability for a point to be part of a continuous blood vessel wave-like form. We retained in a new matrix all the spikes that, based on the cross-correlation analysis, are determined to be blood vessels. All connected spikes, which, through the spike identification algorithm are unique within a minimum window size X, were considered to be a continuous part of those blood vessels even if the cross-correlation result was not as expected. The disturbed cross-correlation value in a real blood vessel spike can occur when the blood vessel is not continuous in the image or is about to branch. The downside of this approach is that will identify also the healthy blood vessels in the background.

Table 1. Cross correlation values for rows 81–85 on *x* axis, for the 4 peaks between *y_min:y_max* window for dataset *s2*

	P1					P2					P3,4					P5				
X	81	82	83	84	85	81	82	83	84	85	81	82	83	84	85	81	82	83	84	85
Y	45:55					80:90					105:125					140:160				
Corr	.96	.96	.95	.94	.98	.78	.78	.85	.86	.90	.95	.92	.89	.88	.90	.83	.86	.81	.86	.94

Method$_2$. *Searching for connected components.* Aiming to keep the connected components, two algorithms were tested: a Depth-First search and a Flood Fill algorithm that considers components with up to X length and Y position gaps as connected. The method takes two parameters, the minimum length for connected peaks to be considered a blood vessel, and a maximum empty spaces for how many empty (missing peaks) spaces can a blood vessel have in a 5×5 matrix. Both algorithms provide average results.

$Method_3$. Recursively analysing the neighbours. The method consists of recursively examine, for each point marked as a spike, the neighbouring points that fall within the corresponding window for each peak, extending the network as long as new spikes are encountered or up to a number X of empty spaces, which will then continue with the same or a new blood vessel if the vessel shows an interruption. This is a good technique that also reconstructs the blood vessel but does not accurately mark the junction areas between multiple blood vessels.

$Method_2$ and $Method_3$ show good results, where, on average, 9 of 10 blood vessels are marked, and 9 of 10 areas with high noise are identified. Due to the fact that, in a certain area, a blood vessel is continuous and presents approximately the same characteristics, by correlating the waveform from X consecutive points, we preferred $Method_1$, as it provides a metric for the shape of a blood vessel.

Find the Vessel Depth in Volume

Identifying all peaks and their corresponding peak width, two kernels were tested for both 2D and 3D processing. The first kernel as a 2D circle and 3D sphere, with high values in the centre that decrease toward the edges and continue with negative values outside. The second kernel, 2D mask and 3D mask, extracted as it is from the resulted mask illustrated in Fig. 4 column *e*. Figure 5 shows a sample from the founded depths, while Table 2 compares them.

Reconstructing the 3D Volume and Volume Approximation

Based on identified peaks, their width and depth, we get the surrounded volume and keep the most intensive points. The 3D rendering was done with the Vedo library [7], a Python module for scientific analysis and visualisation of 3D objects and point clouds, where the input was a 3D array representing a dense point cloud image data. Volume estimation was calculated on the resulted 3D matrix. Based on image scale provided by the OCTA device with the following defaults by manufacture units in micrometers for one pixel: *5.54(width) * 5.54(height) * 25(thickness, default: 3.87)* and thus one square pixel, or a voxel, will have $767.29\ \mu m^3$.

4 Running Experiments

We tried different methods and techniques to obtain the clearest image before applying the wave peak selection. This was tricky, because filtering the image also results in the loss of thin blood vessels. Firstly, we attempted to eliminate as much noise as possible using various filters. We started with a basic threshold filter based on intensity percentile and employed different techniques that consider area intensity, including both custom methods and established methods like the Sauvola filter [9]. Afterward, we tried the flood fill algorithm and various histogram-based techniques, but all these methods also removed parts of the blood vessels. Therefore, we decided to use only a simple image

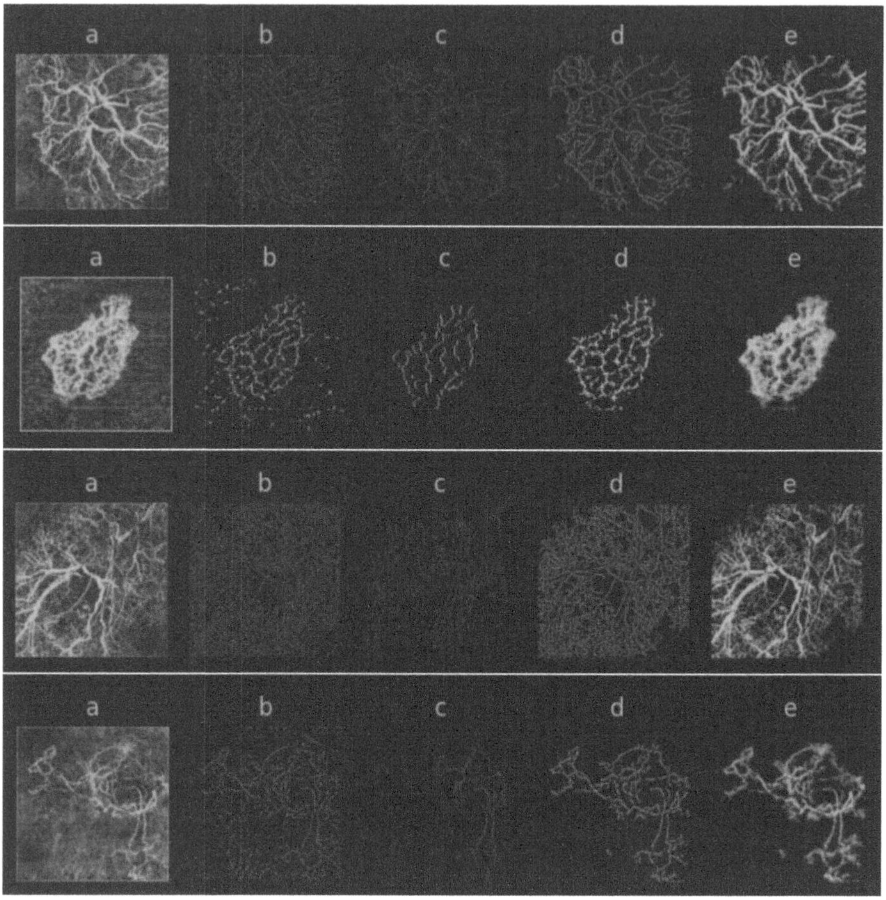

Fig. 4. Identify blood vessels, multiple examples: (a) the main dilated image; (b) find peaks algorithm on filtered waves for 4 axes (rows, columns and two diagonals); (c) filtered points by length and cross correlation; (d) all connected neighbours between 'b' and 'c' images; (e) all connected neighbours expanded to the corresponding width. Datasets: row 1: *s2*; row 2: *d1*; row 3: *ie*; row 4: *s3*

dilation to make everything smoother which was also the most effective and keeping three iterations for main image and two for inner sections. We keep the number of iterations low to avoid removing thin or discontinuous blood vessels (Fig. 6).

In the first step, for each row, column and the first two diagonals, from upper right and upper left, we extracted and processed the resulting wave. Since we did not have a training dataset, the parameters for the spike detection algorithm were tested manually to find the effective values. The spike detection algorithm was used from *scipy.signals.find* peaks method [12]. The most important parameters were the *prominence* and *rel_height* which chooses the relative height at which the peak width is measured as a percentage of its prominence. There is a big difference (±20% accuracy) by using both rows/columns

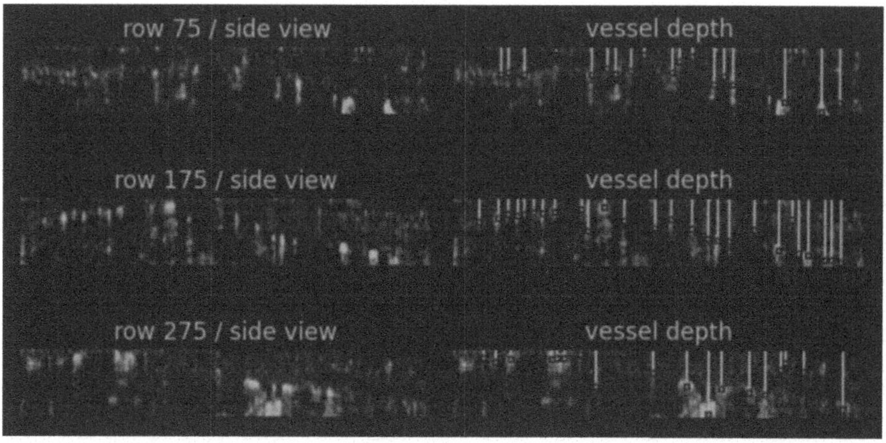

Fig. 5. Side view to row 78, 175 and 275 founded depths using *3d_mask* for blood vessel centre on dataset *ie*. Mapping accuracy is described in Table 2

Table 2. Mapping accuracy for depth measured on side views, Fig. 5, with kernels: *2d_circle / 3d_sphere* - high values in middle decreasing to edges and *2d_mask / 3d_mask* on dataset *ie*

Row / Kernel	2d_circle	3d_sphere	2d_mask	3d_mask
73	9.5	8.5	9	9.5
74	9	9	9.5	9.5
75	9	9	9.5	10
173	9.5	8.5	10	9.5
174	9	8	9.5	10
175	8	8	9	10
273	8	8	8.5	9
274	8	8	9	9.5
275	8	8	9	10
Avg	8.89	8.22	9.22	9.56

and two diagonal comparing to rows/columns only. A reference set of parameters can be seen in Table 3, prefixed with *FP*.

To keep only the good spikes we traversed the image recursively, following neighbouring rows to create the longest possible path in the given window and compute the cross correlation between each point in path and 4 neighbours, two in front and two behind. An example of the wave similarity between the selected point and its neighbours can be seen in Fig. 2(d) and the cross-correlation results in Table 1. The algorithm can accept gaps in the founded path, and we have set a maximum gap value of 7 pixels. This approach will also rebuild the blood vessel if there are discontinuities in the captured image. The results are illustrated in Fig. 7(c). Then we keep all closed neighbours to the founded points. The result is illustrated in Fig. 7(d) where we got all the blood vessels in the form of a skeleton. A reference set of parameters can be seen in Table 3, prefixed with *CC*.

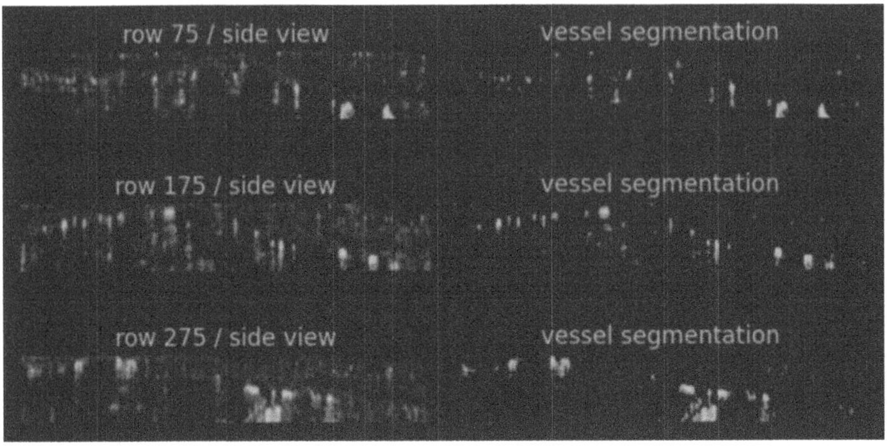

Fig. 6. Side view for vessel segmentation on row sections, dataset *ie*

Table 3. Sample output for three datasets

Exp.	High-Res	Img. No.	width × height	No. of voxels	Volume (μm^3)
s2	Yes	57	509 × 509	1,025,665	786982497.85
d1	No	42	141 × 141	43,169	33123142.01
ie	Yes	84	513 × 513	992,367	761433275.43

With all the blood vessels represented as a skeleton within a matrix and a width reference for each peak, we were able to apply a convolution kernel to each peak to determine the depth within the volume at which they are positioned. For each point, we extracted from the main image a $2{*}width \times 2{*}width$ section around the point and applied the resulting kernel in the image depth on an identical section that includes all the images from the volume. We have tried other kernels, a 2D circle and a 3D sphere but the results were poor and even after two or three iterations the mapping was not as expected. A comparison between the kernels can be seen in Fig. 8 and Table 2.

Algorithm 1 The main sequences

$OCTAimg \leftarrow preprocess()$
$peaks[direction, width, cross\_correlation, depth, intensity] = np.array$
$peaks[spikes, widths] \leftarrow find\_peaks(img)$ ▷ Find all spikes across 4 axes
$peaks \leftarrow filter\_peaks(peaks)$ ▷ Good cross-correlation and length spikes
$peaks[depth] \leftarrow find\_depth(peaks)$ ▷ Get the vessel depth
$result \leftarrow get\_segment(peaks)$ ▷ Get the final section from volume

In the final step, we extracted a 3D rectangular shape from the initial volume corresponding to the width and depth dimensions of each peak, and retained only the pixels

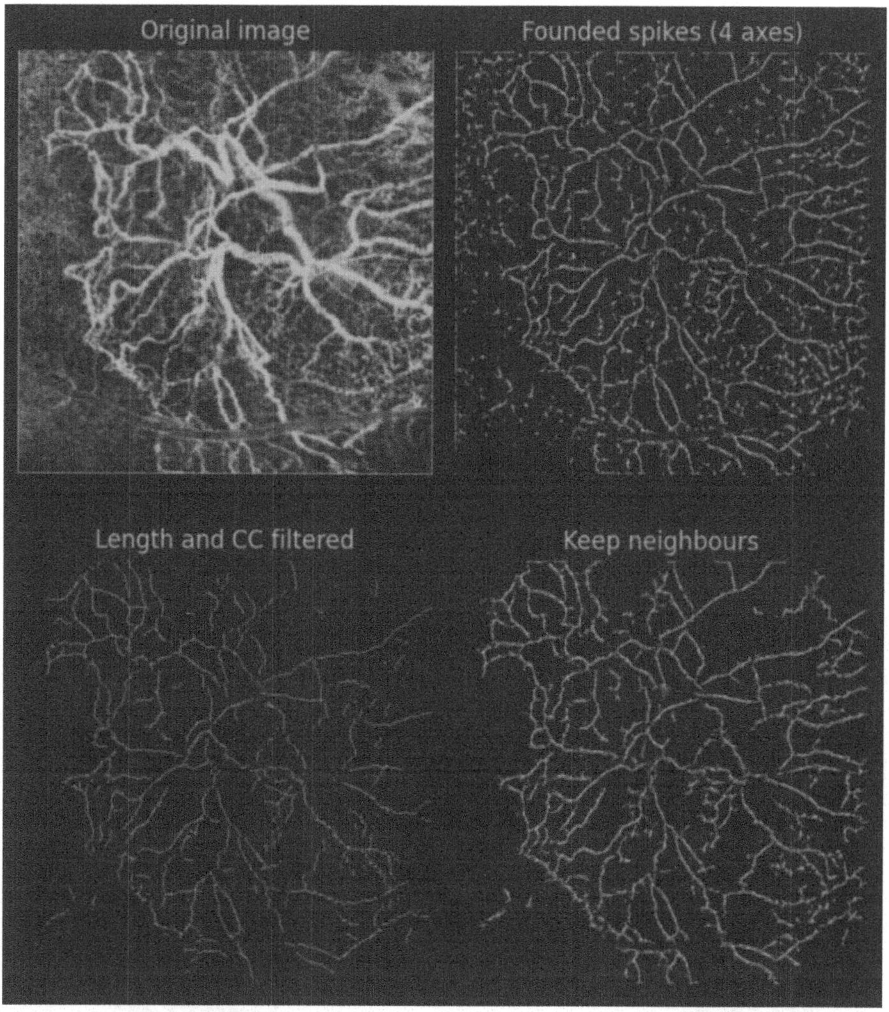

Fig. 7. Identify blood vessel shape: (a) *Original image* - the main dilated image, (b) *Founded spikes* -all founded peaks with filtered waves on 4 axes (rows, columns and two diagonals), (c) *Length and CC* - length and cross correlation filtered points and (d) *Keep Neighbours* - connected neighbours between second and third images

with high intensity. There is a significant difference from getting the corresponding area in 3D and not in 2D. Because getting the most intensive points from a section volume is a step with impact over the final result, we tried to filter each section volume using the DBSCAN clustering algorithm [5] or other similar approaches done manually. The DBSCAN algorithm, which theoretically should work best because it allows us to retain only the cluster with the highest intensity for each extracted volume, was computationally costly, so we set this computation aside for the next stage. For retaining only the intense points around the middle point we use the idea of Babin et al. [1] with a

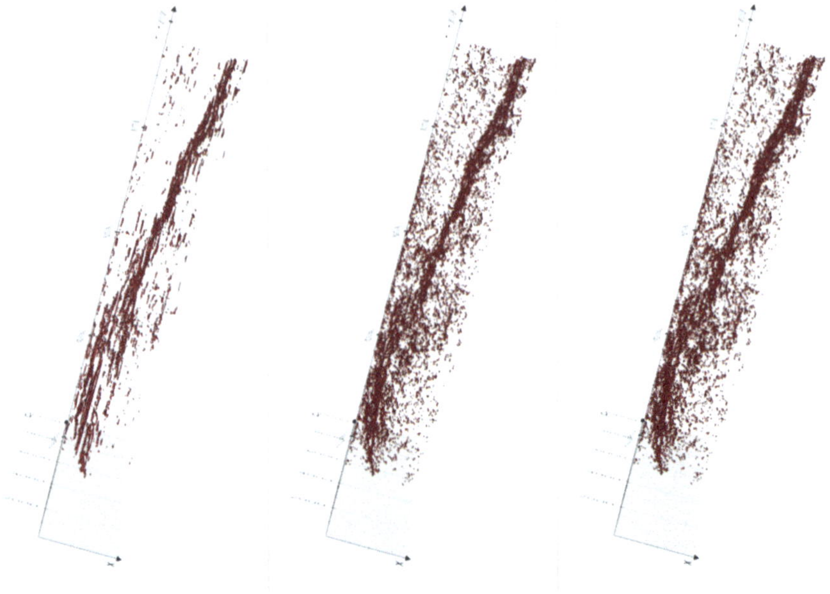

Fig. 8. 3D side view for founded depths with *2d_mask / 2d_circle / 3d_sphere* kernels for dataset *ie*

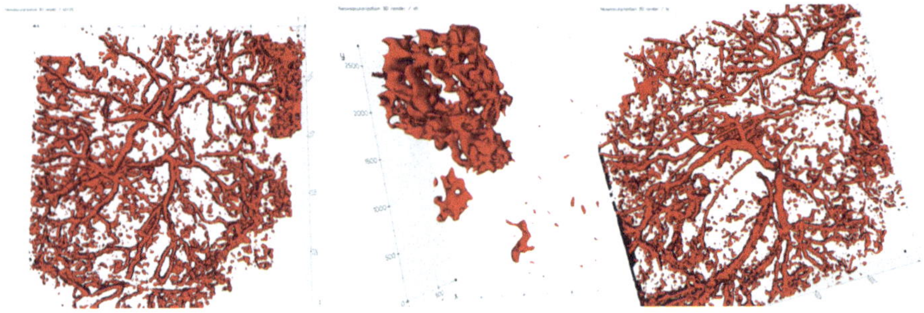

Fig. 9. 3D output for dataset *s2, d1, ie*

combination of pixel neighbourhood structure and intensities of the object intended for segmentation. For rendering, the Vedo [7] library was used to plot the 3D visualisation. A sample of 3D output for 3 images (*s2*, *d1* and *ie*) in Fig. 9 and Fig. 10. The code is available on GitHub (Table 4).

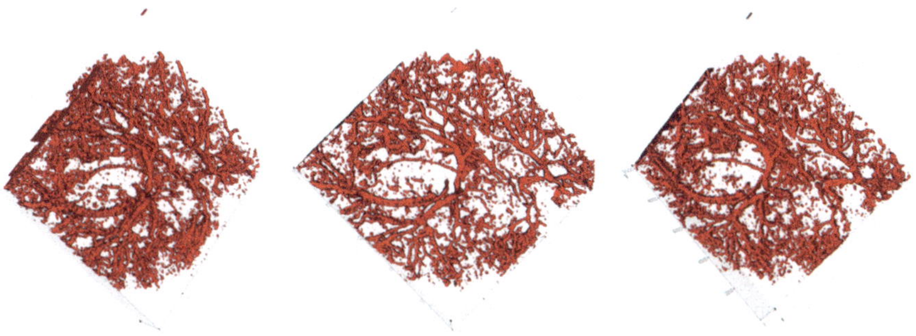

Fig. 10. 3D output in three angles for dataset *ie*

Table 4. Reference values used for algorithm parameterisation

	FP			CC	
Exp.	Prom.	Wlen	Rel_{height}	$Path_{length}$	Tresh
s1	[35, 200]	30	0.8	5	0.7
s2	[50, 200]	40	0.8	15	0.8
s3	[60, 200]	40	0.8	10	0.85
u4	[50, 200]	40	0.7	10	0.8
ie	[40, 200]	45	0.7	10	0.85
d1	[40, 200]	45	0.7	10	0.8

5 Conclusion

Blood vessels were identified by detecting and marking spikes from four different directions—row, column, and two diagonals—treated as a wave form, combined with cross-correlation among five neighbouring points. When the parameters are manually optimised for the available dataset, the method can go up to 100% accuracy. Furthermore, using a kernel extracted from the main image, the depth of a blood vessel was determined with an accuracy of 95.6%. By extracting a 3D volume centered around the point of interest, based on previously calculated coordinates, and retaining only high-intensity points, a 3D reconstruction of the neovascularization area in AMD pathology was achieved. For OCTA scans containing AMD biomarkers, the application of filters is effective only if the filtering of the image or the waves to be processed is minimal. Ongoing work regards replacing the manual tuning of the parameters with a semiautomatic one, possible guided by genetic programming or deep learning.

Acknowledgments. This study is a part of project « New optical coherence tomography biomarkers identified through deep learning for risk stratification in patients with age-related macular degeneration » project code PN-III-P2-2.1-PED-2021-2709.

Disclosure of Interests. The authors have no competing interests to declare that are relevant to the content of this article.

References

1. Babin, D., Vansteenkiste, E., Pizurica, A., Philips, W.: Segmentation of brain blood vessels using projections in 3-d ct angiography images. pp. 8475–8478 (2011)
2. Bilc, S., Groza, A., Muntean, G., Nicoara, S.D.: Interleaving automatic segmentation and expert opinion for retinal conditions. Diagnostics. **12(1):22** (2021)
3. Bressler, N.M., Bressler, S.B., Fine, S.L.: Neovascular (exudative) age-related macular degeneration. In: Volume 2: Medical Retina, pp. 1075–1113. Elsevier Inc. (2005)
4. Choi, M., Han, S., Kim, S.-W., Yun, C., Jaeryung, O.: Volume-rendering three-dimensional image analysis of macular neovascularization in age-related macular degeneration. Eye. **38**(6), 1125–1132 (2024)
5. Ester, M., Kriegel, H.-P., Sander, J., Xiaowei, X., et al.: A density-based algorithm for discovering clusters in large spatial databases with noise. Inkdd. **96**, 226–231 (1996)
6. Hoover, A.D., Kouznetsova, V., Goldbaum, M.: Locating blood vessels in retinal images by piecewise threshold probing of a matched filter response. IEEE Trans. Med. Imaging. **19**(3), 203–210 (2000)
7. Musy, M., et al.: marcomusy/vedo: 2024.5.1 (2024)
8. Tchinda, B.S., Tchiotsop, D., Noubom, M., Louis-Dorr, V., Wolf, D.: Retinal blood vessels segmentation using classical edge detection filters and the neural network. Inform. Med. Unlocked. **23**, 100521 (2021)
9. Sauvola, J., M. Pietik¨ainen.: Adaptive document image binarization. Pattern Recogn. **33**(2), 225–236 (2000)
10. Savastano, M.C., Rizzo, C., Gambini, G., Savastano, A., Falsini, B., Bacherini, D., Caputo, C.G., Kilian, R., Faraldi, F., De Vico, U., Rizzo, S.: Choriocapillaris vascular density changes: healthy vs. advanced exudative age-related macular degeneration previously treated with multiple anti-vegf intravitreal injections. Diagnostics. **11**(11) (2021)
11. Staal, J.J., Abramoff, M.D., Niemeijer, M., Viergever, M.A., van Ginneken, B.: Ridge based vessel segmentation in color images of the retina. IEEE Trans. Med. Imaging. **23**(4), 501–509 (2004)
12. Virtanen, P., et al.; SciPy 1.0: Fundamental algorithms for scientific computing in python. Nat. Methods. **17**, 261–272 (2020)
13. Wang, C., Zhao, Z., Yu, Y.: Fine retinal vessel segmentation by combining nest u-net and patch-learning. Soft. Comput. **25**(11), 5519–5532 (2021)
14. Wang, Y., et al.: Optical coherence tomography angiography-based quantitative assessment of morphologic changes in active myopic choroidal neovascularization during antivascular endothelial growth factor therapy. Front. Med. **8**, 657772 (2021)
15. Abdul-Ameer, H.K.: 3D-shape formation of blood vessels based on computer aided design system. Adv. Mech. Eng. **15**(6) (2023). https://doi.org/10.1177/16878132231181552
16. Van Rossum, G., Drake, F.L.: Python 3 Reference Manual. CreateSpace, Scotts Valley, CA (2009)

Artificial Intelligence in Pediatric Pneumonia Diagnosis

Evelin H. Dulf[1], Raul-P. Banut[2], and Alexandru G. Berciu[2](✉)

[1] Iuliu Hatieganu University of Medicine and Pharmacy, Cluj-Napoca, Romania
[2] Technical University of Cluj-Napoca, Cluj-Napoca, Romania
Alexandru.Berciu@campus.utcluj.ro

Abstract. Pneumonia is a common occurrence in children under the age of five, particularly in those with underlying conditions. It is the leading cause of death in children worldwide. Chest radiography is currently the most effective method for identifying pneumonia. Several research papers have discussed the potential of computer-aided diagnostic tools for pneumonia, but none of them have been designed for pediatric patients. This work addresses the lack of such a tool by proposing a solution based on supervised learning. A comparative study was conducted to ascertain which convolutional neural network performed optimally. The obtained results are comparable with those reported in the literature for adults. The final model is included in a website created for clinical use. The proposed application has the potential to assist clinicians in interpreting data sets and to facilitate accurate diagnoses, particularly in underdeveloped regions where clinicians may lack the necessary expertise.

Keywords: Pneumonia · Artificial Intelligence · Pediatric Patients

1 Introduction

Pneumonia is a common acute respiratory infection affecting the alveoli and airways. The disease can be divided into community-acquired pneumonia (CAP) or hospital-acquired pneumonia (HAP, including ventilator-associated pneumonia VAP). The microorganisms that can cause CAP and HAP differ greatly. The most common microorganisms for CAP are Streptococcus pneumoniae, Haemophilus influenzae, Mycoplasma pneumoniae and Chlamydophila pneumoniae. In the case of HAP, it is Staphylococcus aureus [1].

Treatment differs depending on the type of micro-organism that has infected the individual. Streptococcus pneumoniae is treatable and often seen in children. It can occur following influenza and can be dangerous for the elderly or people with various medical conditions. People aged between 15 and 35 can acquire Mycoplasma pneumoniae. This can sometimes pass without treatment, sometimes requiring antibiotics. An effective vaccine is available for Haemophilus influenzae. Chlamydophila pneumoniae is acquired in the community and is transmitted by droplets of saliva during a coughing fit. Pneumonia caused by Staphylococcus aureus is uncommon but has a high mortality

S. Vlad and N. M. Roman (Eds.): MEDITECH 2024, IFMBE Proceedings, 130, pp. 129–139, 2025.
https://doi.org/10.1007/978-3-031-95671-3_15

rate. It can occur through direct contact with a sick person's skin or personal belongings [2].

Pneumonia often occurs in children up to 5 years old and in the elderly with various chronic conditions. Certain factors and comorbidities can predispose patients to developing the disease. These include comorbidities such as asthma, heart failure, diabetes, chronic kidney disease, and factors such as smoking, alcohol/drug use, poor dental hygiene and nutritional status, overcrowded premises or contact with carriers (birds, cats, rodents) [2].

In children, it is considered the leading cause of death worldwide. According to the World Health Organization, every year, an estimated 1.4 million children under the age of 5 die, accounting for about 18% of all under-5 deaths worldwide. Pneumonia affects children and families everywhere but is most common in areas of South Asia and South-Saharan Africa. Although diagnosed in the early stage, it can be prevented with simple interventions and treated with low-cost, low-tech medicines and care. Worldwide, statistics show that about two million people suffer from pneumonia each year [3].

Today, chest radiography is one of the best methods of identifying pneumonia, and is often chosen instead of a Computer Tomography (CT) scan. In underdeveloped regions where high-quality CT scans may not be available, X-rays are used. This technique is the most common and widely available identification imaging technique, which has an important role in patient care and epidemiological studies. Also, the lack of radiologists in underdeveloped areas and their predictability of disease can affect the lives of some classes of patients. Consequently, an Artificial Intelligence (AI) -based identification solution would assist physicians in diagnosis, as it is a primary focus of numerous researchers in the field [4].

Artificial intelligence (AI) is a relatively new and rapidly evolving field. Its applications are becoming increasingly sophisticated, enabling it to perform tasks that were previously thought to be beyond its capabilities. AI is now a valuable tool that brings numerous benefits to our lives, simplifying many processes and reducing the need for repetitive tasks. Its utility has been demonstrated in a wide range of areas, including the reduction of repetitive tasks for employees and the automation of decision-making in fast-changing contexts. It is anticipated that in the near future, artificial intelligence will play an even more significant role in people's lives, as well as in the field of medicine. While it is unlikely to replace a doctor entirely, it will be able to assist them in various ways. It will be able to suggest personalized treatments or identify specific cases of illness.

In the paper [5], the authors presented five types of Neural Network architectures that could be used to classify Pneumonia and discussed different approaches of other researchers. The dataset consists of 5240 images, from which 25.7% of images are classified as normal, 65% are labelled as bacterial pneumonia and the remainder are labelled as viral pneumonia. It is concluded that the best results were achieved by the ResNet18 model, which had 94.23% accuracy and 91.58% precision.

Wang et al. [6] contributed to the field with a dataset containing 108,948 frontal images from 32,717 patients.

In the paper [7], the authors discuss the challenges faced with databases in such image-processing-based diagnostics. The first one is the large number of images needed.

The common trend to mitigate this problem is to modify a small dataset into a larger one using augmentation techniques. It is reported that data augmentation can improve classification accuracy. In this study, the authors used three augmentation strategies to generate new datasets (Rotation, Scaling and Translation). The second discussed challenge lies in the image sizes, as each algorithm needs different dimensions. Thus for AlexNet, they needed images of 227x227 pixels, and for DenseNet and ResNet they used images with dimensions of 224x224 pixels. The best results reported by these researchers were obtained by the DenseNet network. It had the highest accuracy for both training and testing. The accuracy of the test was between 93.2% and 95% for different types of pneumonia.

Paper [8] discusses the computer-aided diagnosis possibilities in COVID-19 pneumonia and typical bacterial pneumonia in patients from the emergency department. The best-reported performances have an accuracy of 93.8%, with a misclassification rate of 6.2% and a weighted F1 of 93.8.

Automated tools to assist individuals are widely discussed in the literature [9–12]. However, to the best of our knowledge, software for automated diagnosis of pediatric patients with pneumonia is not published in the literature.

The present work addresses the lack of computer-aided diagnostic tools for pediatric patients. It proposes a solution based on supervised learning, using a labelled dataset. This application has the potential to assist clinicians in interpreting data sets and to facilitate accurate diagnoses, particularly in underdeveloped regions where clinicians may lack the necessary expertise.

The paper is structured into four sections. Following this introductory section, section 2 outlines the materials and methods employed. Section 3 presents the results obtained, which are then discussed. The paper concludes with a summary of the key findings.

2 Materials and Methods

The lung imaging set used as the database for training the neural network came from Guangzhou Children's and Women's Medical Center from pediatric patients aged one to five years [8]. For X-ray image analysis, all chest radiographs were initially checked for quality by eliminating all low-quality scans. Imaging diagnoses were graded by two expert physicians before being approved for network training. To account for possible classification errors, the grading set was also checked by the third physician. The dataset is organized into 3 folders (train, test, wave) and contains subfolders for each category (Pneumonia/Normal). Within the set, there are 5683 x-ray images in .jpeg format and 2 image categories (Pneumonia/Normal).

The features utilized for diagnosis are illustrated in Fig. 1, which displays a typical image from the database along with its corresponding annotations. The key anatomical landmarks identified in the diagnosis include the aortic pivot, costocardiac angle, costophrenic angle, carina, clavicle, and trachea. The aortic pivot is seen in the chest radiograph around the T3/T4 level or lateral to the carina. In patients with aortic aneurysm, this may be the area contributing to the "widened mediastinum". The costocardial angles, as well as the costophrenic angles, should be fairly sharp and well-defined if the patient has no significant effusions or pulmonary edema. If they are blurred or lost, it could be a sign of fluid in the lungs or a mass obstructing vision. The carina is the point that divides the trachea into the main right and left bronchus. It lies on the midline. The carina is also the location used by clinicians when assessing the correct position of an endotracheal tube after intubation. The clavicle is attached to the lateral surface of the sternum. The position of the clavicular heads concerning the trachea can assist in determining whether the patient was correctly positioned during radiography. The two clavicular heads should be on either side of the trachea, and the spinous apophyses should be in the middle. The trachea should be placed on the midline and lie between the ends of the right and left clavicular heads. Any deviation from the midline could suggest the presence of a mediastinal mass or the presence of a pneumothorax tension.

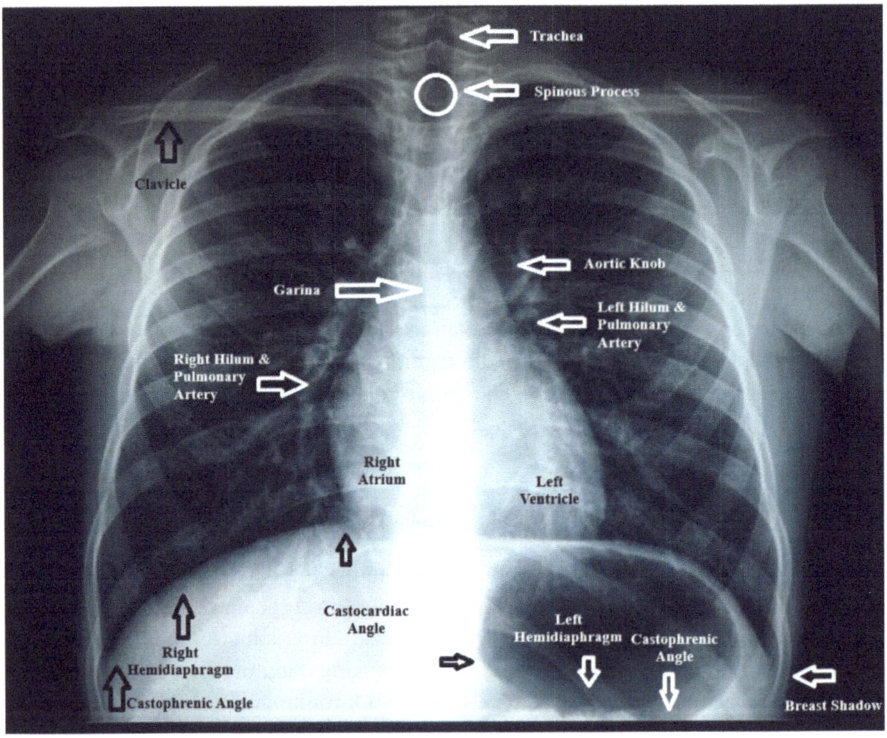

Fig. 1. Common landmarks on chest radiography

Based on the state of the art presented in the previous section, transfer learning is addressed to develop a model that can identify Pneumonia. The model was trained using Google Colab. Prior to implementation, a comparative study was conducted to identify the most suitable architecture for classifying lung images. The objective of the study was to compare the performance of different architectures. The same settings were used in the training process to ensure consistency.

It is not possible to calculate neural network weights by analytical methods; they must be discovered by optimising the chosen algorithm. The chosen optimisation method is called ReduceLROnPlateau. For the trained model, it was chosen to decrease the learning rate after 2 epochs of no changes occurring.

The training of the neural networks was conducted on a graphical processing unit (GPU), necessitating the identification of an optimal balance between the subsets utilized, the values assigned to periods, resources, learning rate, and network depth to prevent memory errors. The number of epochs utilized was 12, with each epoch encompassing forward transitions and backward transitions through all training samples. To avoid overtraining, data were shuffled at the beginning of each epoch.

The learning rate interacts with many aspects of the optimization process. It found that low learning rates require more time and more epochs for training, whereas high rates require less time and fewer epochs. The chosen factor with which to calculate the new learning rate is equal to 0.3, and the lower bound for the learning rate has been set as 10^{-6}.

The amount of data included in each training epoch is an important parameter and is chosen as 32 frames. This is a subset of the training batch (miniBatchSize), which is used for gradient evaluation and modification of the parameters involved in the training process. The determination process of the exact batch size is experimental, as it depends entirely on the data to be kept. It has been observed that a small size of a subset not only leads to faster training but also exhibits generalization of the test dataset compared to a large batch size.

Image classification was performed using a generally developed algorithm capable of automating its development for the comparative study to take place. For image classification, the final classifier and the final training layer of the neural network use the extracted features. In all networks, the last two or three layers usually contain detailed information about how features are combined. The images have been resized to match the input size of each network. The chosen size is 150 x 150 pixels.

The dataset has been artificially expanded in order to eliminate the problem of overfitting. The basic idea is to modify the training data with small transformations in order to reproduce the variations. In order to augment the data, the present work uses the following techniques:

- randomly rotate some training images by 30 degrees;
- randomly enlarge some training images by 20%;
- randomly crop images horizontally by 10% of their width;
- randomly shift images vertically by 10% of height;
- randomly shift images horizontally.

Among the networks tested was the GoogLeNet model, the one developed by Google [14]. This architecture aims to learn with a reduced number of parameters. The data are divided into 3 subfolders, respectively train, test and wave (validation). The train folder contains 88% of the data, the test folder 8% and the validation folder only 4%. Validation was performed at every 230th iteration. This model was used as a standard for the network created from scratch. The structure of the created network can be seen in Fig. 2. The activation function used for the Conv2D layers was ReLU. Value strides were equal to 1, specifying how large to make the displacement step on a data set, and the padding was set to "same". The output size was equal to that of the inputs, in the case of MaxPool2D layers, matrices of 2x2 size. The first layers of the network learned the general features of images, such as contours, and texture, and the deeper layers dealt with extracting the features of each image.

Layer (type)	Output Shape	Param #
conv2d_1 (Conv2D)	(None, 150, 150, 32)	320
batch_normalization_1 (Batch	(None, 150, 150, 32)	128
max_pooling2d_1 (MaxPooling2	(None, 75, 75, 32)	0
conv2d_2 (Conv2D)	(None, 75, 75, 64)	18496
dropout_1 (Dropout)	(None, 75, 75, 64)	0
batch_normalization_2 (Batch	(None, 75, 75, 64)	256
max_pooling2d_2 (MaxPooling2	(None, 38, 38, 64)	0
conv2d_3 (Conv2D)	(None, 38, 38, 64)	36928
batch_normalization_3 (Batch	(None, 38, 38, 64)	256
max_pooling2d_3 (MaxPooling2	(None, 19, 19, 64)	0
conv2d_4 (Conv2D)	(None, 19, 19, 128)	73856
dropout_2 (Dropout)	(None, 19, 19, 128)	0
batch_normalization_4 (Batch	(None, 19, 19, 128)	512
max_pooling2d_4 (MaxPooling2	(None, 10, 10, 128)	0
conv2d_5 (Conv2D)	(None, 10, 10, 256)	295168
dropout_3 (Dropout)	(None, 10, 10, 256)	0
batch_normalization_5 (Batch	(None, 10, 10, 256)	1024
max_pooling2d_5 (MaxPooling2	(None, 5, 5, 256)	0
flatten_1 (Flatten)	(None, 6400)	0
dense_1 (Dense)	(None, 128)	819328
dropout_4 (Dropout)	(None, 128)	0
dense_2 (Dense)	(None, 1)	129

```
Total params: 1,246,401
Trainable params: 1,245,313
Non-trainable params: 1,088
```

Fig. 2. The structure of the created network.

Evaluation metrics play an important role in finding the optimal model. Thus, their choice is an important step in obtaining the classifier.

One of the key concepts in performance calculation is the error matrix. Each row of the matrix represents the predicted classes, and each column represents the actual classes. The elements on the main diagonal are the correct predictions for each class, while the other elements indicate the wrong predictions. From the matrix are extracted the values of the terms true positive (TP), true negative (TN), false positive (FP) and false negative (FN). Using these values, are evaluated the performance measures called sensitivity, specificity, accuracy and F1-score, defined in Eqs. (1)–(4).

$$\text{sensitivity} = \frac{TP}{TP + FN} \tag{1}$$

$$\text{specificity} = \frac{TN}{TN + FP} \tag{2}$$

$$\text{accuracy} = \frac{TP + TN}{TN + TP + FN + FP} \tag{3}$$

$$\text{F1} - \text{score} = 2\frac{sensitivity \times specificity}{sensitivity + specificity} \tag{4}$$

In this case, the greatest attention will be paid to sensitivity, since it is the most accurate identification of patients with pneumonia. However, the F1 score will also be taken into account as a second main criterion, as it represents sensitivity versus accuracy.

For the final application, a website was created for clinical use. It was developed in the programming language React. Flask was used for communication between the client application and the created artificial intelligence model.

3 Results and Discussions

Table 1 shows the results obtained by the mentioned two models. Both models obtained good results, but the second model outperforms in the F1 score and accuracy. The corresponding confusion matrix is presented in Fig. 3.

Table 1. Performance measures obtained with the two models.

Model		Accuracy [%]	Specificity [%]	Sensitivity [%]	F1 score [%]
GoogLeNet	Pneumonia	94	90	89.5	93.3
	Normal	88	90.2	88	90
The proposed model (Fig. 2)	Pneumonia	92.3	91	90.1	94
	Normal	92	90	89.8	90.3

Based on the results it was chosen to integrate the second model into the final computer-aided diagnosis tool. These results outperform the results presented in the literature.

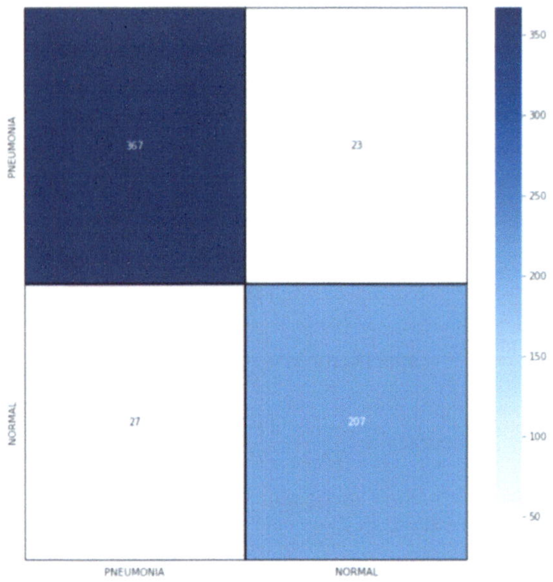

Fig. 3. The confusion matrix

Figure 4. shows the final application. The clinician can upload the image data, Fig. 5, which is classified by the created model and offers the diagnosis after classification, Fig. 6. The layout was kept as simple as possible to be easy to use by any user.

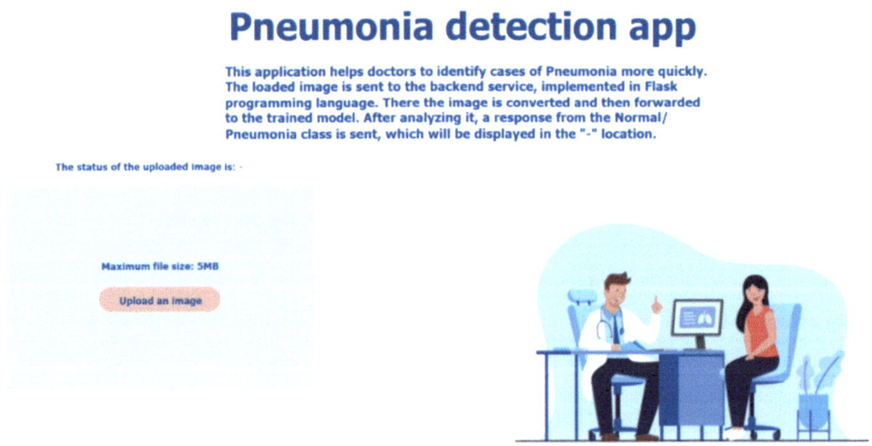

Fig. 4. The computer-aided diagnosis tool main page

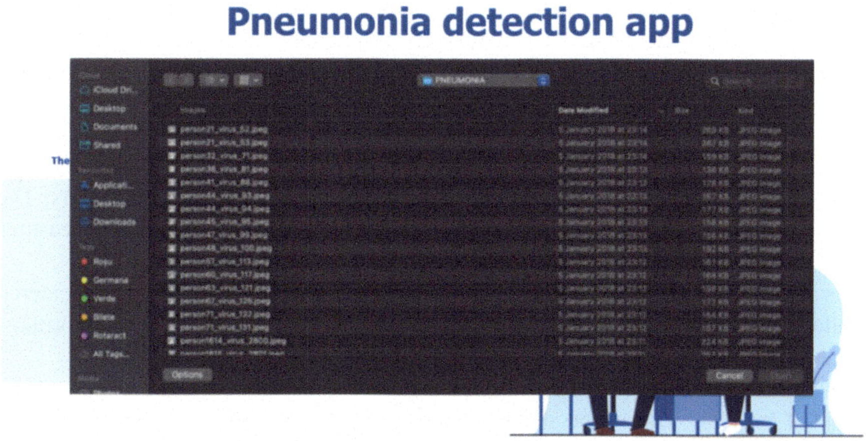

Fig. 5. Uploading an image

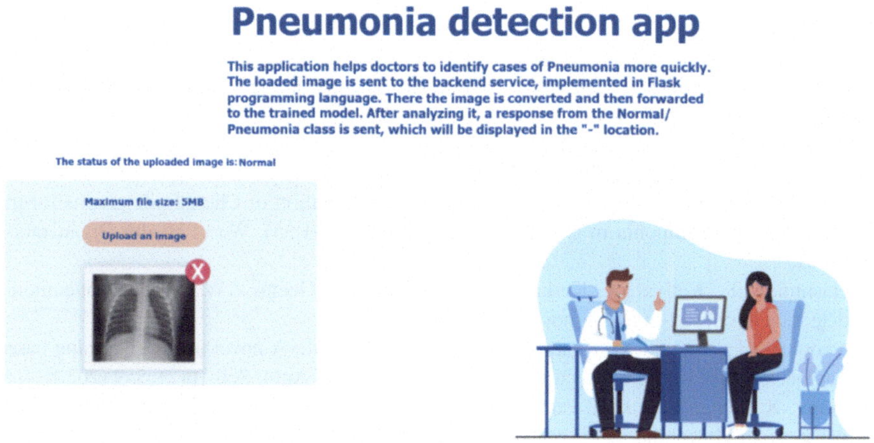

Fig. 6. The diagnosis, which can be "Normal" or "Pneumonia"

4 Conclusions

The app developed in this study is designed to assist doctors in diagnosing pediatric pneumonia, particularly those in underdeveloped regions where they are confronted with a considerable number of patients. It is capable of interpreting a chest X-ray and providing a diagnosis regarding the patient's condition. It is specifically tailored to pediatric patients, where only a few tools have been identified in the literature. Frequently, when under pressure, there can be a mistake, which in medicine can result in the loss of a patient's life. The final medical outcome can be influenced by several factors, including inattention to some problems, which can lead to misdiagnosis. Thus, an application that can provide a doctor with a second opinion is helpful.

The present proposal uses artificial intelligence to implement a diagnostic system for pneumonia from chest X-rays. The transfer learning method is approached using the Python programming language. Supervised learning techniques are also used and image classification. A comparative study was also conducted to ascertain which convolutional neural network performed optimally. This resulted in a network comprising five convolutional layers, five batch normalization layers (Batch Normalization), five aggregation layers (Max Pooling), four layers for Dropout and two densely connected layers. The obtained results are comparable and even outperform in some performance measures similar to tools presented in the literature.

As a further development, a database will be included that collects the images entered and allows doctors to track their patients. Images can be used to train learning algorithms supervised by Naïve Bayes or K-Nearest Neighbors.

Conflict of Interest. The authors declare no conflicts of interest. The funders had no role in the design of the study; in the collection, analyses, or interpretation of data; in the writing of the manuscript; or in the decision to publish the results.

References

1. Torres, A., Cilloniz, C., Niederman, M.S., et al.: Pneumonia. Nat. Rev. Dis. Primers. **7**, 25 (2021)
2. Blasi, F., Aliberti, S., Pappalettera, M., Tarsia, P.: 100 years of respiratory medicine: pneumonia. Respir. Med. **101**(5), 875–881 (2007)
3. World Health Organization: Standardization of Interpretation of Chest Radiographs for the Diagnosis of Pneumonia in Children (No. WHO/V&B/01.35). World Health Organization (2001)
4. Hashmi, M.F., Katiyar, S., Keskar, A.G., Bokde, N.D., Geem, Z.W.: Efficient pneumonia detection in chest x-ray. Diagnostics. **10**(6), 417 (2020)
5. Chouhan, V., Singh, S.K., Khamparia, A., Gupta, D., et al.: A novel transfer learning based approach for pneumonia detection in chest X-ray images. Appl. Sci. **10**(2), 559 (2020)
6. Wang, X., Peng, Y., Lu, L., Lu, Z., Bagheri, M., Summers, R.: Hospital-scale chest x-ray database and benchmarks on weakly-supervised classification and localization of common thorax diseases. IEEE CVPR. **7**, 46 (2017)
7. Rahman, T., Chowdhury, M.E., Khandakar, A., Islam, K.R., Islam, K.F., et al.: Transfer learning with deep convolutional neural network (CNN) for pneumonia detection using chest X-ray. Appl. Sci. **10**(9), 3233 (2020)
8. Labeled Optical Coherence Tomography (OCT) and Chest X-Ray Images for Classification: https://data.mendeley.com/datasets/rscbjbr9sj/2?__hstc=25856994.0f51a158ec6ac3f06ba3c9d5204b29b7.1717352772850.1717352772850.1717352772850.1&__hssc=25856994.1.1717352772850&__hsfp=1325817387. Accessed Oct. 2023
9. Berciu, A.G., Dulf, E.H., Stefan, I.A.: PillCrop: the solution for the correct administration of medicine. IFAC-PapersOnline. **54**(15), 328–333 (2021)
10. Stoleru, C.A., Dulf, E.H., Ciobanu, L.: Automated detection of celiac disease using machine learning algorithms. Sci. Rep. **12**(1), 4071 (2022)
11. Danku, A.E., Dulf, E.H., Banut, R.P., Silaghi, H., Silaghi, C.A.: Cancer diagnosis with the aid of artificial intelligence modeling tools. IEEE Access. **10**, 20816–20831 (2022)
12. Dulf, E.H., Bledea, M., Mocan, T., Mocan, L.: Automatic detection of colorectal polyps using transfer learning. Sensors. **21**(17), 5704 (2021)

13. Chest Radiograph – Normal: https://www.ebmconsult.com/articles/radiology-chest-xray-norma. Accessed Sept. 2023
14. Militante, S.V., Dionisio, N.V., Sibbaluca, B.G.: Pneumonia detection through adaptive deep learning models of convolutional neural networks. In: 2020 11th IEEE control and system graduate research colloquium (ICSGRC), pp. 88–93. IEEE (2020)

Accurate MRI-Based Alzheimer's Disease Detection Using CNNs and Encoding Techniques

Florina-Luminița Rostaș, Raluca-Cosmina-Maria Ardelean, Laura-Nicoleta Ivanciu(✉), and Emilia Șipoș

Bases of Electronics Department, Technical University of Cluj-Napoca, Cluj-Napoca, Romania
laura.ivanciu@bel.utcluj.ro

Abstract. Alzheimer's disease is a well-known neurological condition that is the most common cause of dementia. It affects older adults leading to a decline in cognitive function, and memory which affects daily activities. This study investigates which combination between encoding technique (One-Hot, Label, Binary, Ordinal, and Leave-one-out) for the dataset and activation functions (ReLU, LeakyReLU) in the CNN model delivers the best performance, when Precision, Recall and F1score are evaluated. The different CNN models were trained over 10, 50 and 100 epochs. By using a publicly available MRI scans dataset, with four categories of Alzheimer's disease, *Non_Demented, Mild_Demented, Moderate_Demented,* and *Very_Mild_Demented*. The study shows that an accuracy of 100% can be obtained when OneHot and LeakyReLU are employed. The results outperform similar research, where the same dataset, encoding techniques and activation functions are used.

Keywords: Alzeimer's disease · encoding technique · CNN · activation function · MRI

1 Introduction

Discovered by Alois Alzheimer in 1906 [1], Alzheimer's disease (AD) is the most common progressive neurodegenerative pathology among older people; the disease kills healthy cells in the brain causing memory loss and dementia. Although non-treatable, early detection can slow down its progress and ensure accurate diagnosis. There are millions of AD cases worldwide and the number is expected to increase significantly in the coming years. Imaging techniques are used to study, determine and diagnose Alzheimer's disease, such as computed tomography (CT), and positron emission tomography (PET), the most used method being magnetic resonance imaging (MRI). MRI is a non-invasive brain imaging method, without ionized radiation, that provides high resolution scans of soft tissues. According to the Alzheimer's Association, about 1 in 9 people aged 65 and above has AD [2]. The different stages of the disease are cognitively normal (CN), significant memory concern (SMC), early mild cognitive impairment (EMCI), mild cognitive impairment (MCI), late mild cognitive impairment (LMCI) [3].

In recent years, various methods of detecting AD using Convolutional Neural Networks (CNNs) have been employed. The method presented in [4] is based on an 8layer

S. Vlad and N. M. Roman (Eds.): MEDITECH 2024, IFMBE Proceedings, 130, pp. 140–150, 2025.
https://doi.org/10.1007/978-3-031-95671-3_16

CNN and a dataset of 186 MRI images which include 98 healthy patients and 98 AD patients; 3 activation functions are used: ReLU, LeakyReLU and sigmoid and 3 pooling techniques: medium, max and stochastic. They demonstrated that the most effective combination is the LeakyReLU activation function and max pooling technique, achieving an accuracy of 97.65%, sensitivity of 97.96%. To expand the dataset, they applied an augmentation method that increased the number of imagines to 17.738. A 50-layer residual neural network (ResNet) is proposed in [5], where abandon layers were used inside the fully connected layer to avoid overfitting. Results indicate that the LeakyReLU, max pooling technique, backpropagation algorithm and Adam optimizer is suitable for AD diagnosis, exhibiting an accuracy of 97.49%. An architecture for a CNN using MRI data from the Alzheimer's Disease Neuroimaging Initiative is presented in [6]; the network has two CNN models with different pooling layers and filter sizes using multi-class classification. The proposed model achieves an accuracy of 99.57%. In [7], two convolutional neural networks are used; the first CNN consists of 6 convolutions and 6 max-pooling layers with kernel filters of different sizes. The second CNN used 5 convolution and 3 max-pooling layers, having a kernel filter with the same size. The accuracy of the proposed methods is 90.74%, and 86.69% respectively. The activation function was ReLU. A review of research initiatives that use deep learning to detect AD is available in [8]. Data from three public Kaggle datasets (ADNI, OASIS and MRI) [9] is combined to obtain three representations of the brain: axial, sagittal and coronal. The proposed CNN model is uses 20 different architectures, such as ResNet01, AlexNet, DeepConvVet and others, with accuracies between 85% and 99.89%.

The purpose of this study is to investigate which combination between encoding technique (One-Hot, Label, Binary, Ordinal, and Leave-one-out) and activation function (ReLU and LeakyReLU) delivers the best accuracy, when used in the same CNN, and with the same dataset.

The remainder of the paper is organized as follows: Sect. 2 describes the materials and methods employed in the study, Sect. 3 presents the results obtained using several test scenarios and performance metrics, Sect. 4 analyzes the results and states the best combination, while Sect. 5 concludes the paper.

2 Materials and Methods

To analyze the best combinations of encoding techniques and activation functions, the proposed methodology follows the diagram in Fig. 1. First, the dataset containing the MRI images is imported and split; encoding techniques (One-Hot, Label, Binary, Ordinal and Leave-one-out) are used one at a time, then the dataset is divided for training, validation and testing.

The images are fed into the CNN, with either ReLU or LeakyReLU as activation functions for the convolutional layers. Each convolutional layer is paired with a MaxPooling layer, then come the Flatten, SoftMax and output layers. The model is compiled and trained, and the results are analyzed via performance metrics: F1 Score, Recall, Precision, Accuracy and Loss, based on the confusion matrix.

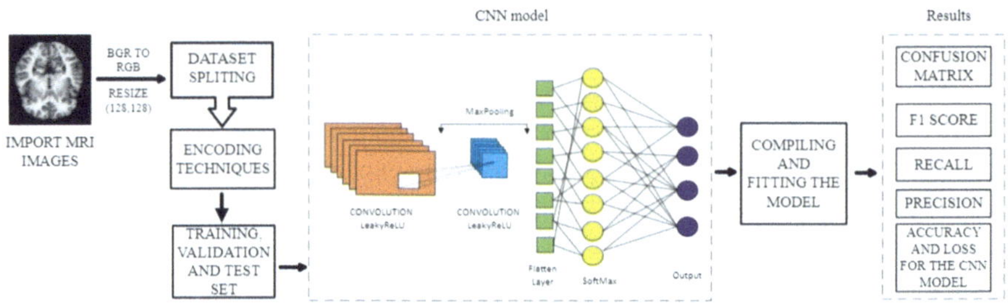

Fig. 1. Block diagram of the proposed methodology.

2.1 The Proposed CNN Model

Convolutional Neural Networks are a class of deep neural networks, commonly applied to analyzing visual imagery. The CNN works by extracting features directly from images. In this case, the CNN model (Fig. 2) is divided into two sections: the convolutional and pooling part, and the second, fully connected part. For each convolutional and pooling part, the LeakyReLU activation function is used, and a MaxPooling layer to reduce the dimensionality. A kernel filter is applied to preserve the most important value of the area. The fully connected part uses the characteristics extracted for the final classification. The output layer consists of four neurons, corresponding to the four classes: *Non_Demented, Mild_Demented, Moderate_Demented* and *Very_Mild_Demented.*

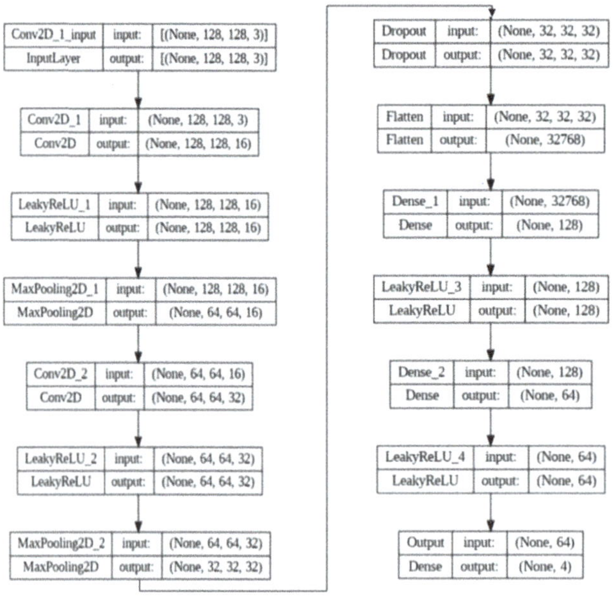

Fig. 2. Summary of the CNN model.

2.2 MRI Dataset

The dataset used for this paper has 6400 MRI transverse views of the brain pictures, preprocessed MRI images, black and white, available in a public repository [9]. The MRI images are classified into *Non_Demented, Mild_Demented, Moderate_Demented, and Very_Mild_Demented*. Sample images for each class are available in Fig. 3, while Fig. 4 shows the distribution of images across classes.

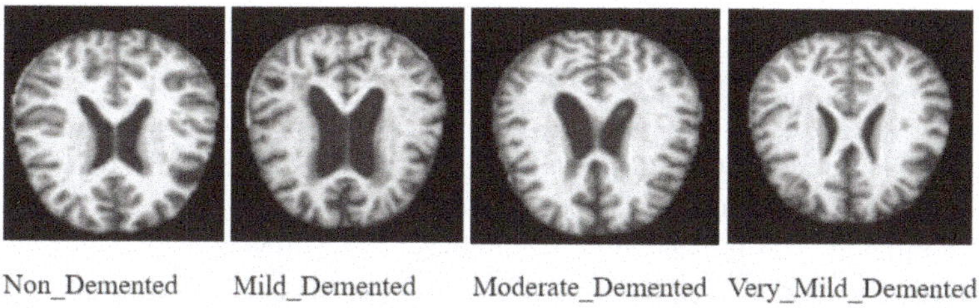

Fig. 3. Sample images of MRI Alzheimer dataset.

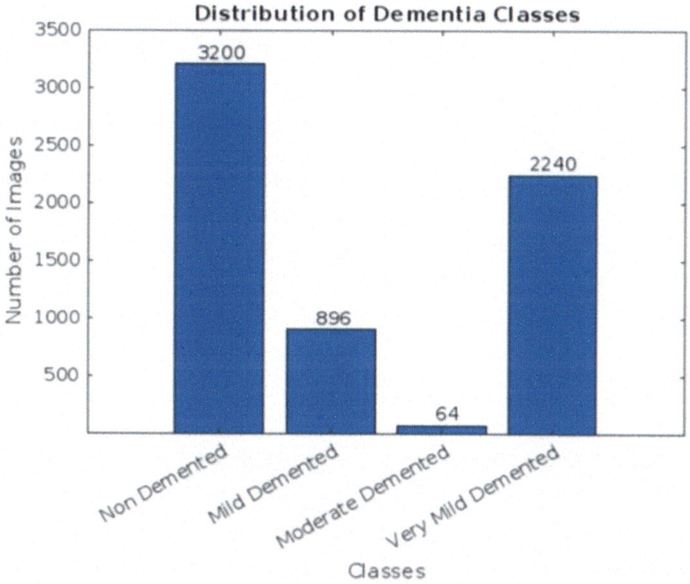

Fig. 4. Distribution of Dementia Classes.

The dataset is randomly split into 80% for training (5120 images), 10% for validation (640 images), and 10% for testing (640 images), using the *train_test_split* function (Table 1).

Table 1. Number of images across each set.

Class	Train	Test	Validation
Non demented	2565	319	316
Mild demented	712	93	91
Moderate demented	54	6	4
Very mild demented	1789	222	229

2.3 Activation Functions

In this research three activation functions are used: ReLU, LeakyReLU and Softmax. The first two activation functions are compared to each other to obtain the best performance metrics, combined with the encoding techniques. The SoftMax activation function is mandatory for the output layer. ReLU activation function sets all negative inputs to zero, and can lead to "dying ReLU", while LeakyReLU introduces a small slope, to keep training the neurons, thus solving the "dying ReLU" issue.

$$\text{ReLU}(\text{x}) = \begin{cases} x, \text{x} > 0 \\ 0, \text{x} \leq 0 \end{cases}$$

$$\text{LeakyReLU}(x) = \begin{cases} x & , x \geq 0 \\ 0.1 \cdot x, & x < 0 \end{cases}$$

where x denotes the neuron's input, and α represents a constant, usually 0.01 or 0.1.

SoftMax acts as an activation function, and it turns logits (numeric output of the last linear layer of a multiclass classification neural network) into probabilities. by taking the exponents of each output and then normalizing each number by the sum of those exponents [11].

$$f(s_i) = \frac{e^{s_i}}{\sum_j e^{s_j}}, \tag{3}$$

where s_i represents logits scores for a class i and s_j represents all the logits scores for all classes j.

2.4 CNN Layers

The CNN layers are used to extract and process the features of an image to reach a goal, in this case image classification. The layers used for CNN model are MaxPooling, Dropout, Flattening and fully connected layers.

MaxPooling is an essential process in Convolutional Neural Networks, that significantly contributes to reducing the dimensions of feature maps while preserving key information. Once the feature map is rectified by the ReLU/LeakyReLU activation function, the next operation is to down-sample the images to reduce the dimensionality through a pooling layer [11]. Max pooling is the most applied pooling operation in CNN. The major advantage of this process is dimensionality and overfitting reduction, also a kernel filter is applied, to save the most important value of the area selected.

The dropout layer represents a regularization technique that prevents overfitting by randomly disabling some neurons during the training process. This type of layer is useful in CNNs where the number of parameters is very high. Once the dimensionality of the data volume is reduced to a manageable size, by means of the flattening layer, a fully connected layer connects all inputs to all outputs.

2.5 Encoding Techniques

Converting categorical variables is an essential step in preparing data for machine learning tasks. When handling categorical data, which consists of non-numerical values like text or categories, it is important to transform these into a numerical format to ensure compatibility with machine learning algorithms. The encoding techniques used are *One-Hot, Label, Binary, Ordinal* and *Leave-One-Out*, the following notations are used S-small, M-medium, L-large (Table 2) [16–18].

Table 2 Description of encoding techniques [16–18]

Encoding Techniques	Description	Advantages	Disadvantages	Dataset		
				S	M	L
One-hot	– Columns with binary data	– Simple – Useful for nominal variables	– Increases dimensionality with many categories	x	x	
Label	– Each category has a unique integer	– Simple – Efficient – Suitable for most models	– Misleading for nominal variables	x	x	x
Binary	– Data as binary digits	– Balance of interpretability and efficiency	– Less intuitive			x
Ordinal	– Categorical to numerical preserving the order	– Simplicity – Time conservation – Reduce dimensionality	– Sensitivity – Overload – Normalization challenges with large data		x	x
Leave-one-out	– Categorical to target average minus the current instance	– Reduces overfitting – Use all available data	– Complexity – Risk of over-adaptation	x	x	x

3 Results

The simulations for five encoding techniques: One-Hot, Label, Binary, Ordinal and Leave-One-Out and LeakyReLU/ReLU activation functions were completed using Python as a programming language, in the Google Colab environment, with libraries such as Keras, OpenCV, Seaborn, Matplotlib, TensorFlow, scikit-learn, and numpy. For this study, the ADAM optimizer is used with a learning rate of 0.001, a categorical cross-entropy loss function, a mini-batch size of 32, a validation batch size of 32, and a validation frequency of once per epoch. To compare the simulations with all five encoding techniques and activation functions, Recall, F1 score and Precision were computed. The values for: True Positive - number of correct predictions, False Positive- number of instances incorrectly classified as positive, False Negative- number of instances incorrectly classified as negative are taken from the confusion matrix.

Precision indicates how often the model is correct, when predicting the target class.

$$Precision = \frac{True\ Positive}{True\ Positive + False\ Positive} \tag{4}$$

Recall indicates the success rate of predictions of positive values [10].

$$Recall = \frac{True\ Positive}{True\ Positive + False\ Negative} \tag{5}$$

F1 score is a combined metric, usually used for classification tasks, mostly for imbalanced datasets, which is the case for the data used in this study.

$$F_1 Score = 2 \cdot \frac{Precision \cdot Recall}{Precision + Recall} \tag{6}$$

To choose the appropriate combination for detecting AD, we use two activation functions, LeakyReLU and ReLU, for all five encoding processes. The number of epochs varied from 10 to 50 to 100, and the comparison was made by the F1 Score (Table 3). LeakyReLU achieves a better F1 score.

Table 4 represents all the performance metrics that were calculated, over 100 epochs. Combining the activation functions and encoding techniques, the best performance metrics for 100 epochs were obtained in 30% of cases for ReLU activation function and in 70% for LeakyReLU activation function.

Finally, since the combination between One-Hot encoding technique and LeakyReLU activation function proved to be the best choice for this CNN, with validation loss of 0.2% and validation accuracy of 100% (Fig. 5).

The confusion matrix (Fig. 6) was generated from the validation set and it shows the true positive (correct classification) rate is 100%, all values outside the main diagonal are null.

4 Discussion

Table 5 shows the performance of other similar research to detect Alzheimer's disease. In [10], where the same dataset with data oversampling is used, the stages of Alzheimer's disease are split into four categories and an accuracy of 96.35% is obtained. Paper [4]

Table 3. The F1 score values for different numbers of epochs.

Encoding Technique	Number of epochs	F1 LeakyReLU	F1 ReLU
One-hot	10 epochs	98.18%	98.44%
	50 epochs	99.61%	99.27%
	100 epochs	100%	99.54%
Binary	10 epochs	95.53%	90.37%
	50 epochs	97.93%	95.21%
	100 epochs	99.85%	96.29%
Ordinal	10 epochs	98.69%	97.61%
	50 epochs	99.90%	99.71%
	100 epochs	99.63%	99.84%
Label	10 epochs	99.44%	97.63%
	50 epochs	99.54%	99.53%
	100 epochs	99.80%	99.71%
Leave-one-out	10 epochs	98.61%	98.85%
	50 epochs	99.34%	99.53%
	100 epochs	99.54%	99.61%

Table 4. Comparison between ReLU and LeakyReLU activation functions for 100 epochs.

	ReLU				LeakyReLU			
	F1	Acc	Recall	Precision	F1	Acc	Recall	Precision
One-hot	99.54%	99.53%	99.35%	99.73%	**100%**	**100%**	**100%**	**100%**
Binary	96.29%	96.35%	95.51%	96.31%	**99.85%**	**97.50%**	**99.83%**	**99.87%**
Ordinal	**99.90%**	**99.84%**	**99.89%**	**99.91%**	99.63%	99.53%	99.56%	99.70%
Label	99.71%	99.53%	99.70%	99.72%	**99.80%**	**99.84%**	**99.73%**	**99.88%**
Leave-one-out	**99.61%**	99.37%	**99.57%**	99.66%	99.54%	**99.53%**	99.39%	**99.70%**

uses the same MRI scans and CNNs, applying a binary classification and three activation functions:sigmoid, ReLU and LeakyReLU. In paper [5], ResNetF with LeakyReLU activation function and data augmentation technique achieves an accuracy of 97.49%. In [6], the 99.57% accuracy is obtained using an MRI dataset with CNNs in 5-way classification.

Using an imbalanced dataset ensures that the CNN model is trained on data that closely resembles real-life cases; the model becomes more effective and appropriate to be used in actual medical diagnosis protocols. Moreover, the imbalanced datasets often force the model to learn to identify the most important cases, such as the Moderate_Demented

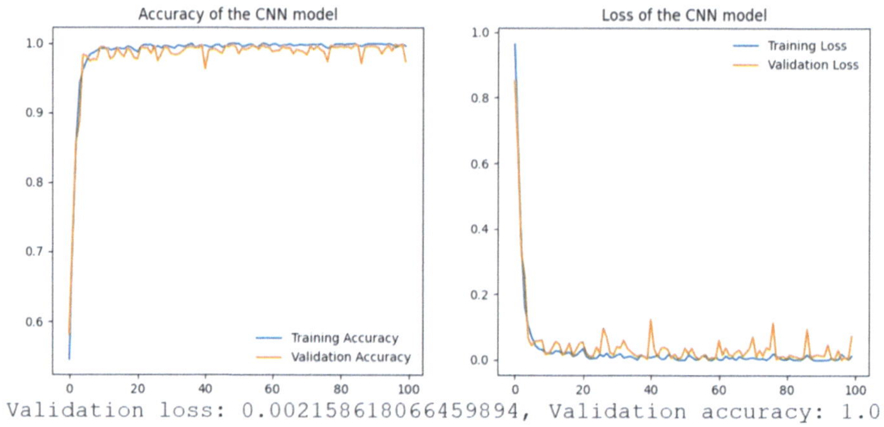

Fig. 5. Accuracy and Loss rate in the train and validation set for One-Hot and LeakyReLU.

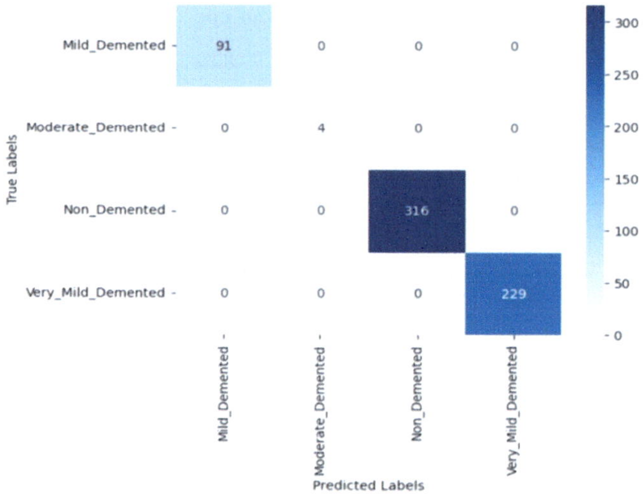

Fig. 6. Confusion Matrix for One-Hot and LeakyReLU.

class, thus enabling prevention mechanism against the possible worsening of Alzheimer's disease.

5 Conclusions

Alzheimer's disease is a well-known neurological condition that is the most common cause of dementia. It affects older adults leading to a decline in cognitive function, and memory which affects daily activities. This study investigates which combination between encoding techniques for the dataset and activation functions in the CNN model delivers the best accuracy. By using a publicly available MRI scans dataset, with four

Table 5. Comparison of accuracy values for similar work and proposed model.

Paper	Accuracy
Murat et al., [10]	96.35%
Shui_Hua Wang et al., [4]	97.65%
Faten Salim Fanoon et al., [5]	97.49%
A.M. El-Assy et al., [6]	99.57%
Proposed model	**100%**

categories of Alzheimer's disease, the study shows that an accuracy of 100% can be obtained when OneHot and LeakyReLU are employed. The different CNN models were trained in 10, 50 and 100 epochs. Possible future developments revolve around adding other types of medical images or exploring different CNN architectures.

Conflicts of Interest. The authors have no conflicts of interest to declare that are relevant to the content of this article.

References

1. Scharchter, A.S., Davis, K.L.: Alzheimer's disease. Dialogues Clin. Neurosci. **2**(2), 91–100 (2000)
2. A. Association: Alzheimer's Disease Facts and Figures, U.S. (2024)
3. ADNI | Background & Rationale
4. Wang, S.H., Phillips, P., Sui, Y., Liu, B., Yang, M., Cheng, H.: Classification of Alzheimer's disease based on eight-layer convolutional neural network with leaky rectified linear unit and max pooling. J. Med. Syst. **42**, 1 (2018)
5. Hanoon, F.S., Alasadi, A.H.H.: A modified residual network for detection and classification of Alzheimer's disease. Int. J. Electr. Comput. Eng. **12**(4), 4400–4407 (2022)
6. El-Assy, A.M., Amer, H.M., Ibrahim, H.M., Mohamed, M.A.: A novel CNN architecture for accurate early detection and classification of Alzheimer's disease using MRI data. Sci. Rep. **14**(1), 3463 (2024)
7. Mehmood, A., Abugabah, A., AlZubi, A.A., Sanzogni, L.: Early Diagnosis of Alzheimer's Disease Based on Convolutional Neural (2021)
8. Saikia, P., Kalita, K.S.: Alzheimer Disease Detection Using MRI: Deep Learning Review (2024)
9. Kaggle: https://www.kaggle.com/datasets/sachinkumar413/alzheimer-mri-dataset. Accessed 2024
10. Avşar, M., Polat, K.: Classifying Alzheimer's Disease Based on a Convolutional Neural Network with MRI Images. pp. 46–57 (2023)
11. Sadiku, M., Ashaolu, T.J., Ajayi-Majebi, A., Musa, S: Artificial Intelligence in Education, iUniverse, (2022)
12. Rallabandi, S.: (26 March 2023). https://medium.com/@sreeku.ralla/activation-functions-relu-vs-leaky-relub8272dc0b1be. Accessed 9 July 2024
13. Kingma, D.P., Ba, J.: Adam: a method for stochastic optimization (2015)

14. Huseynov, I..: Medium.com. (28 February 2023). https://medium.com/@ismailhuseynov6666/early-detection-of-alzheimersdisease-deep-learning-approach-b96378e4a904. Accessed 10 July 2024
15. Hand, D., Christen, P., Kirielle, N.: F*: an interpretable transformation of the F-measure. Mach. Learn. **110**, 451–456 (2021)
16. Gnat, S.: Impact of categorical variables encoding on property mass. ScienceDirect. **192**, 3542–3550 (2021)
17. Baruah, I.D.: Medium. https://medium.com/anolytics/all-you-need-to-know-about-encoding-techniques-b3a0af68338b. Accessed 2024
18. Roy, B.: Towards Data Science. (16 July 2019). https://towardsdatascience.com/all-about-categorical-variable-encoding-305f3361fd02. Accessed 2024

Health Technology Assessment; Education and Teaching

Prototype for a Telemedicine Applications within the Framework of Energy—Positive Homes to Enhance Healthcare Accessibility

Roxana-Valentina Briscan[1(✉)], Ciprian Forț[2], Simona Vlad[1], and Denisa Șteț[1]

[1] Energy Transition Research Center, Technical University of Cluj-Napoca, Cluj-Napoca, Romania
roxana.briscan@ethm.utcluj.ro

[2] National Institute for Research and Development of Isotopic and Molecular Technologies, Cluj-Napoca, Romania

Abstract. Although technology is continuously developing, telemedicine is still not yet utilized to its full potential into a positive home environment. Many patients lack home-based equipment to monitor their health condition. To address this issue the authors of this paper propose the development of a small and low-cost wearable Electromyograph prototype with surface electrodes. The device is specifically designed for patients diagnosed with Myasthenia Gravis. The purpose of this device is to enhance patients' comfort and to reduce the energy, time, and costs associated with visiting specialist doctors. Another benefit of the device is that it provides doctors with easy access to real-time data sent by the patient.

Keywords: EMG device · Myasthenia Gravis · wearable device · telemedicine · home-based health monitoring · real-time health data · electromyography · medical wearable technology · digital health solutions

1 Introduction

1.1 Context and Motivation

The modern technology used in the last century influences telemedicine which follows an upward slope. Although in 1970 Bird referred to an instrument necessary for "healing at a distance" (from Latin "medicus" and Greek "tele") [1], nowadays, the term has a more comprehensive meaning. According to the European Commission, telemedicine can be defined as "healthcare services in situations where the health professional and the patient are not in the same location" and "it involves secure transmission of medical data and information" [2]. This service is mostly used for patients diagnosed with cardiovascular, neuromuscular, and respiratory diseases. Although Yellowlees [3] considers the tarnished reputation of telehealth an even greater expense with the waste of equipment and human resources, telemedicine offers numerous advantages. A major benefit of monitoring from distance is the elimination of unnecessary movement of patients and their companions

S. Vlad and N. M. Roman (Eds.): MEDITECH 2024, IFMBE Proceedings, 130, pp. 153–161, 2025.
https://doi.org/10.1007/978-3-031-95671-3_17

[4]. In addition to improving the quality and comfort of a patient's life, this technology helps healthcare professionals to monitor vital signs in real time, while also having permanent access to a database of all previous results. In accordance with [5], with telehealth devices, the emergency room visits decrease and simpler symptoms in low risk were resolved without clinic visits. These factors support the hypothesis that telemedicine is essential in healthcare services.

Neuromuscular diseases affect approximately 14 million people worldwide, with Myasthenia Gravis (MG) categorized as a rare disease within this group [6]. MG is classified as a "rare" disorder because its prevalence rates range from 150 to 200 cases per million globally [7]. When a patient is suspected of having this illness, a series of diagnosis tests are necessary to confirm the diagnosis. The process typically starts with the Testilon test, followed by the measurement of acetylcholine receptor antibodies in the blood. Electromyography (EMG) studies are then conducted as the final steps to conclusively diagnose the disease [16].

During this period when humanity's focus is increasingly on developing energy-positive and sustainable environments, integrating telemedicine into a positive home environment enhances physical health, emotional stability, mental well-being, and overall quality of life. It improves healthcare accessibility, reduces stress, supports chronic disease management, and promotes a supportive and balanced home environment.

Drawing from the above context, the present study aims to introduce a low-cost wearable EMG device equipped with surface electrodes, designed to monitor the EMG signals of patients diagnosed with MG from the comfort of their homes thereby reducing the energy, time, and costs associated with a visit to a specialist doctor.

1.2 Research Purpose

The presented work serves four primary objectives: (i) to offer a solution for remote health monitoring, (ii) to introduce a cost-effective wearable device tailored for patients with MG, (iii) to underline the non-invasive method of monitoring through surface electrodes and (iv) to leverage the latest advancements in telemedicine technology. Analyzing the current monitoring solutions, such as "eSense Muscle" and "Allengers Scorpio EMG", it can be observed that the price is not accessible to an ordinary patient who needs monitoring. Starting from this motivation, this low-cost EMG fulfills the consumer's need to purchase at a more affordable price. Subsequently, in the long term, it can contribute to increasing life expectancy as well as decrease the costs, energy, and time wasted on routine doctor visits.

Through these objectives, the presented work attempts to contribute to the progress in improving patient care and advancing the field of telemedicine.

1.3 Structure of the Paper

After introducing the context and motivation, and outlining the main purposes of the research, along with the reasons for selecting the application area in the preceding subsections, Sect. 2 introduces the proposed framework for designing and developing the EMG device. This section begins with a summary of relevant literature, followed by a brief description of the phases involved in the design and development of the proposed

EMG. Subsequently, a dedicated section for results and discussions will emphasize the validity of the work through practical tests and comparisons with existing similar equipment. Finally, the conclusions section summarizes the key aspects of the paper's findings and suggests potential future work for device improvements.

2 Methodology

The methodology that led to the development of the proposed EMG device involves five main stages, as highlighted in Fig. 1: literature review, design, development, optimization and deployment.

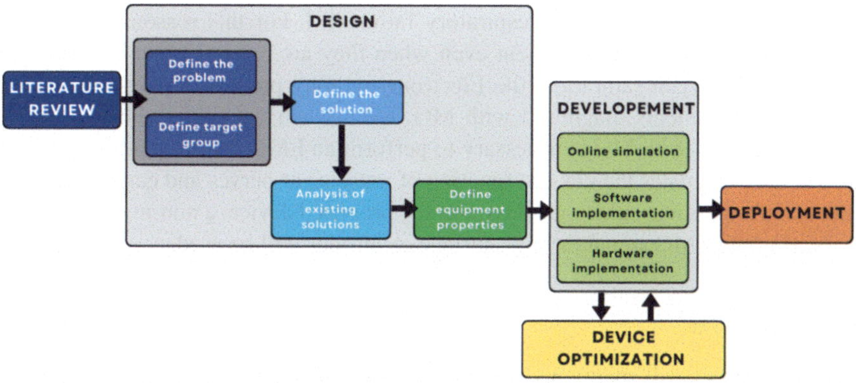

Fig. 1. Flowchart of the development process for the proposed EMG device

As an initial step, a comprehensive **literature review** has been conducted on MG pathology, EMG concepts, and experimental EMG prototypes to gain insights into the current state of knowledge. Additionally, the requirements of the proposed technical application were carefully analyzed, considering the device's objectives, the target audience (patients), and the technical constraints.

The subsequent step involved the **design phase**, wherein the solution for the target problem and group was defined and constraints were identified. Following this, the process advanced to establish the specific components of the EMG device.

During the **development stage**, tasks comprised conducting online simulations in Proteus, programming the software in Arduino IDE, and assembling the hardware components. The **testing phase** focused on validating the results, while completion marked the finalization of the prototype.

2.1 Literature Review

Myasthenia Gravis is a neuromuscular disease and an autoimmune disorder that affects the neuromuscular junction resulting in reduced force of muscle contraction and difficulties in movement, speech, or vision. The first MG related historical records dates to more than 350 years ago, when an Oxford physician named T. Willis described patients who experienced wellness in the morning but fatigue towards noon, often leading to speech difficulties. Later, in 1868, the first detailed case of a myasthenic patient was reported by Hérard followed by Erb in 1879 describing similar symptoms in three patients. Today, these symptoms characterize MG disorder [9].

Depending on the geographic location, the prevalence of MG varies from 1.5 to 17.9, or from 2.19 to 36.71 cases per 100.000 population [8]. This suggests that there are approximately 56,000 to 123,000 patients in Europe and around 60,000 in the US. The precise cause of its occurrence is still unknown, however, studies conducted so far suggest that the thymus plays a crucial role in the manifestation of this dysfunction, as it is one of the organs involved in regulating the immune system [13].

Currently, mortality rates from MG range from 5–9%, with an overall in-hospital mortality rate of 2.2% rising to 4.7% during myasthenic crisis. The most important predictors of death are age and respiratory failure [8]. For this reason, is crucial to monitor the vital signs of the patient even when they are not in hospital. Besides the heart rate and oxygen saturation, the Electromyography procedure plays an essential role in helping patients diagnosed with MG. Even when the subject is suspected or already diagnosed with MG is necessary to perform an EMG test. In a graphical form, the procedure evaluates the electric function of muscles or nerves and can establish if a subject confronts a muscular weakness. For the proposed device, a non-invasive method of capturing the EMG signal through surface electrodes will be used.

2.2 Design Stage

Based on the previously mentioned literature review, the importance of home monitoring equipment for patients diagnosed with MG and its technical requirements were defined.

An EMG measurement is usually taken by sensing the conduction of the electrical potential between two skin electrodes that are connected to a muscle [10]. To accomplish this, it is necessary to use a favorable sensor, with two skin electrodes and a reference electrode. For this reason, a MyoWare muscle sensor was selected [12]. Whit small dimension, the sensor requires a supply voltage (V_S) between 3.3 and 5 [V], a supply current of approximately 9 [mA], and provides an output signal voltage between 0 and $+V_S$ [V]. The sensor offers the possibility to output the raw EMG signal or an amplified, rectified, and integrated signal that will work well with a microcontroller's analog-to-digital converter [12].

The "healing at a distance" equipment requires data transmission technology. For the proposed device, an HC-10 Bluetooth Module [14] was selected first for a short distance. This module is a friendly electronic component, easy to use and integrated with Arduino board. It is notable that it can be connected with a mobile application for free. Arduino Nano is an acquisition board used in development. For constructive reasons, it improves the circuit's structure by reducing the number of connection cables and having a small dimension, for material considerations it is a low-cost board. Because this application intends to collect a biological signal is essential to possess a display for the measurement, for that an LCD was selected. The Serial Bluetooth Terminal application was selected for wireless connectivity between the prototype and mobile devices, enabling numerical data display. An alarm was implemented to activate when the signal exceeds the prescribed minimum and maximum limits or when the electrodes are improperly positioned for accurate measurement.

2.3 Development Stage

It's well-known that assembling a circuit is a complex task with numerous challenges. To make this process more efficient, simulations are invaluable. For this application an experimental circuit diagram with all pins was implemented in Circuit Simulator Software—Proteus. The simulation objective was to verify the implementation of the circuit connections, with a particular focus on detecting short circuits as shown in Fig. 2.

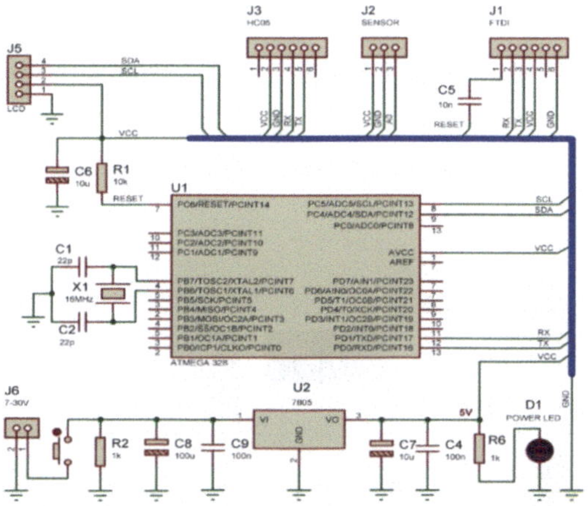

Fig. 2. The circuit designed in Proteus

Following the simulation, it can be concluded that the circuit has been constructed accurately, allowing to proceed to the next implementation phase.

Figure 3 emphasizes the role of each component within the proposed EMG device, providing a more suggestive representation of its functionality.

Element no.1 illustrates the connection between adapter and power supply module while no.2 describes the connection between Power Supply Module and Arduino Nano board. Element no.3 highlights the placement of the sensor on the Biceps muscle. No.4 illustrates the connection between sensor and board, no.5 illustrates the connection between the board and the LCD and no.6 describes the connection between board and Bluetooth Module. No.8 presents the interdependence with mobile application. Similarly, no.7 shows the boundary between the board and the Buzzer. No.9 describes the connection between the board and the Arduino IDE on the computer. The direction of the arrows indicates the flow of information within the circuit.

Based on the proposed structure (Figs. 2 and 3), the hardware implementation of the circuit was made on a breadboard. The software implementation includes a series of instructions for acquiring, transmitting and displaying data to the user and on the mobile application. After running some functional tests on breadboard, the entire circuit was

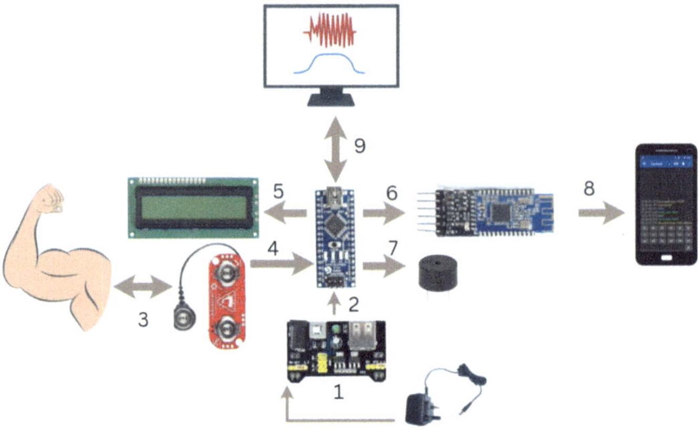

Fig. 3. Representation of the EMG device's hardware components

encapsulated into a plastic case. The final configuration of the prototype is visible in Fig. 4.

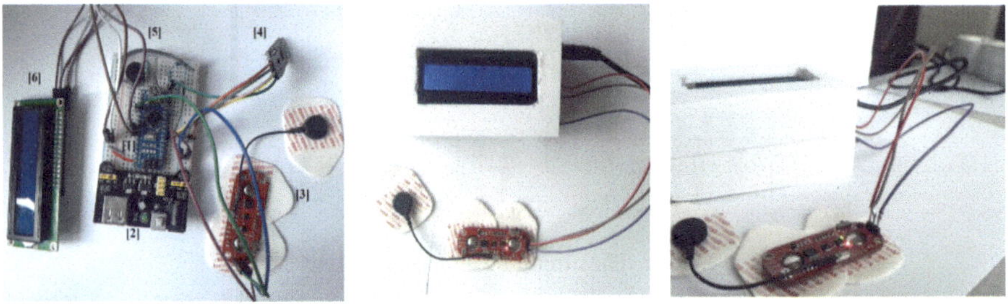

Fig. 4. Final prototype: (a) components connected on breadboard; (b) top view; (c) side view

Figure 4. illustrates the comprehensive structure of the electronic circuit, including all components; respectively provides views of the final physical prototype from different perspectives.

3 Results and Discussions

The data collected using the low-cost EMG were obtained from 8 subjects, both female and male, who tested negative for Myasthenia Gravis. These subjects were within the age ranges of 20–30 and 40–50. Measurements were taken at three reference intervals, each lasting 10[s].

The first interval analyzed the muscle potential when the muscle was relaxed. In the second interval, the potential was analyzed while the muscle supported a weight of 1[Kg], and in the third interval, the muscle underwent repetitive contractions as it shows in Fig. 5. All the measurements presented in Fig. 5. were obtained with 500[ms] delay and 4 measurements/second with a step of 0.25[s]. The muscle sensor collected the measurements, which were then processed by the Arduino board and transmitted to the mobile application through the Bluetooth Module. The average values of EMG envelope for each interval are: 0.924[mV], 1.161[mV] and 1.342[mV].

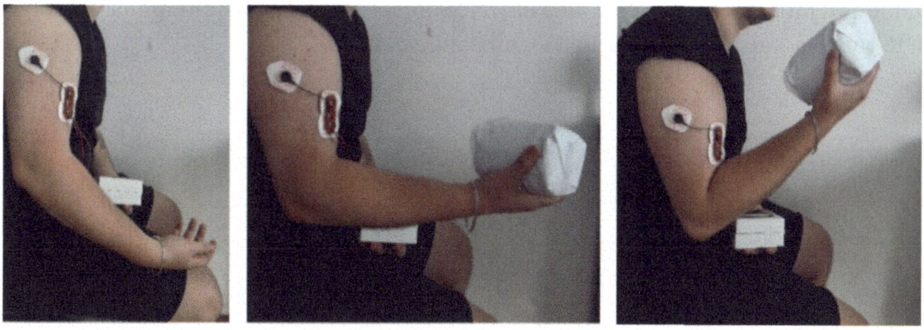

Fig. 5. EMG signal acquisition methods

To validate the accuracy of the measurements collected with the EMG prototype another set of measurements was taken using the Vernier laboratory kit [15] from the same subject. The data was collected following the same structure: three reference intervals, each lasting 10[s], a step of 0.25[s]. (See Fig. 6.) The average values obtained with the Vernier kit of EMG envelope are: 0.941[mV], 0.986[mV] and 0.106[mV]. These are similar with the average values from the first graph which describes measurements from low-cost EMG prototype. It is visible that both envelopes follow the same upward slope.

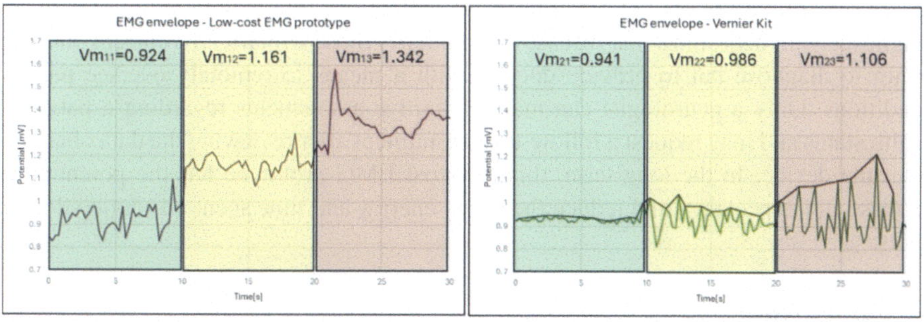

Fig. 6. Comparative measurements (vm_{xy} -where vm denotes the average potential value across the interval x for the instrument, and y represents the interval)

Another set of measurements emphasizes the variation in muscle activity among different male subjects (see Fig. 7). One subject, positive diagnosed with MG by a neurologist and currently under medical treatment, is compared to a subject without the disease. Despite having similar physiological characteristics, notable similarity in muscle activity can be observed. In the first interval, the average values for both subjects fall within the same range, around 0.9 [mV]. In the second interval, where the Biceps muscle supports a weight of 1 [kg], the average measurement from the diagnosed subject is 1.028 [mV], which is comparable to the 1.232 [mV] recorded from the healthy subject. Similarly, in the third interval, the averages of 1.143 [mV] and 1.362 [mV] from the diagnosed and healthy subjects, respectively, are also comparable. These findings point out that the muscle activity measurements of the patient under treatment present a notable resemblance to those of the subject without MG, demonstrating that treatment can normalize muscle response to a significant extent.

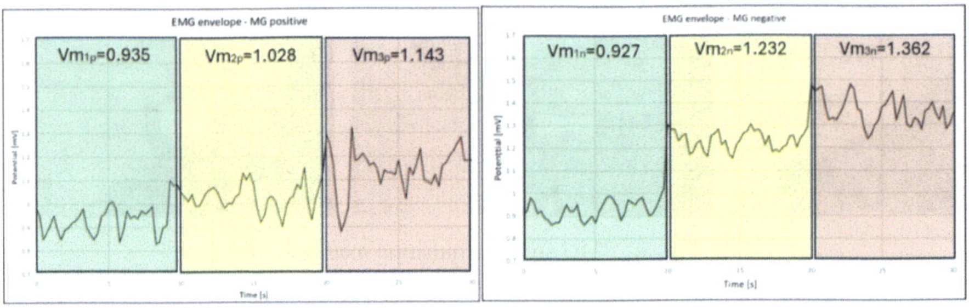

Fig. 7. Comparative measurements from different subjects (where: v_m is the average value of the potential over the interval, p is the positive subject, n is the negative subject, respectively $x = 2.3$ is the number of the interval)

Analyzing the measurements acquired by low-cost EMG device and the Vernier kit, was possible to underscore the resemblance in the data collected through both methods as well as the notable similarity between measurements acquired with EMG device from a healthy subject and a patient under treatment. At the same time, the proposed device offers an affordable and non-intrusive method for remotely monitoring patients diagnosed with neuromuscular diseases, particularly those with MG. Its current purpose is not to diagnose but to provide doctors with a means to remotely oversee patient conditions. Only a neurologist can make definitive assessments regarding a patient's health status and may request a follow-up appointment after reviewing the data obtained from the device. In the long term, this low-cost EMG prototype has the potential to increase life expectancy and reduce the costs, energy, and time spent on routine doctor visits.

4 Conclusion

The proposed EMG prototype, besides the fact that it focuses on the EMG signal monitor, also enhances the quality and comfort of life's patients by reducing the energy, time, and costs associated with visiting specialist doctor. With a low price is more affordable and with valid measurements this device satisfies patients' needs, compared to other equipment already on the market.

Nevertheless, the device presents various limitations such as potential questions regarding the quality of measured values during body movement and the HM-10 module's restriction on data transmission beyond 100 [m] but as a future development the solutions of these limitations will be prioritize.

In this regard, future studies can focus on enhancing the device by: (i) making it more robust and portable, (ii) ensuring compatibility with an integrated application capable of monitoring, saving data, displaying alert messages, and transmitting data to professionals using Wi-Fi, and (iii) adapting it for the monitoring of other types of diseases as well.

By connecting energy-positive homes with telemedicine, we can create a sustainable, efficient, and health-promoting living environment that leverages the best of both worlds. This integration not only supports the health and well-being of residents but also contributes to a more sustainable and resilient healthcare system.

Acknowledgments. This paper was supported by the grant "Renewable Energy- based Positive Homes", no. 52/PHE/2024, project no PN-IV-P8-8.1-PRE-HE-ORG-2023-0147 awarded by CNCS/CCCDI -UEFISCDI.

Conflicts of Interest. The authors declare no conflicts of interest.

References

1. Strehle, E.N., Shabde, N.: One hundred years of telemedicine does this new technology have a place in pediatrics. Arch. Dis. Child. **91**(12), 956–959 (2006). https://doi.org/10.1136/adc.2006.099622
2. Raposo, V.L.: Telemedicine: the legal framework (or the lack of it) in Europe. GMS Health Technol. Assess. **12**, Doc03 (2016). https://doi.org/10.3205/hta000126
3. Dyk, V.L.: A review of telehealth service implementation frameworks. Int. J. Environ. Res. Public Health. **11**(2), 1279–1298 (2014). https://doi.org/10.3390/ijerph110201279
4. Krishnan, G.: Telemedicine and neurosciences. J. Clin. Neurosci. **12**(8), 851–862 (2005). https://doi.org/10.1016/j.jocn.2005.07.004
5. Hwei, L.R.Y., Octavius, G.S.: Potential advantages and disadvantages of telemedicine: a literature review from the perspectives of patients, medical personnel, and hospitals. J. Commun. Empowerm. Health. **4**(3), 180–186 (2021). https://doi.org/10.22146/jcoemph.64247
6. UCL Homepage, About Neuromuscular Diseases
7. Dresser, L., Wlodarski, R., Rezania, K., Soliven, B.: Myasthenia gravis: epidemiology, pathophysiology and clinical manifestations. J. Clin. Med. **10**(11), 2235 (2021). https://doi.org/10.3390/jcm10112235
8. Bubuioc, A.M., Kudebayeva, A., Turuspekova, S., Lisnic, V., Leone, M.A.: The epidemiology of myasthenia gravis. J. Med. Life. **14**(1), 7–16 (2021). https://doi.org/10.3390/jcm10112235
9. Deymeer, F.: History of myasthenia gravis revisited. Noro Psikiyatr. **58**(2), 154–162 (2022). https://doi.org/10.29399/npa.27315
10. Reaz M.B.I, Hussain M.S., Mohd-Yasin F.: Techniques of EMG signal analysis: detection, processing, classification and applications. Biol. Proced. Online, 8: 163 (2006). doi: https://doi.org/10.1251/bpo115
11. Bulter, N.A.: Development of a myoelectric detection circuit platform for computer Interface applications. DigitalCommons@CalPoly. (2019). https://doi.org/10.15368/theses.2019.21
12. MyoWare™Muscle Sensor (AT-04-001) Datasheet by SparkFun Electronics
13. NORD Homepage: Myasthenia Gravis Report (2022)
14. HC-10 Bluetooth Module, Datasheet by Components101 Homepage
15. Vernier Science Education Homepage
16. Baylor College of Medicine Homepage. Myasthenia Gravis

Modern Methods of Visual Training for Children with Strabismus and Amblyopia Through Assisted Serious Games

Barbu Cristian Braun(✉), Corneliu Nicolae Drugă, Ionel Şerban, and Mitu Leonard

Transilvania University of Brasov, Brasov, Romania
braun@unitbv.ro

Abstract. This paper describes how a solution was developed in case of visual problems that frequently occur in childhood could be largely avoided or drastically reduced. It presents a research stage focused on the prevention of the most common vision pathologies in children. The development of the method proposed at this stage of the research was based on the following hypotheses: the first of these is that, unfortunately, not enough importance is given to screening and, as the case may be, visual training among children. The second one is that, according to some statistical studies, more than 20% of the children tested have a visual problem, more or less severe, and pathologies such as strabismus, amblyopia or ametropia have the highest frequency. A third hypothesis is that non-invasive, short-term and interactive methods (as serious games) of visual investigation and training, have a positive impact on both children and optometrists. Thus, a non-conventional method has been developed, complementary to classical methods, whereby vision testing and game training constitute the main activity in some therapy sessions. It was proceeded to the conception, design and testing of visual testing and training methods in as serious assisted games, dedicated to amblyopia and strabismus among school and preschool children. The method was repeatedly tested and improved throughout its development and finally it was tested on a sample of 10 children. It could be concluded that the procedure could be implemented from the point of view of screening and training vision type testing in schools and kindergartens.

Keywords: Amblyopia · Squint · Children · Interface · Serious Games · Visual · Training

1 Amblyopia and Squint in Children

It is known that the visual function provides the most information, more precisely approximately 80% of the amount of information coming from the environment, arrives via visual function [1, 2].

On the other hand, it is known that vision in children develops the latest of all the 5 senses, with the visual sense continuing to develop until approximately 14 years of age.

S. Vlad and N. M. Roman (Eds.): MEDITECH 2024, IFMBE Proceedings, 130, pp. 162–179, 2025.
https://doi.org/10.1007/978-3-031-95671-3_18

In general, until this age there is a physiological Hyper-metropia that gradually resorbs, along with the physical development of the eyeballs in the orbits. However, there are also cases in which this hyper-metropia cannot be completely resorbed, there being what is called a residual hyper-metropia [1, 3].

But from the point of view of visual function in children, the most common problems are related to pathologies such as amblyopia or strabismus. Both pathologies can have physiological causes, as well as causes determined by the environment in which children carry out their daily activities (at home, at nursery school, at kindergarten or at school). The good part is that, especially the two pathologies mentioned above, detected from an early age can be drastically improved or even completely remedied thanks to conventional and non-conventional visual training procedures [3, 4].

The first of the two most common eye pathologies in children, strabismus is characterized by the lack of synergy in the coordination of both eyes and, implicitly, by fixation and fusion problems of the two images on visual cortex, providing from both optical ways [1, 2, 7]. Amblyopia, also called the pathology of the lazy eye, is characterized by the fact that the specific eye makes the processes of fixation and accommodation much more difficult and in a much longer time than in the case of the healthy eye. In addition, for the fixation and accommodation processes of an amblyopic eye to ensure, a sufficiently strong contrast against a target background is necessary [3, 4, 7].

But the main problems related to this aspect are on the one hand in many situations the negligence or ignorance of the parents and, on the other hand, they can be related to the fear or not taking the procedures for visual testing and training seriously by the children. In order to be able to resolve such impediments as best as possible, on the one hand, large-scale campaigns are currently being carried out to raise awareness (primarily by parents and educators) of the importance of control and, if necessary, of visual training from as young an age as possible (even under 2 years old) [3, 5].

On the other hand, to reduce as much as possible the reluctance of parents and also the fear or boredom of children towards these screening or visual training procedures, in recent years the development of friendly non-conventional methods has taken on an unprecedented scale in this sense [6, 7].

2 Proposed Non-conventional Vision Testing Method

Knowing that it is extremely important, at the national and international level, to carry out visual screening procedures from the earliest possible age, in schools, kindergartens and even nurseries, the question arose of motivating parents, educators and children for this purpose. The importance of this aspect is due the fact that, unfortunately, a significant percent from the people (meaning especially the parents or educators) does not pay enough attention to this aspect. On the other hand, children's fear related to the idea of going to the doctor is not to be taken into account, a fact that complicates the solution of such a problem even more. The same problems are encountered when it comes to visual training in special cases, until it is necessary to correct through repeated exercises and procedures a certain vision problem in the case of a school or preschool child [8].

Solving these types of problems could consist of two distinct aspects: on the one hand, campaigns to promote and clearly explain the importance of regular eye testing for children at the earliest possible ages and, on the other hand, finding strategies through which testing to be more transparent and attractive. The research stage described in the paper focuses mainly on the second aspect. Thus, in the context where the concept of serious gaming is currently becoming more and more widespread in personality development procedures, in healing therapies, in the development of skills and intelligence, especially in children, it is currently being implemented at ever wider scale and in the matter of visual function. Related to this aspect, the purpose of the research focused not only on the development of methods for testing and training the visual function in this context, but also on finding simple, effective methods with costs as affordable as possible even for families with financial problems [10, 11].

As a result of the fact that in the development of vision in children there is an increased frequency of the appearance of certain hereditary or acquired pathologies, such as amblyopia, myopia, strabismus, the methods proposed for this purpose focused on two of these pathologies: amblyopia and strabismus. The main goal of the research is to develop methods by which visual testing and training in recovery in the case of these pathologies can successfully complement traditional methods. But what is specifically aimed at is the considerable increase in the attractiveness of children in such methods of play testing, increasing the degree of attention and involvement from the child, but also from the parent or educator [11].

Concretely, for this purpose, in this stage of the research two software interfaces were successfully designed, programmed and tested in order to test and train the visual function for each of the two pathologies [12]. The development of the two game-assisted tests was based, on the one hand, on the generally known attraction of children for games on smart-phones, tablets or laptops and, on the other hand, the recommendations of optometrists specific to strabismus and amblyopia. Specifically, in the case of strabismus, the recommendations are that training should be done in binocular vision, but the strabismus eye should be forced to fixate on a target in an opposite direction. For example, if the strabismus eye is the left and the strabismus manifests outward (towards the temporal area), then the subject will fixate a target located towards the central part of a display, and if the strabismus manifests itself towards the nasal area, then the fixated target should be located on the left side of the display, etc. On the other hand, in the case of amblyopia, the specialists' recommendations are that the healthy eye should be occluded, and the amblyopic eye be forced to fixate and accommodate both on a fixed and on a dynamic target [13, 14].

Thus, the software interface *Assisted serious games for Amblyopia.vi* was developed for detection, but especially for visual training in the case of amblyopia, and in the case of strabismus, the interface *Gaming tests for squint.vi.* Both software interfaces were developed in LabVIEW, then describing how they were designed, programmed and tested.

For amblyopia play testing and training, the LabVIEW software interface *Assisted serious games for Amblyopia.vi* was designed to run through four stages of the play test, all of which are accessible to children, yet as attractive as possible. Besides, the graphic aspect, of the chromatics, was also followed, so that all the fundamental colors appear in the graphic interface, on a strong contrast, this, in turn, is intended to arouse curiosity and the desire to solve the tests through the game. Moreover, through the tasks that must be solved at each stage of the test, the motivation of the child in question was followed, the idea being that of rewarding the correct and timely solution, through the examiner (parent, optometrist or educator) [13, 14].

Thus, the four stages designed for this specific play-testing interface for amblyopia training procedures are as follows: stage 1: step-by-step filling (in distinct primary colors) of four entities familiar to children (house, plane, car, flower); 2nd stage: following a predetermined route (by coloring some dots that make up the route) from a START zone to a motivating entity, namely a candy (following the route with as few mistakes as possible and in good time to the entity candy attracts rewarding the child with real candy from the examiner); 3rd stage: extinguishing some dots of different sizes from a certain category from a chromatic point of view (e.g. extinguishing red dots) from a multitude of differently colored and different sized dots; 4th stage: selective coloring of certain square-type entities located next to some virtual switches arranged in a certain orientation (e.g. coloring of squares located next to switches oriented to the right).

Obviously, the entire procedure of play testing through this software interface must be carried out under the following conditions: the child must be continuously assisted throughout the duration of the test. This is necessary both to be assisted in case he has doubts about what to do in the immediate next moment. On the other hand, the assistance of the subject is necessary in order to follow several aspects: to have a correct posture during the solving of the test, not to cheat, to be able to be evaluated by the examiner the way in which the test is solved (the way he looks towards display, gesture mimicry, etc.). In other words, with regard to the conditions for carrying out the test, the child in question must have occluded the healthy eye, solving the test in monocular vision, using the amblyopic eye, to force him to fixate on the graphic entities in order to solve the tasks.

Another particularly important aspect is the fact that, in the case of amblyopia, visual training therapy must involve a longer period, for example for 6 months or even 1 year, including at least 3–4 weekly training sessions. A training session, according to the recommendations of optometrists and ophthalmologists, should include several activities, all in a 1–2 hour session. These activities should involve following stimuli through conventional methods (for older children) but also through non-conventional methods such as physical or virtual play testing. In the present case, we are talking about the virtual game testing method, but obviously this method must be interspersed with the other methods mentioned above. It is also recommended that the sequence of procedures for testing in a training session be done alternately, to avoid boredom and to avoid the risk of memorizing work steps.

For amblyopia, the software interface programming for assisted screening and visual training was based on the following logic diagram:

Resuming to the software interface for training and game testing in the case of amblyopia, the following describes briefly how the graphical programming of the interface was done, specific to each stage of the testing procedure.

Thus, for the first stage of the test, namely the coloring of the virtual objects, each of them was constituted by a series of status LED sub-entities in Boolean logic, with the off and on states. For example, for the house object, 10 Boolean status LEDs were defined, which were graphically modeled in different shapes and sizes and having distinct colors in the logical on state. An example can be seen in Fig. 1 for the house object, before and after coloring in the first stage of solving the test [13] (Fig. 2).

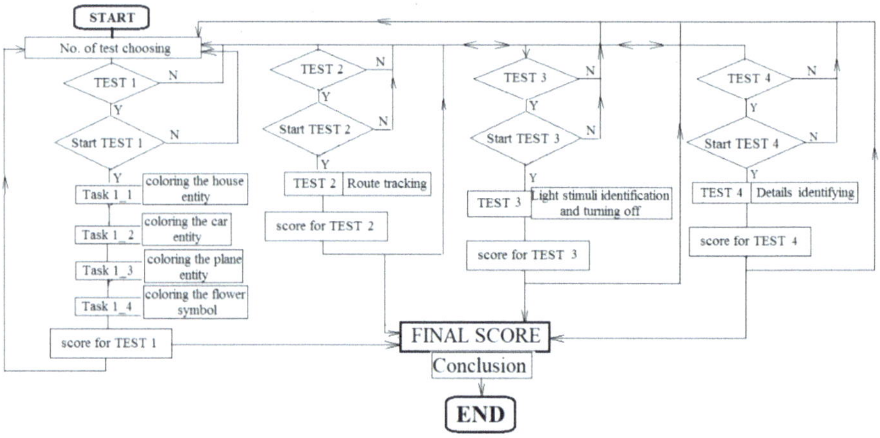

Fig. 1. Logic diagram for software interface programing for amblyopia assisted training

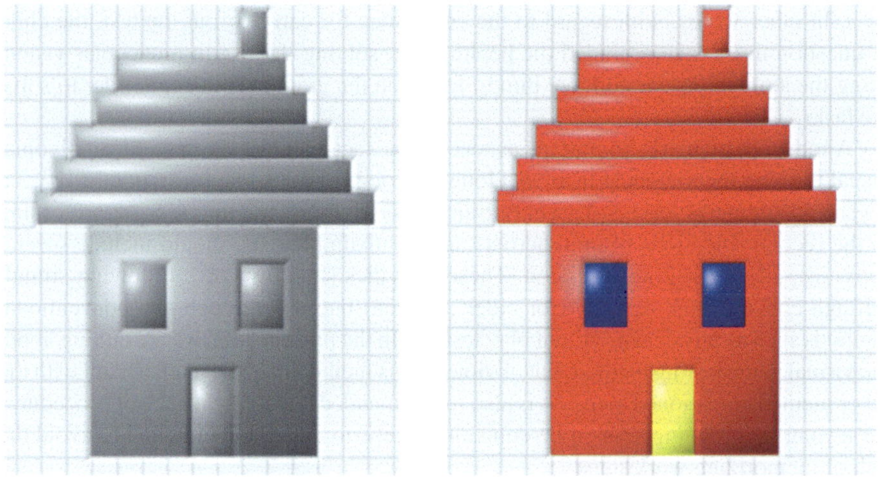

Fig. 2. Example of a virtual object before and after coloring in solving the first stage of the test in the case of the amblyopia training procedure.

From the aspect of graphic programming, the 10 Boolean entities were related to a True - False structure, specific to the two states off and on, with manual actuation, the entities being of input type. These, together with the structure, were included in another repetitive structure of the While-Loop type, which, by running, imposes the condition that the test can be run further only after finishing the coloring of the current object. Obviously, this structure was also provided with a timer by which the coloring time of the current object could be counted automatically. A sequence of graphical programming specific to the coloring of the house entity in the first stage of testing is exemplified in Fig. 3 [13].

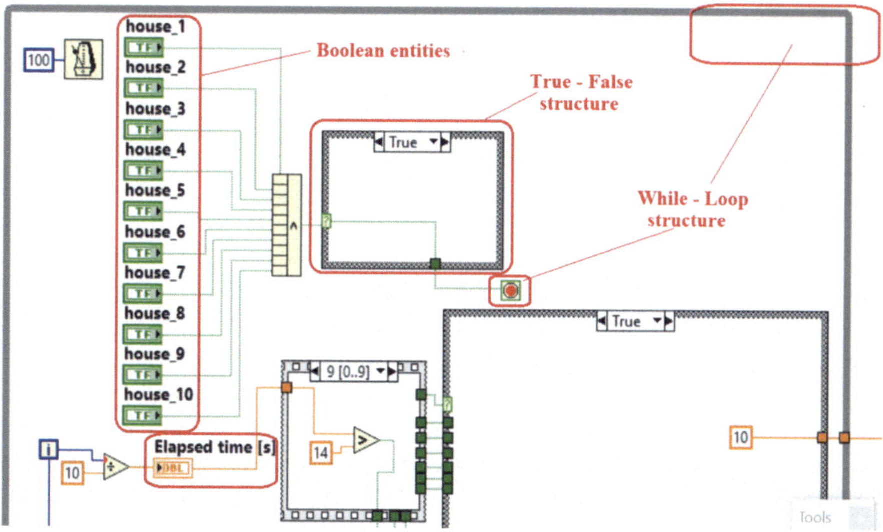

Fig. 3. Example of graphical programing specific to the first task for the first testing step (filling the virtual house).

For each virtual object coloring in this first step, string input and output variables were defined and related in the programming algorithm, through which the current message specific to each task in it could be displayed stage. For example, for the first task, the message to be displayed is *Fill color the house*, for the 2nd task the current message is *Fill color the car*, etc. These string variables were each related by a *True - False* structure, conditioning the appearance, in successive order, of the messages after different time intervals during the course of the first stage of solving the test (Fig. 4). The programming algorithm specific to the first stage included another aspect, namely the quantification of the answer according to the duration in which the tasks are solved. For this, an output variable of numeric type was defined, related to the index of the *While-Loop*, through which the duration in which each task was solved in this first stage of the test could be displayed. Thus, depending on the solving time, the programming algorithm assigns a grade for that task. For example, if the first task (coloring the house) was done in less than 6 seconds (time assessed by optometrists as normal for a subject without visual problems), then the grade given is 10 [13]. If the duration is between 6 and 7 seconds, then the grade awarded is 9, if the duration is between 7 and 8 seconds, then the grade awarded is 8, etc.

For this first stage, as for all stages, the algorithm was programmed to display as a numerical output variable in decimal format the grade specific to the stage. In the case of the first stage, the algorithm consisted in calculating the arithmetic mean of the 4 intermediate grades, for each of the 4 tasks of this stage.

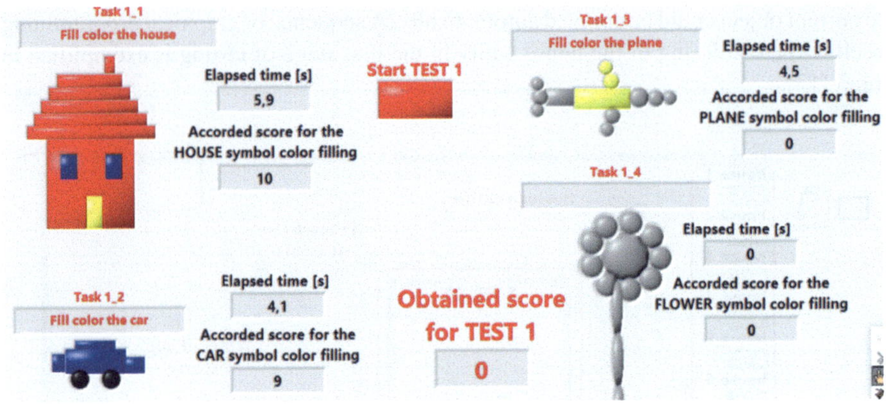

Fig. 4. The first test step during its current solving.

Figure 4 shows the graphical interface during the first stage of the test [12, 13].

For each virtual object that must be filled, the established time reported to the accorded scores was different depending on their complexity, due to the specialist's requirements. The programing algorithm was applied considering the average between all stages (Eq. 1):

$$S_1 = (1/n)^* \left(\left(v^* C_{ts}\right)/\tau_i\right)\text{--}v^* W_{ts} \tag{1}$$

where S_1 means the score for the 1st test ($S_1 = 1 \div 10$), n – the number of stage for the 1st test, C_{ts} – the number of correct turned on entities, W_{ts} – number of wrong turned on entities, v – the value coefficient for each correct or wrong turned on entity.

For the second stage, the programming algorithm assumed, in the first phase, the definition of a set of virtual Boolean LED, strung together, constituting the route. From a graphical point of view, both the starting area of the route and its end, represented by the candy (as a reward), were also made up of virtual LEDs with the two logical states (off and on). Thus, in fact, traversing the route means nothing more than switching on the lit state of all the virtual LEDs between START and END. All the virtual status LED entities constituting the path in the 2nd stage, from the point of view of graphics programming, were related by a logical *AND* multiplexing function linked to a *True-False* structure. In this way, the condition was imposed that only after the completion of the route is the completion of this current stage of the test validated. Obviously, in order to receive a reward (a candy), the course must be completed meaning maximum of 1–2 errors in a time small enough that the grade given is at least 8. Figure 5 shows the second stage of the test run during its solving [13, 14]. For each stage of the test, its start was conditioned by pressing a virtual Boolean button called *Start Test i* (i = 1 to 4), as can be seen in Figs. 5 and 6.

In terms of accorded score for the 2nd stage, it was establish depending on the necessary time to finish all track, by successively entities composing the route turning on. For example, if the necessary time is up to 75 seconds the maximum score (10) will be accorded, if the necessary time is between 75 and 80 seconds than the accorded score will be 9 of 10, if the time for response is between 80 and 85 seconds the accorded score

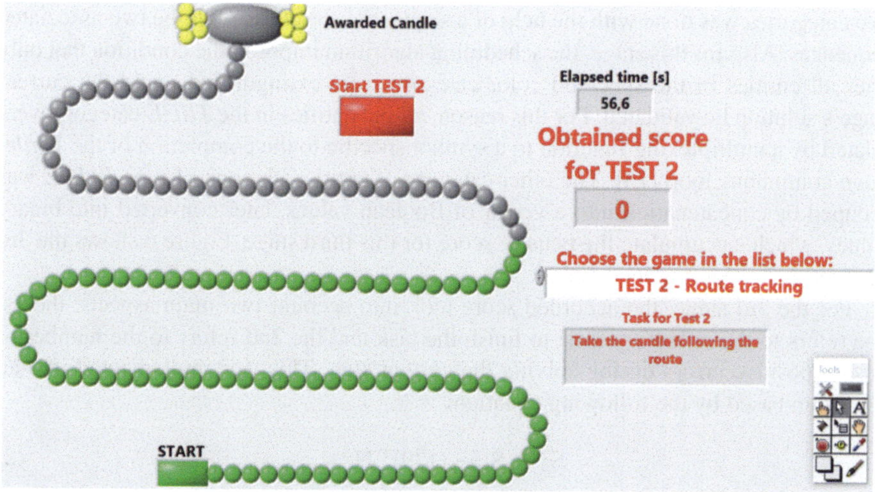

Fig. 5. The second test step during its current solving, meaning to follow the route.

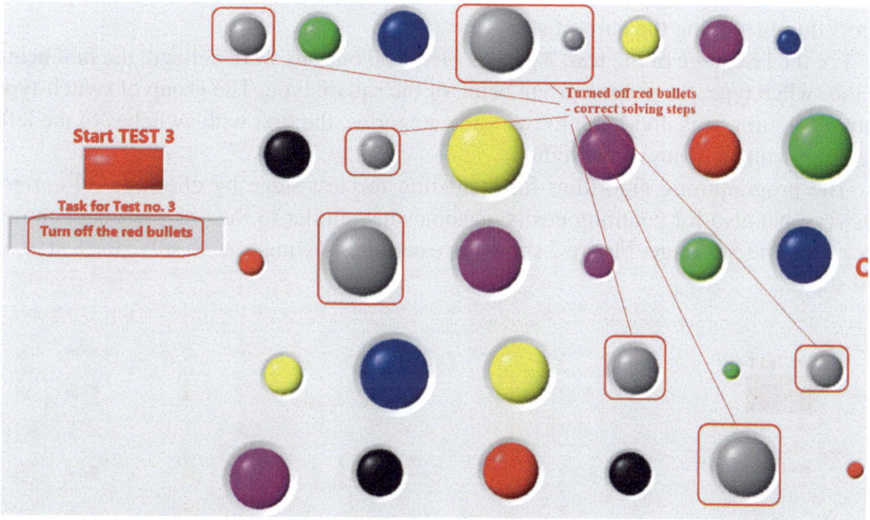

Fig. 6. The 3rd test step during its current solving, meaning turning off the specified colored category of entities (i.e. red bullets).

will be 8 of 10 and so on. All these considerations for the programing algorithm have been considered due to different optometrist requirements.

For the 3rd stage of the test, which requires entities from a single color group to go out, the status LEDs in the form of dots have been defined and grouped into two broad categories: Status LEDs from the TRUE category, namely those that, by their extinction, imply a correct solution of the current task and those from the FALSE category, meaning false-positive errors, i.e. a wrong solution of the current task. Their grouping into the two categories was done with the help of a sequential structure, with the two associated sequences. Also for this stage, the scheduling algorithm imposed the condition that only after all entities in the specified color category were extinguished could the current stage's solution be validated. For this reason, all the entities in the *TRUE* category were related by a multiplexing function to a symbol specific to the completion of the *While-Loop* continuous loop [13]. The other category of entities, the one of type *FALSE* was grouped by concatenation into a vector of Boolean values, later converted into binary values, which, accumulate the penalty score for this third stage. Figure 6 shows the 3rd stage of the test during its solution.

For the 3rd stage, the accorded score took into account two main aspects: the 1st one refers to the necessary time to finish the task and the 2nd refers to the number of false – positive errors during solving the current step. This was implemented into an algorithm based by the following equation:

$$S_3 = S_\tau - (1/9)^* N_{\varepsilon\_fp} \tag{2}$$

where S_3 means the score for the 3rd testing step, S_τ is the score depending only by the necessary time for solving (i.e if the time is up to 24 seconds this score is maximum (10), else as the time is longer, as the score is lower); N_{ε_fp} – the number of false – positive errors during solving the current step

For the last stage of the test, 2 groups of virtual buttons were defined, the first being of the switch type, the second group being of the square type. The group of switch-type entities, in turn, was divided into two sub-categories, the first with switches to the left, the other with switches to the right.

The programming algorithm for validating the test stage by checking all correct answers, but also for counting errors, is somewhat similar to the one presented above, specific to the 3rd stage. Figure 7 shows an example showing how to solve the test in its last stage.

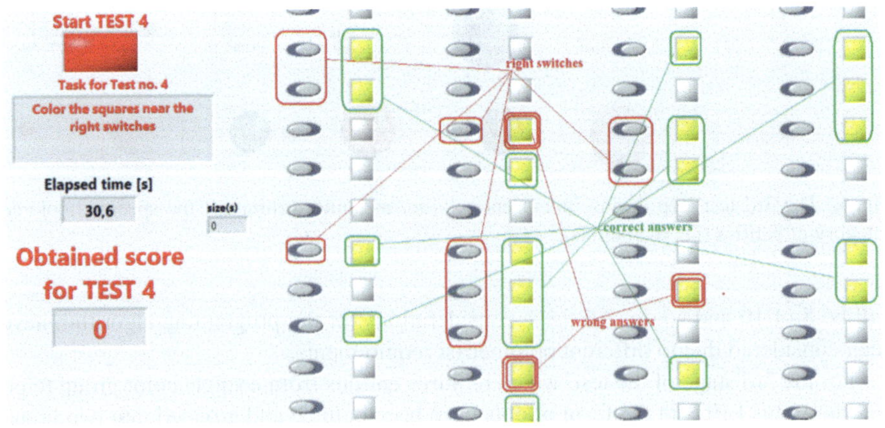

Fig. 7. The 4th test step during its current solving, meaning turning on the yellow square buttons from the right switches to the right.

Regarding the use of the software interface specific to the game-test training in the case of amblyopia, the stages necessary to run it are the following: 1 – reinitialisation of the interface, through the commands *Edit – Reinitialise all values to default*, 2 - selection (one by one) of the current stage for testing from the dialog box, *Choose the game in the list below*, starting the current stage of the test, through the command *Start test 1, 2 3 or 4*, tracking the solution of the test and viewing the final grade after completing the test.

An essential role that the examiner has in assisting the child in the training procedures through this evaluation method is to note and compare the results obtained each time, obviously the testing must take place in identical conditions and at approximately the same times of the day. The subject must definitely be rested, perfectly healthy and not have any kind of physiological problem at the time of the visual training procedures.

The most important aspect is that, at the end of the duration of the visual training procedures, noticeable improvements in the obtained results are observed, this being evidence that the amblyopia in the trained eye has been largely or even completely resolved.

Regarding the other pathology, namely nasal or temporal strabismus, the 2nd software interface was designed so that, with the healthy eye in occlusion, the subject goes through the test games, following three distinct work steps. A logic diagram for the entire programing algorithm is shown in Figs. 8 and 9.

The first stage of the test was designed so that with the free eye (affected by strabismus), the child identifies by switching some virtual 3D buttons the selective lighting of some light stimuli from a string arranged on the display. The arrangement of the stimuli and switch buttons should be such that they are in an antagonistic position to the squinting eye. Thus, if the strabismus manifests itself towards the nasal area of the left eye or towards the temporal area of the right eye, then the entities to be tracked will be located on the left side of the display. If the strabismus manifests itself towards the nasal area of the right eye or towards the temporal area of the left eye, then the entities to be tracked will be located on the right side of the display.

Following the same procedure, from the point of view of the placement of the graphic elements, for the second stage of the testing, the current task that needs to be solved is to complete a Patch-type route, the correctness and duration of its completion being of interest here. The design of the interface assumed that in the third stage the reaction time on the part of the subject was tested at the end of a dynamic cycle simulating the automatic travel of a road to a house-type entity. The subject's reaction must consist of pressing a virtual button that is colored, at which point the time during which he had the reaction is automatically recorded.

For the first stage of testing, the programming algorithm involved defining a string of Boolean output entities, only a few of which would fire for 1–2 seconds at a time during test running.

For this, it was also necessary to use a closed loop structure of the While - Loop type, in relation to which the moment when the selective ignition of those entities can be related. Through other Boolean switch-type input entities, the subject can respond to what it has seen and remembered about the entities lighting up. Figure 10 shows a sequence of this programming algorithm.

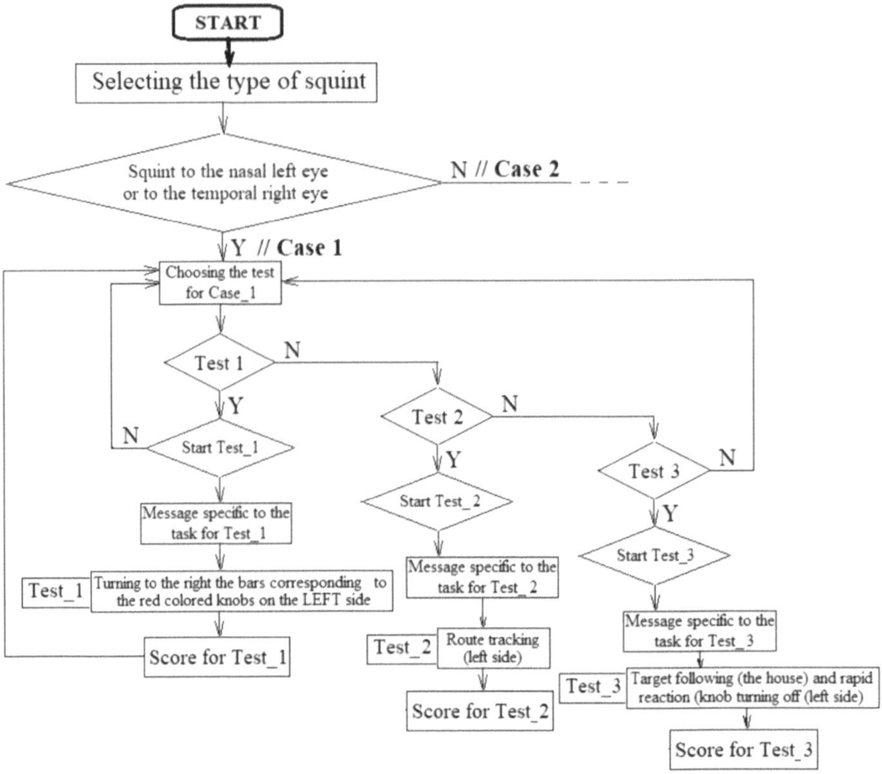

Fig. 8. Logic diagram for software interface programing in case of visual training to correct the eye squint_1st case (squint to the nasal LE or to the temporal RE).

The programming of the algorithm that quantifies the correct and erroneous answers, but also the duration of solving this task is similar to that in the case of the 3rd stage of testing in the case of amblyopia training. An example of solving the test in the first stage is shown in Fig. 11.

The definition of strings of virtual Boolean buttons, was necessary in the programming algorithm for the second step of testing, since the condition was also imposed here (through the *While-Loop*) that only upon completion of the entire route to validate the current work stage. Obviously, in this case too, the duration of the route is of interest, the grade being awarded at this stage depending on this aspect. The scoring algorithm for awarding the grade was similar to that of the 2nd step test for amblyopia training procedure. However, in contrast to the case of the interface for amblyopia, at this stage, a subroutine was included in the programming algorithm through which, every time there is a deviation from the route (a square that constitutes the route is not found and clicked outside, in the background) to count one error each, the score for this stage being affected by this aspect as well.

The Boolean output entities that make up the path that is automatically traveled by simulation, specific to the 3rd stage, were related according to the *While-Loop* structure

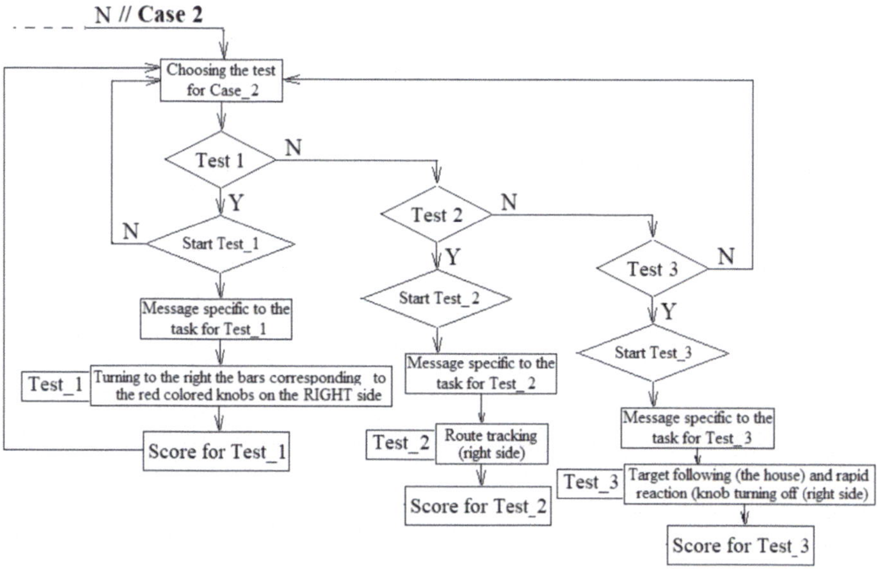

Fig. 9. Logic diagram for software interface programing in case of visual training to correct the eye squint_2nd case (squint to the temporal LE or to the nasal RE).

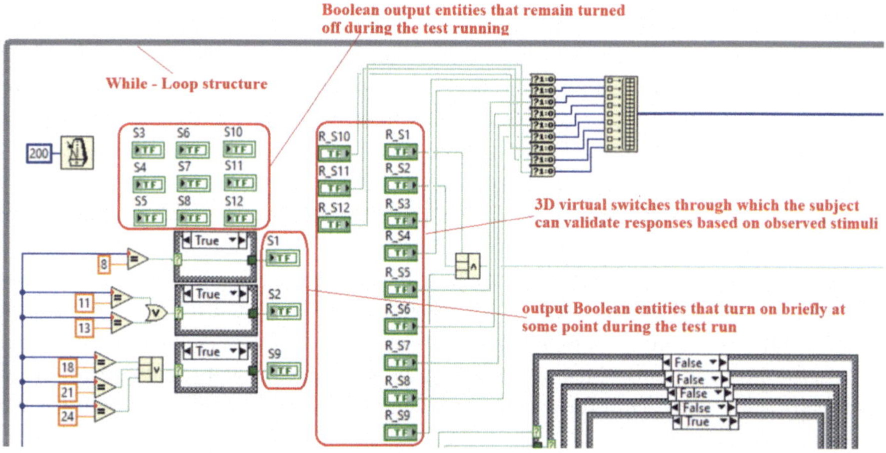

Fig. 10. Sequence of programing algorithm specific for the 1st testing step.

index, so that, when simulating the automatic path to the home entity, these elements to flash automatically at intervals of less than one second. The algorithm for this step has been programmed in such a way that, upon completion of the turning on of all output Boolean entities (completion of the route simulation), the automatic turned on of the house entity (consisting of several virtual Boolean buttons of output). Thus, only after this moment can the manual switching of the input button, namely the subject's reaction,

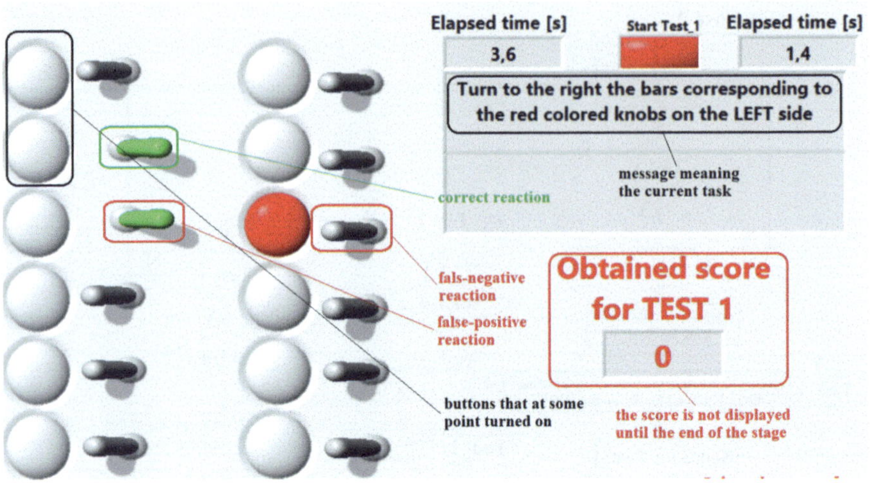

Fig. 11. Example of solving the first step of the testing in case of training procedure for squint.

be taken into account. In these conditions, immediately after switching the input button, the running of the *While-Loop* must be stopped, so that at the end the duration from the turning on of the house-type stimulus can be timed. An example of solving the test in the last its stage is shown in Fig. 12.

As for the procedure for training through assisted testing in the case of strabismus, the working stages for using the software interface described above are the following: 1 - selecting the stage for testing (there is also the possibility of going through selective stages); 2 - choosing the type of strabismus for which the training is done; 3 - starting the test for the current work stage; 4 - passing the test; 5 - display of final results. Figure 13 exemplifies the procedure to be followed specific to the first two steps required to start the test. Testing, as with the amblyopia training software interface, involves manually turning ON the button to start the test in its current step.

For each stage completed within the test, the respective note can be viewed, thus the examiner can get an overview of the evolution of the visual training of the subject in question in terms of combating strabismus. This fact is very important because very useful information is obtained regarding the aspects of visual training that should be emphasized more. The information is particularly useful and effective both for the examiner and the tested subject.

Regarding the training procedure through physical alternative games, complementary to the virtual ones (presented above), in this research, two low-cost physical games, made of recycled materials, were developed [13].

The first of these consists of making a route created from colored lines between hole-type entities that have been drilled into a thin plywood. The image of this physical game can be seen in Fig. 16. This first physical game-test was designed and manufactured for training in case of amblyopia [13].

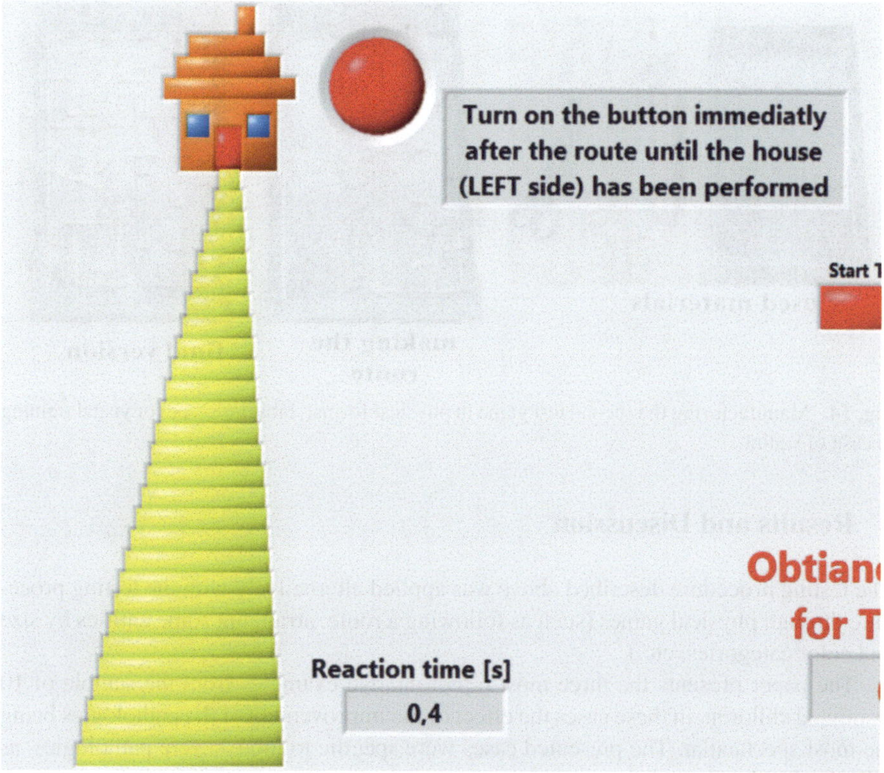

Fig. 12. Example of solving the last step meaning the reaction time.

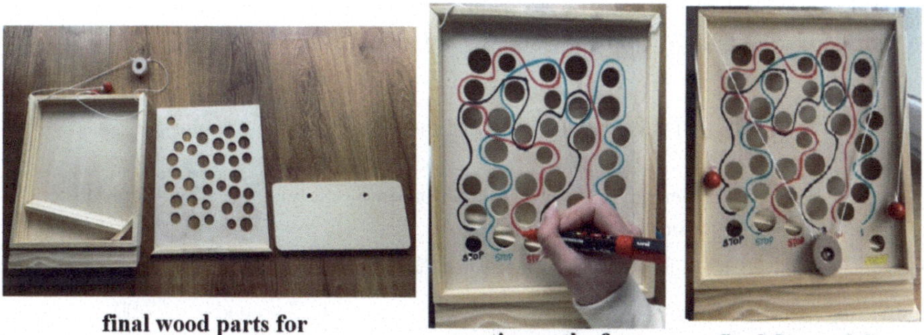

final wood parts for the physical game **creating paths from colored lines** **final form of the physical game-test**

Fig. 13. Manufacturing the first test game in physical format following colorful tracks.

The second game-test, labyrinth type, was designed and made for the purpose of visual training in the case of strabismus and required the following components: a cardboard box, plastic sticks and two wooden balls of different sizes. The stages of practical implementation and the final version of this test are presented successively in Fig. 14.

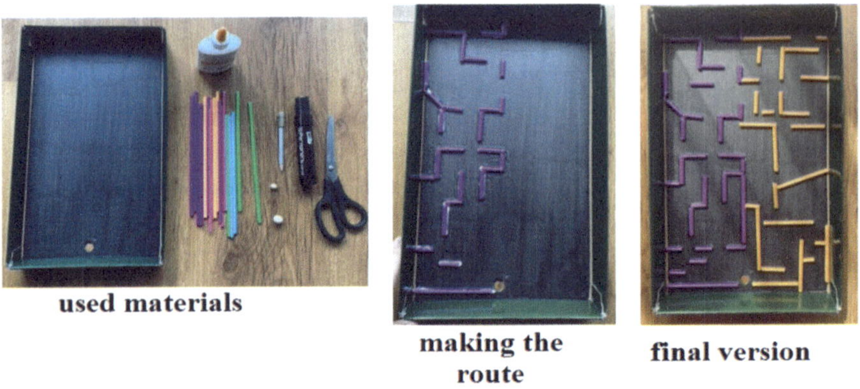

Fig. 14. Manufacturing the second test game in physical format, labyrinth type for visual training in case of squint.

3 Results and Discussion

The testing procedure described above was applied alternatively with the testing procedure through physical games (such as following a route, arranging some entities by size and color categories, etc.).

The paper presents the three most representative examples from the sample of 10 examined children, in these cases the effect of the improvement of the pathologies being the most spectacular. The presented cases were specific to more severe pathologies, as follows: for the case of three children, two of whom were previously diagnosed with amblyopia and one being diagnosed with strabismus [12, 13].

Regarding the first subject, he followed an alternative testing procedure through physical and virtual games, specific to training in the case of strabismus, the results of the tests can be quantified both subjectively and objectively in Table 1.

Table 1. Testing results in case of the 1st child (having squint).

Game type	Understanding game performance	Reaction time	Solving time	Accorded score per test
2nd physical game	Good	Very good	33.38 seconds	10.00
2nd virtual game	Good	Enough	11.2 seconds	7.00
Additional data				
Receptivity	Very good			
Relaxing	Very good			
Coordination	Good			

The second subject, diagnosed with amblyopia, was obviously tested by test-games specific to the training to combat amblyopia, i.e. by the first test-game in physical format, respectively by the first software interface for assisted testing in the virtual environment. Test results of subject no. 2 are presented in Table 2 [13].

Table 2. Testing results in case of the 2nd child (amblyopia).

Game type	Understanding game performance	Reaction time	Solving time	Accorded score per test
1st physical game	Good	Enough	87 seconds	8.00
1st virtual game	Good	Not enough	128.9 seconds	3.00
Additional data				
Receptivity	Good			
Relaxing	Good			
Coordination	Enough			

Subject no. 3, having amblyopia like the 2nd subject, followed the same procedure of physical game and virtual game testing as in the case of subject no. 2. The results can be viewed in Table 3.

Table 3. Testing results in case of the 3rd child (amblyopia).

Game type	Understanding game performance	Reaction time	Solving time	Accorded score per test
1st physical game	Good	Good	65 seconds	8.00
1st virtual game	Good	Enough	43.6 seconds	5.00
Additional data				
Receptivity	Very good			
Relaxing	Very good			
Coordination	Good			

4 Conclusion on the Proposed and Applied Method

The objective of the study was to evaluate the effect of visual training on people with strabismus and amblyopia through a procedure that included virtual test games and physical test games. The results and conclusions were obtained from the analysis of the monitored data in Excel for each subject.

It was found that for the children having amblyopia it was more difficult to pay concentration on the tests solving, but even these have been attracted by the idea of testing due to physical and virtual games.

The study on the 10 examined children represents only one stage in the research regarding the development and implementation of this new method of evaluation and training of the visual function [13]. To complete the study, but also to be able to certify to a much better extent the efficiency and usefulness of this proposed new method, in the next researches it is considered to apply it to much larger samples of children. For this purpose, a strategy has been proposed that in the next year screening actions will be carried out in schools and kindergartens, a number of approximately 100–150 children examined is expected, and in the next 2 years it is desired to expand the sample to over 300 of children [4, 6]. Based on statistical information that approximately 2–3 children out of 10 have visual problems such as strabismus or amblyopia, an application of the case study on assisted visual training procedures is estimated for approximately 60–80 children. In this way, it will be possible to prove to a much better extent the degree of efficiency and the usefulness of the method under the aspect of visual assisted and alternative training [12, 13].

The further researches on this theme will be focused on the proposed non-conventional testing by serious games method implementation as visual screening and training procedure in children having sight diseases especially in case of squint or amblyopia.

Resuming the obtained results it can be observed that the alternative testing methods of squint and amblyopia have given some important information representing a very good base for the furtherly study. It could be found the high potential of the gaming tests as innovative solution not only in the Optometry, but even in Medical Engineering domains [13].

Acknowledgments. The research described in the paper was largely due to the contribution of Optometry graduate students, 2022-2023 promotion, within the Diploma Project, the practical activity being carried out in the laboratories of Transilvania University.

Conflict of Interest. The authors declare they have no conflict of interest.

References

1. Baritz, M., Braun, B.: Modelling synergistic eye movement in the visual field, 6th International Conference Engineering Graphics Design, ICEGD 2015, by SORGING, Brasov, Romania, 11th—13th June, 2015, Fascicle 3, Section Engineering Computer Graphics, ISSN: 2344–4681, pp. 1–4 (2015). http://www.sorging.ro/ro/revista/volume-10-special-issue-fascicle-3/modelling-synergistic-eye-movements-in-the-visual-field
2. Baritz, M., Braun, B.: The correlation between ocular physio-pathological manifestations and bipodal postural reactions, the 6th IEEE International Conference on E-Health and Bioengineering - EHB 2017 Grigore T. Popa University of Medicine and Pharmacy, Sinaia, Romania, June 22–24, 2017, pp. 567–570 (2017)
3. Baritz, M.I.: Evoluţia funcţiei vizuale, Notiţe de curs, Platforma elearning, Universitarea Transilvania din Braşov pp. 5–8 (2022)
4. Jennifer, H.: Vision Screening (04/2020). https://aapos.org/glossary/vision-screening-description

5. Wright, K.W., Spiegel, P.H., Thompson, L.S.: Handbook of Pedriatic Strabismus and Amblyopia, Library of Congress Control Number: 2005932932 ISBN 10: 0–387–27924-5 e-ISBN 0 387–27925-4 ISBN 13: 978–0387–27924-4
6. Barbu, D.: Analysis and Modeling of the Visual Function. Transilvania University Publishing House, ISBN:973-635-130-0 (2003)
7. Baritz, M.: Analysis of human bio-behaviour and eye movements correlated with induced positive/negative emotional level. In: The International Scientific Conference e-Learning and Software for Education, pp. 38–44. ELSE, Bucharest (2016)
8. Consilium Ophthalmologicum Universale: Visual acuity measurement standard. Int. Council Ophthalmol. **15**, 16 (1984)
9. Pop, R.: Oftalmologie, Casa Cărţii de ştiintă (1995)
10. Mufarriq, S., Mirzaman, K., Muhammad, K.; Mohammad, Y., Saeed, N.: Causes of visual impairment in children with low vision, J. Coll. Phis. Surg., Pakistan, JCPSP 21 (2). pp 88–92 (2011). https://www.researchgate.net/publication/49847509_Causes_of_visual_impairment_in_children_with_low_vision
11. Talepasand, S.; Eskandaripour, M.; Taghinezhad, A.: Comparison of Working and Visual Memory in Children with and Without Dyslexia, Zahedan J Res Med Sci, Semnan University, Iran, November 2018, pp. 3–5 (2018).
12. Braun, B., Baritz, M.: Flexible method for "serious" games category for assisted visual training for children with visual impairments. In: 16th International Scientific Conference e-Learning and Software for Education, Bucharest, April 23-24, 2020. ELSE (2020)
13. Şoldan, C.: Proiectarea şi dezvoltarea unei metodologii alternative pentru antrenamentul vizual al copiilor strabici şi ambliopici, Diploma Project, coordinator: Lecturer Dr. Eng. Braun Barbu (2022)
14. National Instruments: Introduction to LabVIEW 8.6, site: www.ni.com (2008)

Comparative Analysis of Machine Learning Algorithms on Kidney Transplant Medical Data

Gabriel Cristian Dragomir-Loga(✉)

Computer Science Department, Technical University of Cluj-Napoca, Cluj-Napoca, Romania
gabriel.dragomir@cs.utcluj.ro

Abstract. In this work, we present the results obtained after using some of the most modern Machine Learning (ML) algorithms: K-Means (KM), Decision Trees (DT), Logistic Regression (LR), Bernoulli Naive Bayes (BNB), Gaussian Naive Bayes (GNB), Random Forest (RF), Support Vector Machines (SVM), K-Nearest Neighbors (KNN), Gradient Boosting (GB). We worked with real clinical data that refer to kidney transplantation. Features include age, gender, Killer-cell immunoglobulin-like receptor (KIR) characteristics and patient status. The conclusion depends on the small volume of data available and tells us that on average, using the above characteristics to classify the patient's status, the accuracy does not exceed 70% for any of the algorithms we used. GB and GNB outperformed all the other estimators (accuracy 70%) with penalty in execution time for GB—32.5 min, compared to other estimators where the execution time was under 1 min. GNB was the best to precision. SVM was the best to specificity. GB was also the best at recall and F1 score.

Keywords: Machine-Learning · KIR · kidney-transplant

1 Introduction

1.1 Killer-Cell Immunoglobulin-Like Receptors (KIRs) Role in Diseases

Natural killer NK cells play a critical role in the immunological response mechanism obtained by transplantation.

Genotypes that are inhibitory KIR receptor dominant are likely susceptible to infection and reproductive disorders but protective against autoimmune diseases, whereas activating KIR receptor dominant genotypes are likely susceptible to autoimmunity but protective against viral infection and cancer [1].

Figure 1 shows a schematic of KIR protein structural organization [2].

The data set for the study in the current paper consists of transplant recipients who received a kidney transplant between January 2010 and December 2019 at ICUTR Cluj-Napoca [3]. The control group consisted of 97 voluntary blood donors of the National Registry of Hematopoietic Stem Cells Voluntary Donors (RNDVCSH). All samples collected had the written consent of the patient. There were 16 patients who have been diagnosed with acute rejection from 84 patients which have received a kidney transplant.

S. Vlad and N. M. Roman (Eds.): MEDITECH 2024, IFMBE Proceedings, 130, pp. 180–194, 2025.
https://doi.org/10.1007/978-3-031-95671-3_19

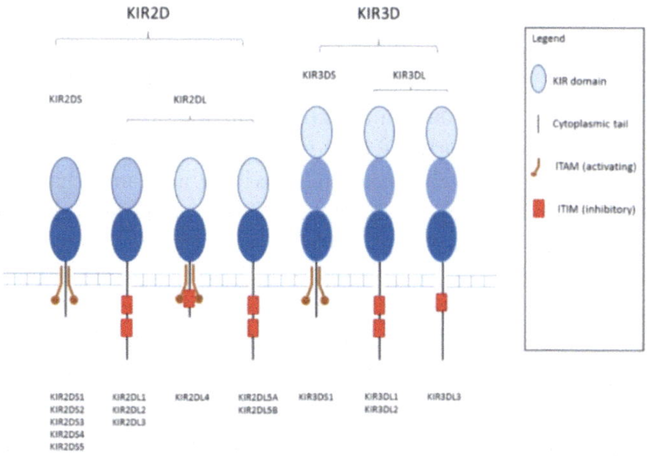

Fig. 1. Structural organization of KIR proteins [2].

KIR genotyping was performed for the presence or absence of the 16 KIR genes: 2DL1, 2DL2, 2DL3, 2DL4, 2DL5, 2DS1, 2DS2, 2DS3, 2DS4, 2DS5, 3DL1, 3DL2, 3DL3, 3DS1, 2DP1, 3DP1.

1.2 The Importance of AI (Artificial Intelligence) in Current Life

Artificial Intelligence (AI) is a field in computer science dedicated to creating machines that can perform tasks typically requiring human intelligence [4]. These tasks include reasoning, learning, perception, language understanding, and problem-solving.

AI systems are designed to mimic human cognitive functions and can be divided into narrow AI, which is tailored for specific tasks, and general AI, which aspires to understand and reason across a wide range of activities. AI has seen substantial growth with the advent of deep learning and neural networks, which enable machines to process vast amounts of data, recognize patterns, and make decisions with minimal human intervention.

Machine Learning (ML) is a subset of AI focused on developing algorithms that enable computers to learn from and make predictions based on data [5]. Rather than being explicitly programmed to perform a task, ML algorithms build models from sample data, known as training data, to make data-driven predictions or decisions. ML encompasses various methods, including supervised learning, unsupervised learning, and reinforcement learning, each suitable for different types of tasks.

Supervised learning has two different flavors: classification (the algorithm learns to assign input data into predefined categories or classes) and regression (involves predicting a continuous-valued output variable based on input variables). Applications of ML span numerous fields, from natural language processing and computer vision to finance and healthcare, where it facilitates the automation of complex decision-making processes. In the following we will present some widely used techniques in the domain of Machine Learning.

Logistic Regression (LR), part of supervised learning, is a statistical method used for binary classification problems, where the outcome variable is dichotomous, such as yes/no or true/false [6]. Despite its name, it is a linear model that predicts the probability of a particular class by fitting data to a logistic function, also known as the sigmoid function. This model is particularly effective when the relationship between the features and the outcome is approximately linear and is widely used in scenarios like disease diagnosis, customer churn prediction, and credit scoring. Logistic Regression is appreciated for its simplicity, interpretability, and robustness to overfitting when used with regularization techniques.

Linear Regression is a fundamental and widely used statistical method in supervised learning (especially regression) for modeling the relationship between a dependent variable and one or more independent variables [7]. It operates under the assumption that there is a linear relationship between the input features and the continuous output variable. The goal of linear regression is to find the best-fitting straight line, known as the regression line, that minimizes the sum of squared differences between the observed and predicted values. This method is particularly effective for understanding the impact of several variables on a single outcome and is widely applied in various fields such as economics, finance, biology, and social sciences for tasks like predicting sales, assessing risk, or forecasting trends. Despite its simplicity, linear regression serves as a powerful baseline for regression analysis and is often the starting point for more complex modeling techniques, providing insights into data trends and relationships while being easy to interpret and implement.

Naive Bayes is a family of simple, yet powerful probabilistic classifiers based on Bayes' theorem, with a fundamental assumption that the features are independent of each other given the class label [8]. Despite this "naive" assumption of independence, which rarely holds true in real-world scenarios, Naive Bayes models perform remarkably well in many practical applications, particularly when dealing with high-dimensional datasets.

There are different variants of Naive Bayes classifiers, such as *Bernoulli Naive Bayes* (BNB), which is suitable for binary or Boolean feature vectors, and *Gaussian Naive Bayes* (GNB), which is used when the features are assumed to follow a normal distribution. The simplicity and computational efficiency of Naive Bayes make it ideal for large-scale text classification tasks, spam filtering, sentiment analysis, and real-time prediction systems. Its interpretability and robustness to noise in the data further contribute to its widespread use in various domains.

Support Vector Machines (SVMs) are supervised learning models that are particularly effective for classification tasks in high-dimensional spaces [9]. The core idea of SVM is to find the hyperplane that best separates the data into different classes with the maximum margin between the nearest data points of any class, known as support vectors.

SVMs are versatile, capable of using different kernel functions to handle linear and non-linear separable data, making them suitable for a wide range of applications, from text categorization and image classification to bioinformatics and handwriting recognition.

Decision Trees (DT) are a type of supervised learning algorithm used for both classification and regression tasks [10]. They model decisions and their possible consequences in a tree-like structure, where each internal node represents a test on an attribute, each branch represents the outcome of the test, and each leaf node represents a class label or a continuous value.

Decision Trees are intuitive and easy to interpret, making them a popular choice for many applications, including risk assessment, credit scoring, and medical diagnosis. However, they can be prone to overfitting, especially when they grow too complex, necessitating pruning techniques to enhance generalization.

Random Forest (RF) is a supervised learning technique that constructs multiple decision trees during training and merges their outputs to improve the model's accuracy and control overfitting [11]. Each tree in the forest is trained on a random subset of the data and features, promoting diversity among the trees and enhancing the robustness of the model.

Random Forest is highly versatile, capable of handling both classification and regression tasks, and is used extensively in various domains, including financial forecasting, bioinformatics, and environmental science. Its strength lies in its ability to handle large datasets with high dimensionality and provide estimates of feature importance.

K-Nearest Neighbors (KNN) is a simple, yet powerful non-parametric algorithm used for classification and regression tasks [12]. It operates by finding the 'k' closest data points to a query point in the feature space and assigning a class or predicting a value based on the majority class or average value of the neighbors.

KNN is highly intuitive and easy to implement but requires careful consideration of the choice of 'k' and distance metric. It is widely used in recommendation systems, pattern recognition, and real-time applications where the data distribution is unknown or dynamic.

Gradient Boosting (GB) is an advanced supervised learning technique that builds a model in a stage-wise fashion from a series of weak learners, typically decision trees [13]. The algorithm focuses on correcting the errors of the previous models by fitting new models to the residuals of the predictions, effectively minimizing the loss function used.

Gradient Boosting is highly effective for both classification and regression tasks and is known for its accuracy and ability to handle complex datasets with mixed types of features. It has been successfully applied in areas such as credit scoring, ranking problems, and various prediction challenges in machine learning competitions.

K-Means (KM) is a widely used unsupervised machine learning algorithm designed for clustering data into a predefined number of groups, or clusters, based on their similarity [14].

The algorithm aims to minimize the variance within each cluster while maximizing the variance between different clusters. It begins by selecting random centroids for each cluster and iteratively refines these centroids by minimizing the squared Euclidean distance between the data points and their respective cluster centers. K-Means is particularly useful in exploratory data analysis, customer segmentation, and image compression due to its simplicity, scalability, and efficiency.

The motivation behind the analysis of ML algorithms in the context of renal transplantation starts from the intention to discover a treatment path that prevents *rejection based on KIR genes*. Despite advances in surgical techniques and patient follow-up, immunologic rejection continues to have adverse effects on graft function and survival. Therefore, if it can be predicted, it would mean saving lives.

2 Description of the Analysis Method

Figure 2 presents the processing pipeline for classification algorithms. For KM algorithm the processing pipeline does not have fragmentation step into train data and test data and the step which determines optimal parameters (k) is a visual one (*elbow* method). We used the KM algorithm to observe how the data set can be divided into groups.

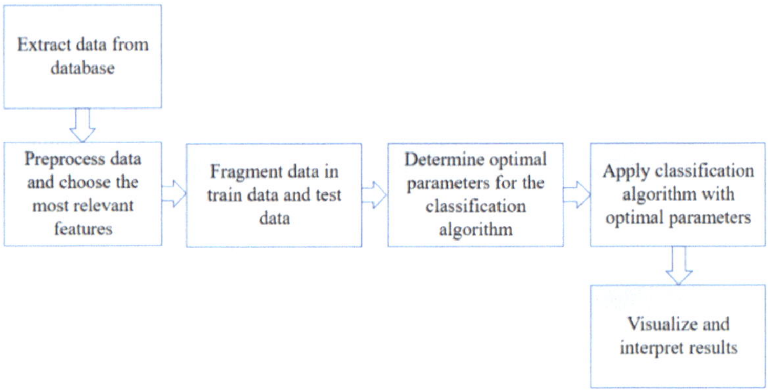

Fig. 2. Processing pipeline for classification algorithms.

2.1 Extract Data from Database

In this step the raw data were obtained by initiate query on the database (see [3] for the structure of our database).

Database is of data warehouse type. We used KIR DataMart to obtain dataset for ML algorithms. The dataset has the following details: id, gender, age, 16 KIR characteristics and status.

2.2 Preprocess Data and Choose the Most Relevant Features

The input data were processed to adapt them as best as possible for the ML algorithms.

For each patient, the age at the time of the analysis (healthy patient) or the age at the time of admission (patient with kidney disease) was transformed into an age group. Thus, we obtained the following values: "18–40", "41–55", ">55" which result in 1, 2, 3 after coding.

For the KIR characteristics (16 in number) 0 (characteristic absent) or 1 (characteristic present) was considered and a 16-bit string was formed which was transformed into a natural number. For instance, the sequence K2DL1, K2DL2, K2DL3, K2DL4, K2DL5,

K2DS1, K2DS2, K2DS3, K2DS4, K2DS5, K3DL1, K3DL2, K3DL3, K3DS1, K2DP1, K3DP1: 1, 1, 1, 1, 1, 0, 1, 1, 1, 0, 1, 1, 1, 1, 1, 1 was transformed into natural number 64447.

The patients' gender was coded binary (female 0 and male 1).

The patient's status was also coded binary (healthy 0 and transplant recipient 1) The data about the patients also contained information about acute rejection. In the preliminary stage of our experiments with the ML algorithms, this characteristic was also considered (multiclass classification). Since the number of patients with rejection was small (16 patients belong acute AKAR group out of 84 total recipients), the results were unconvincing and only binary classification was still used.

In the data preparation phase before using the ML algorithms, the features were normalized (except for the classification criterion). Standardization of a dataset is a common requirement for many machine learning estimators. They might behave badly if the individual features do not more or less look like standard normally distributed data.

The standard score z of a sample x is calculated like in formula (1).

$$z = (x - u)/s \tag{1}$$

In formula (1) u is the mean of the data samples and s is the standard deviation of the data samples.

For example, an entry in dataset with the features: Age Group, Gender, KIR, Patient Type has original values: 1, 1, 64,447, 1 and after standardization has the values: –0.796337, 0.850770, 0.975858,1.

2.3 Fragment Data in Train Data and Test Data

Learning the parameters of a prediction function and testing it on the same data is a methodological mistake. Overfitting appears when applied a model of supervised machine learning too closely or exactly to a particular set of data and may therefore fail to fit to additional data or predict future observations reliably.

It is common practice to divide dataset in 80/20 train data/test data. In our experiments the number of training examples was 144 and the number of testing examples was 37 chosen randomly with seed 123.

2.4 Determine Optimal Parameters for the Classification Algorithm

To find optimal parameters, an exhaustive search over specified parameter values for each estimator was done.

In Table 1, we can see the search domains corresponding to classification algorithms.

The parameters of the estimator are optimized by cross-validated (CV) grid-search over a parameter grid. In the basic approach, called *k*-fold CV, the training set is split into *k* smaller sets. For each of the k "folds" a model is trained using k-1 of the folds as training data and the resulting model is validated on the remaining part of the data. Performance measure is accuracy. The performance measure reported by k-fold cross-validation is the average of the values computed in the loop.

Table 1. The range of values of the parameters for each estimator.

Estimator	Parameter	Range of values
KNN	n_neighbors	3, 4, 5, 6
	weights	'uniform', 'distance'
	algorithm	'ball_tree', 'kd_tree', 'brute'
	p	1, 2
BNB	alpha	0.01, 0.1, 0.5, 1.0, 10.0
	fit_prior	True, False
	class_prior	None, [0.1,]* len(n_classes)
	binarize	None, 0.0, 8.5, 10.0
GNB	priors	None
	var_smoothing	np.logspace(0, −9, num = 100)
DT	criterion	'gini', 'entropy', 'log_loss'
	splitter	'best', 'random'
GB	loss	'log_loss', 'exponential'
	learning_rate	0.1, 1, 10, 100
	n_estimators	1, 10, 100, 1000, 10,000, 1,000,000
	criterion	'friedman_mse', 'squared_error'
	max_depth	3, 5, 7, 10, 12
LR	penalty	'l1', 'l2', 'elasticnet', None
	solver	'lbfgs', 'liblinear', 'newton-cg', 'newton-cholesky', 'sag', 'saga'
RF	n_estimators	5, 10, 20, 50, 100
	criterion	'gini', 'entropy', 'log_loss'
	max_depth	10, None
	min_samples_split	2, 3
	max_features	'sqrt', 'log2', None
SVM	C	0.1, 1, 10, 100, 1000
	gamma	1, 0.1, 0.01, 0.001, 0.0001
	kernel	'rbf', 'linear', 'poly', 'sigmoid'

In our experiments, the number of jobs to run in parallel was set to all-processors, that means 4 parallel jobs. The number of folds was set to 5.

The values in Table 1 are motivated by personal experimental results as well as results presented in the field of ML. The detailed description of the parameters of each estimator from Table 1 can be found in [15].

2.5 Apply Classification Algorithm with Optimal Parameters

After obtaining optimal parameters, each classification algorithm is applied.

At this step, the following measurements will be used in the analysis: *accuracy*, *precision*, *recall* (*sensitivity*), *specificity*, *F1 score*. In medicine, individuals who have the condition are considered "positive" and those who do not are considered "negative".

Accuracy measures how close a given set of observations are to their true value. *Precision* measures how close the observations are to each other and is the fraction of relevant retrieved instances among all the retrieved instances. *Sensitivity* (true positive rate) is the fraction of relevant retrieved instances among all relevant instances. *Specificity* (true negative rate) is the probability of a negative test result, conditioned on the individual truly being negative. The *F1 score* is the harmonic mean of the *precision* and *recall*. The highest possible value of an *F1-score* is 1.0, indicating perfect *precision* and *recall*, and the lowest possible value is 0, if *precision* and *recall* are zero. The *Support* is the number of occurrences of each class. In our study testing data set had 20/17 *Support* for 0/1 classes.

2.5.1 Discussion Related to Optimal Parameters

Running classification algorithms with optimal parameters in Table 2, do not lead to the best results.

The GB algorithm used the most resources in the phase of determining the optimal parameters. Suggested optimal parameters did not provide maximum results. Parameters with the best results were the following: criterion = 'friedman_mse', max_depth = 7, learning_rate = 10, loss = 'exponential', n_estimators = 10,000.

In the case of KNN, as the number of neighbors increases, the classification becomes more specific and the accuracy decreases (Table 3). If we change the domain for n_neighbors to 1. 30 than the results of optimal parameters search are: algorithm = 'brute', n_neighbors = 25, p = 2 and weights = 'uniform'. With these values of parameters, at running of KNN estimator, we obtained: *Accuracy* = 0.49, *Precision* = 0.38, *Recall* = 0.18, *Specificity* = 0.75, *F1 score* = 0.24 which is worse than in Table 4.

GB and GNB outperformed all the other estimators in terms of accuracy. GNB was the best for precision. SVM was the best to specificity. GB was also the best at recall and F1 score. Execution time for GB was 32.5 min. The other estimators had the execution time under 1 min.

2.6 Visualize and Interpret Results

In Fig. 3 we represented in 2D space the 4 features of dataset in two images.

At this time, we will talk about the K-means clustering algorithm. We used the *elbow* method to determine the number of clusters. As can be seen in Fig. 4 knee of the curve is approximately at point 5. Next, we applied the KM algorithm to the entire dataset, and we obtained the clusters. In Fig. 5, we can see the centroids for the 5 obtained clusters. We have organized representation of five clusters in 3D form like in Fig. 6.

For example, for the KNN algorithm, the data in Table 5 are interpreted so that the algorithm estimated 15 TP and 8 TN, respectively, 9 FP and 5 FN. We consider positive class 0 (healthy patient, even if in medicine diseased means positive).

Table 2. Optimal parameters for each estimator found through grid-search.

Estimator	Parameter	Optimal value	Best Accuracy Through Grid Search
KNN	n_neighbors	4	0.619
	weights	'uniform'	
	algorithm	'ball_tree'	
	p	1	
BNB	alpha	10.0	0.569
	fit_prior	True	
	class_prior	None	
	binarize	0.0	
GNB	priors	None	0.646
	var_smoothing	0.008111308307896872	
DT	criterion	'gini'	0.604
	splitter	'random'	
GB	loss	'log_loss'	0.709
	learning_rate	0.1	
	n_estimators	1,000,000	
	criterion	'squared_error'	
	max_depth	3	
LR	penalty	'l2'	0.604
	solver	'lbfgs'	
RF	n_estimators	20	
	criterion	'entropy'	
	max_depth	10	0.695
	min_samples_split	3	
	max_features	None	
SVM	C	100	
	gamma	0.001	0.660
	kernel	'rbf'	

Formulas 2, 3, 4 and 5 describe the calculation method for performance parameters from Table 4 starting from the confusion matrices from Table 5.

The confusion matrix is critical in understanding the metrics from Table 4. Confusion matrix for each estimator (Table 5) has class 0 on the first line and class 1 on the second one (true class on X axis). On columns we have predicted classes. So, the main diagonal represents true predicted classes (true positives (TP) in the top left corner and

Table 3. Classification report (healthy 0 and transplant recipient 1) that was obtained by applying the optimal parameters for each estimator.

Estimator	Classification	Precision	Recall	F1 score
KNN	0	0.62	0.75	0.68
	1	0.62	0.47	0.53
BNB	0	0.59	0.80	0.68
	1	0.60	0.35	0.44
GNB	0	0.60	0.75	0.67
	1	0.58	0.41	0.48
DT	0	0.70	0.80	0.74
	1	0.71	0.59	0.65
GB	0	0.71	0.75	0.73
	1	0.69	0.65	0.67
LR	0	0.68	0.75	0.71
	1	0.67	0.59	0.62
RF	0	0.62	0.65	0.33
	1	0.56	0.53	0.55
SVM	0	0.61	0.85	0.71
	1	0.67	0.35	0.46

Table 4. Measure of predictive performance for each estimator.

Estimator	Accuracy	Precision	Recall	Specificity	F1 score
KNN	0.62	0.62	0.47	0.75	0.53
BNB	0.59	0.60	0.35	0.80	0.44
GNB	**0.70**	**0.81**	0.50	0.89	0.62
DT	0.62	0.59	0.59	0.65	0.59
GB	**0.70**	0.69	**0.65**	0.75	**0.67**
LR	0.68	0.67	0.59	0.75	0.63
RF	0.68	0.69	0.53	0.80	0.60
SVM	0.65	0.75	0.35	**0.90**	0.48

true negatives (TN) in the lower right corner). The upper right corner corresponds to false positives (FP) and lower left corner corresponds to false negatives (FN).

$$\text{Accuracy} = (\text{TP} + \text{TN})/(\text{P} + \text{N}) \tag{2}$$

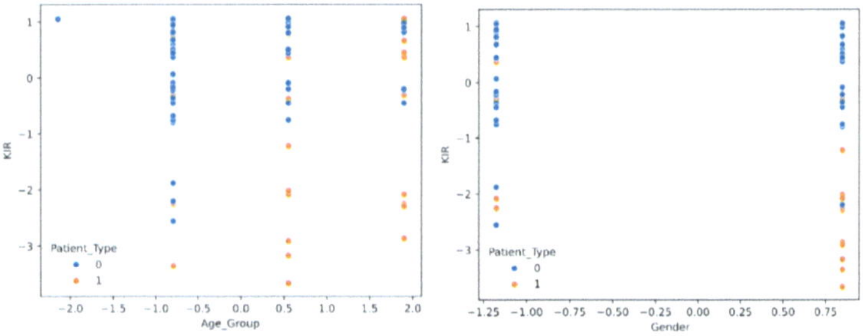

Fig. 3. 2D data view Patient Status (KIR, Age Group) and Patient Status (KIR, Gender).

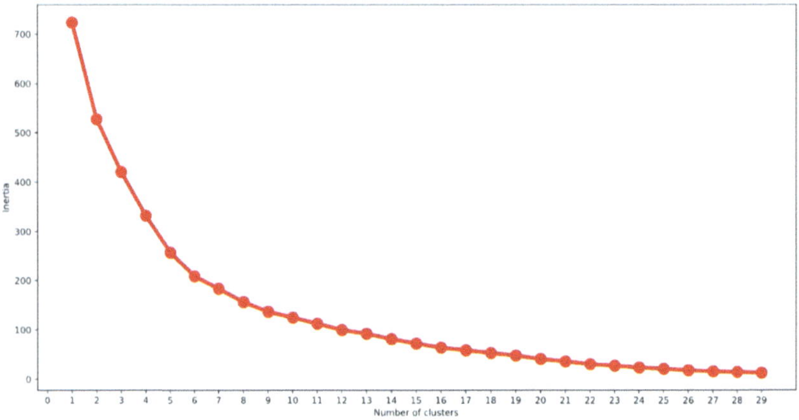

Fig. 4. Inertia = Sum of squared distances of samples to their closest cluster center, weighted by the sample weights (KM).

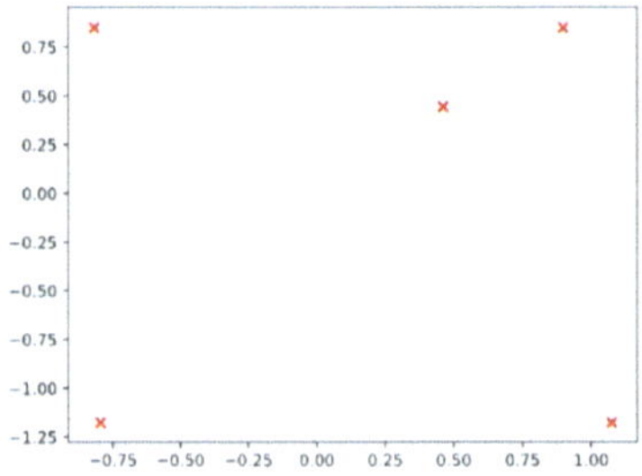

Fig. 5. Clusters centroids KM. Centroids are located at the following points in normalized data set: [0.46, 0.45, −2.47, 0.87], [−0.80, −1.18, 0.09, 0.32], [0.90, 0.85, 0.13, 0.51], [1.07, −1.18, 0.36, 0.56], [−0.82, 0.85, 0.28, 0.36]

In formula (2) P = TP + FP = 20 and N = TN + FN = 17.

$$\text{Precision} = \text{TP}/(\text{TP} + \text{FP}) \tag{3}$$

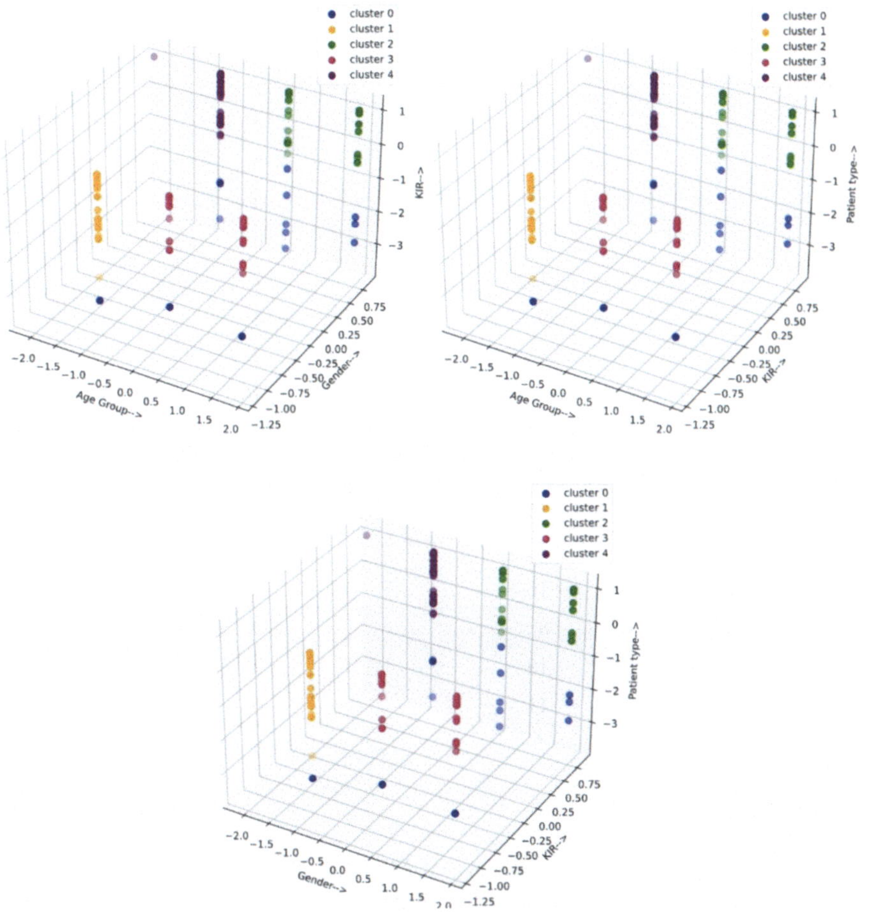

Fig. 6. 3D data view of Clusters KM.

$$\text{Recall or Sensitivity} = \text{TP}/(\text{TP} + \text{FN}) \tag{4}$$

$$\text{Specificity} = \text{TN}/(\text{TN} + \text{FP}) \tag{5}$$

Sensitivity and specificity provide insight into the predictive capabilities of a classification model. While high sensitivity minimizes false negatives, high specificity minimizes false positives.

Table 5. Confusion matrices for each estimator.

Estimator	Confusion matrix	
KNN	15	5
	9	8
BNB	16	4
	11	6
GNB	18	2
	9	8
DT	13	7
	7	10
GB	13	7
	7	10
LR	15	5
	7	10
RF	16	4
	8	9
SVM	18	2
	11	6

3 Conclusion

In the current paper, we used 8 supervised ML algorithms and 1 unsupervised ML algorithm for the analysis of medical data in the context of kidney transplantation. In the specialized literature for renal transplantation, we did not find any studies analyzing the use of ML. The results obtained (Table 4) can predict with accuracy 70%, if the GB algorithm is used (the percentage is not very good - a good percentage would mean somewhere over 90%) for features age, gender, 16 KIR components if a person needs a kidney transplant.

In the preliminary stage of our experiments with the ML algorithms, multiclass classification was considered (healthy patient, chronic rejection patient, stable renal function patient). Since the number of patients with rejection was small (16 patients belong acute AKAR group out of 84 total recipients), the results were unconvincing and only binary classification was still used (healthy patient, patient with renal transplant). By unconvincing we mean underfitting, which is because the data set is small, for the test subset there were 0 cases predicted by most algorithms. KNN: Accuracy = 0.76, Precision = 0.86, Sensitivity = 0.86, Specificity = 0.86, F1_score = 0.86. BNB: Accuracy = 0.82, Precision = 0.82, Sensitivity = 1.0, Specificity = 1.0, F1_score = 0.9. GNB: Accuracy = 0.71, Precision = 0.8, Sensitivity = 0.86, Specificity = 0.86, F1_score = 0.83. DT: Accuracy = 0.71, Precision = 0.85, Sensitivity = 0.79, Specificity = 0.79, F1_score = 0.81. GB: Accuracy = 0.71, Precision = 0.85, Sensitivity = 0.79, Specificity = 0.79,

F1_score = 0.81. LR: Accuracy = 0.76, Precision = 0.81, Sensitivity = 0.93, Specificity = 0.93, F1_score = 0.87. RF: Accuracy = 0.71, Precision = 0.85, Sensitivity = 0.79, Specificity = 0.79, F1_score = 0.81. SVM: Accuracy = 0.76, Precision = 0.81, Sensitivity = 0.93, Specificity = 0.93, F1_score = 0.87.

In the analysis stage of the optimal parameters of the algorithms, the obtained values have proved that did not give the best accuracy. For some of the algorithms, such as GB, we searched in the results of the analysis of the optimal parameters to find the parameters that give the highest score, thus improving the performance (accuracy and sensitivity in sense of predict that the person with the given features will not need a kidney transplant).

The best algorithms in terms of accuracy were GB and GNB (0.7).

GB paid a penalty in execution time of 32 minutes. The other algorithms take less than 1 minute.

GNB was the best for precision (0.81). SVM was the best to specificity. GB was also the best at sensitivity (0.65) and F1 score (0.67). SVM was the best to specificity (0.9).

The conclusion is that ML algorithms can offer a solution in the classification of patients in the context of renal transplantation, but to be sure how they can be used in the prevention of AKAR, it is necessary to build the classification model using a larger data set.

Conflict of Interest. The authors declare that they have no conflict of interest.

References

1. Parham, P.: Immunogenetics of killer cell immunoglobulin-like receptors. Mol. Immunol. **42**(4), 459–462 (2005)
2. Downing, J., D'Orsogna, L.: High-resolution human KIR genotyping. Immunogenetics. **74**(4), 369–379 (2022)
3. Dragomir-Loga, G.C., Dican, L.: KIR analysis in context of kidney transplantation. In: 10th IEEE International Conference On E-Health and Bioengineering Conference (EHB), pp. 1–4, Iasi, Romania (2022)
4. Earl, B.: Hunt: Artificial Intelligence. Academic Press. Inc., New York, NY/San Francisco/CA, London (1975)
5. Alpaydin, E.: Machine Learning. Revised and Updated Edition. The MIT Press, Cambridge, MA/London (2021)
6. Nick, T.G., Campbell, K.M.: Logistic Regression. In: Ambrosius, W.T. (ed.) Topics in Biostatistics. Methods in Molecular Biology, vol. 404, pp. 273–301. Humana Press (2007)
7. James, G., Witten, D., Hastie, T., Tibshirani, R., Taylor, J.: Linear regression. In: An Introduction to Statistical Learning. Springer Texts in Statistics, pp. 69–134. Springer, Cham (2023)
8. Eguturi Manjith Kumar Reddy, Akash Gurrala, Vasireddy Bindu Hasitha, Korupalli V Rajesh Kumar: Introduction to Naive Bayes and a Review on Its Subtypes with Applications. 1st Edition. Chapman and Hall/CRC (2022)
9. Steinwart, I., Christmann, A.: Support Vector Machines, 1st edn. Springer, New York, NY (2008)
10. de Ville, B.: Decision trees. Wires Comput. Stat. **5**(6), 448–455 (2013)
11. Rigatti, S.J.: Random forest. J. Insur. Med. **47**(1), 31–39 (2017)

12. Kramer, O.: K-nearest neighbors. In: Dimensionality Reduction with Unsupervised Nearest Neighbors. Intelligent Systems Reference Library, vol. 51, pp. 13–23. Springer, Berlin, Heidelberg (2013)
13. Bentéjac, C., Csörgő, A., Martínez-Muñoz, G.: A comparative analysis of gradient boosting algorithms. Artif. Intell. Rev. **54**(3), 1937–1967 (2021)
14. Ahmed, M., Seraj, R., Islam, S.M.S.: The k-means algorithm: a comprehensive survey and performance evaluation. Electronics. **9**(8), 1295 (2020)
15. Scikit-Learn Homepage: https://scikit-learn.org/stable/api/sklearn.html

Predictive Mathematical Modeling of COVID-19 Pandemic: A Comparative Analysis of ARX and ARMAX Approaches

Cristina-Maria Stancioi(✉), Vlad Muresan, Mihail Abrudean, and Marius Nicolae Roman

Technical University of Cluj-Napoca, Cluj-Napoca, Romania
stancioi_cristina@yahoo.com

Abstract. This study compares ARX and ARMAX models for forecasting COVID-19 case dynamics. Both models have been frequently used in time series forecasting and are considered here for their efficacy to estimate the spread of COVID-19. The ARX model utilizes delayed values of the dependent variable as well as exogenous inputs, whereas the ARMAX model includes a moving average component to account for random disturbances and noise in the data. This study focuses on each model's performance, accuracy, and stability, emphasizing how the presence of external factors like movement patterns, government actions, and vaccination rates influences predictive ability. The work is based on real-world COVID-19 data, and the models are evaluated using important metrics such as root mean square error (RMSE) and goodness of fit. The results show that, while both models are successful, the ARMAX model makes more accurate predictions in the face of noise and changing external factors. This comparison emphasizes the benefits and drawbacks of each method and provides insights on developing time series models for pandemic forecasting.

Keywords: Covid-19 Pandemic · Mathematical models · ARX-model · ARMAX-model · Prediction

1 Introduction

The COVID-19 pandemic, caused by the new coronavirus SARS-CoV-2, has resulted in an unsurpassed globally health disaster since it began in late 2019 [1, 2].

COVID-19 has rapidly spread throughout countries and regions, resulting in high mortality, serious disruptions in healthcare systems, and huge socioeconomic repercussions around the world. Governments, healthcare providers, and academics have significantly depended on predictive models to forecast infection patterns, optimize resource allocation, and inform public health actions such as lockdowns, vaccination campaigns, and movement restrictions [3].

Accurate forecasting of COVID-19 case counts is crucial for timely decision-making, especially as the pandemic progresses with novel variations, shifting infection rates, and

S. Vlad and N. M. Roman (Eds.): MEDITECH 2024, IFMBE Proceedings, 130, pp. 195–200, 2025.
https://doi.org/10.1007/978-3-031-95671-3_20

different degrees of public adherence to health interventions. Traditional time series models, such as ARX and ARMAX mathematical models, have emerged as useful methods for forecasting the trajectory of infectious disease epidemics [4, 5]. These models combine past infection data with external factors such as social distancing measures, testing rates, and vaccination advances to improve forecast accuracy.

2 Implementation and Obtained Results

The article analyzes input-output data pairs to identify transfer functions for the Multiple-Input Single-Output (MISO) mathematical model as described in Fig. 1. The input signals are processed using a black-box system, and the output signal y(t) represents the system's reaction to these inputs [6]. The data utilized for this identification procedure has been filtered ensuring that noise and outliers are reduced in order to establish valid input-output relations.

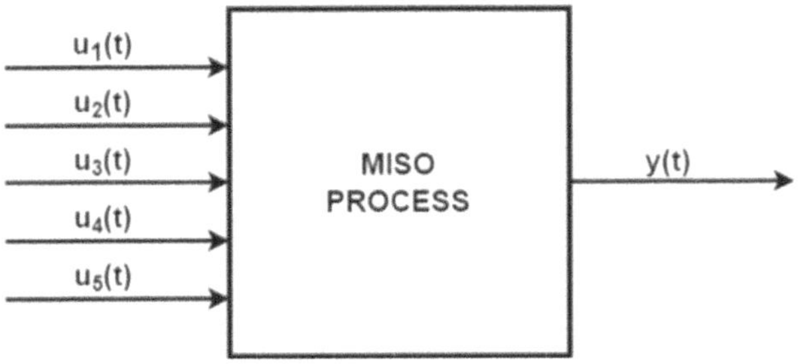

Fig. 1. MISO system

In our system, the five input signals are represented by: number of tests taken every day, temperature in closed spaces, outside temperature, outside humidity and the Covid-19 measures taken by the authorities to limit the spread of the Covid-19 pandemic.

The output signal is represented by the total number of cases.

Given the severely nonlinear and unpredictable nature of virus transmission, particularly with COVID-19, determining the optimal transfer function order is a difficult challenge. Each input signal, such as mobility, vaccination rates, or government initiatives, can impact the output (the number of COVID-19 cases) in different ways and with various delays, making the identification procedure both critical and difficult.

The ARX model depicts the relation between the output (in this case, COVID-19 cases) and its prior values as well as external inputs (interventions or mobility data) [7, 8]. The model may be mathematically represented as follows in Eq. (1):

$$y(t) = \sum_{k=1}^{n} a_k \cdot y(t-k) + \sum_{i=1}^{m} \sum_{j=1}^{p_i} b_{i,j} \cdot u_i(t-j) + e(t) \quad (1)$$

Where:

- y(t) is the output signal (Covid-19 total number of cases);

- $u_i(t)$ are the exogenous inputs (the above mentioned five inputs);
- a_k represents the autoregressive part (past Covid-19 cases)
- $b_{i,j}$ are the coefficients for the exogenous inputs;
- $e(t)$ the error term.

The major difference between ARX-models and ARMAX-models is represented by the additional component in the ARMAX-model, which is the MA(moving average) term [9, 10]. This has the role in noise correlations by simulating the influence of previous errors on actual output. This improves the ARMAX model's ability to handle organized noise or disruptions in data.

The ARMAX-model equation in our case is represented by (2):

$$y(t) = \sum_{k=1}^{n} a_k \cdot y(t-k) + \sum_{i=1}^{m} \sum_{j=1}^{p_i} b_{i,j} \cdot u_i(t-j) + \sum_{l=1}^{q} c_l \cdot e(t-l) + e(t) \qquad (2)$$

Where:

- $c_l \cdot e(t-l)$ is the moving average term which influences the error term e(t) by taking into account the past errors.

Using the ARX-model the daily COVID-19 cases were estimated utilizing prior case counts and exogenous inputs such as movement data, vaccination rates, and public health interventions. However, it presupposes that any noise (such as delays in case reporting or data gathering) is uncorrelated.

In addition to employing the same exogenous inputs as the ARX model, the ARMAX model would take into consideration structural disruptions such as data reporting delays, testing backlogs, or systematic changes in case detection. This leads to more accurate forecasts in a real-world pandemic when such perturbations are widespread.

In the presented work, the estimation was done for a period of 120 days and the prediction for the following 60 days. In the case of the ARX-model, as stated also before, the best fit obtained was 88.74%(see Fig. 2) and regarding the ARMAX-model, the best fit increased in percentage with 7.87% as it can be seen in Fig. 3.

In the case of the ARX-model, when the noise in the data is not completely random, predicting capabilities are limited. If the noise contains temporal correlations, the ARX model may perform poorly since it does not account for the impact of previous mistakes.

The ARMAX-model is more strong in predicting because it incorporates the MA component, which estimates the impact of previous fluctuations on future values. This improves its ability to capture the dynamic behavior of systems including autocorrelated noise.

When predicting COVID-19 case dynamics, the ARMAX model outperformed the ARX model, particularly in cases including noise and disturbances. By including the moving average (MA) term, ARMAX was able to detect correlations in the noise (for example, reporting delays and testing anomalies), resulting in more accurate forecasts.

On the other hand, the ARX model was effective at displaying the relationships between inputs (e.g., mobility data, vaccination rates) and COVID-19 case patterns, but its lack of capacity to model noise properly led to less precise predictions, especially during highly fluctuating periods.

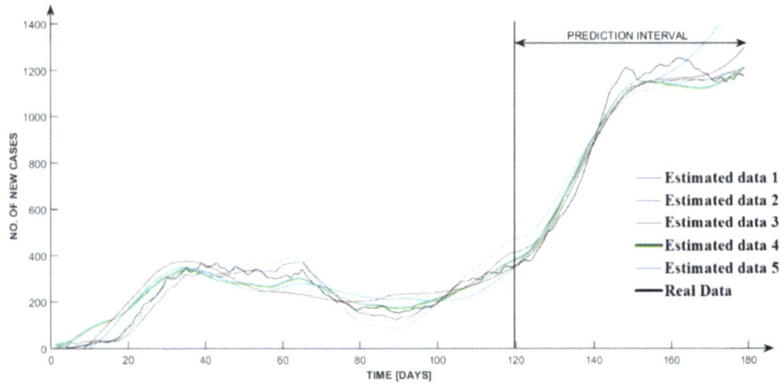

Fig. 2. ARX-model - five estimated transfer functions that model the system.

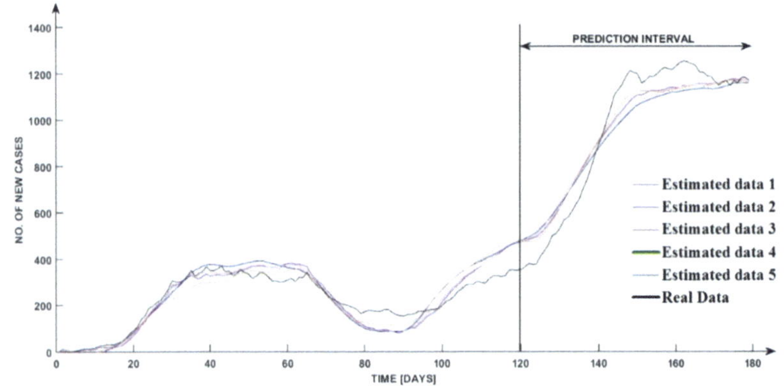

Fig. 3. ARMAX-model - five estimated transfer functions that model the system.

The ARX model provides a structured and interpretable method for predicting COVID-19 instances by combining historical case data with several exogenous variables such as migration, vaccines, and government policies. While the model is linear and may not capture all intricacies, it serves as a strong foundation for short-term forecasting and decision-making throughout the epidemic. Future improvements may include nonlinear extensions or hybrid techniques to increase its accuracy in capturing the dynamic nature of COVID-19 transmission.

The ARMAX model improves COVID-19 case prediction by including a moving average component to manage noise and random disturbances, as well as autoregressive and exogenous input factors. This model provides a thorough framework for understanding how previous case data and external factors affect future case numbers. While the ARMAX model is more robust and adaptive than simpler models, it requires precise parameter estimates and validation to produce accurate and trustworthy projections, especially in the dynamic and complicated environment of a worldwide pandemic.

3 Conclusions

Implementing an ARX-model or an ARMAX-model in a MISO system enables dynamic modeling of several inputs impacting a single output, making it ideal for complicated scenarios such as COVID-19 case prediction. The ARX/ARMAX model may effectively represent the links between inputs and output by finding the appropriate model order, delays, and exogenous variables, hence guiding actions and improving forecasts in public health systems.

Looking ahead, this study highlights growing trends in predictive modeling, illustrating the importance of include increasingly complicated exogenous factors including viral variations, seasonal impacts, and behavioral changes. The incorporation of machine learning approaches into ARX/ARMAX frameworks is proposed as a possible approach for improving model performance and flexibility in forecasting not just pandemics but also other time-sensitive public health problems. These findings highlight the necessity of improving prediction algorithms to forecast future waves of COVID-19 and other infectious disease outbreaks.

Conflict of Interest. All authors certify that they have no affiliations with or involvement in any organization or entity with any financial interest or non-financial interest in the subject matter or materials discussed in this manuscript.

References

1. Marín-Machuca, O., Chacón, R.D., Alvarez-Lovera, N., Pesantes-Grados, P., Pérez-Timaná, L., Marín-Sánchez, O.: Mathematical modeling of COVID-19 cases and deaths and the impact of vaccinations during three years of the pandemic in Peru. Vaccines (Basel). **11**(11), 1648 (2023). https://doi.org/10.3390/vaccines11111648. PMID: 38005980; PMCID: PMC10674587
2. Wu, Z., McGoogan, J.M.: Characteristics of and important lessons from the coronavirus disease 2019 (COVID-19) outbreak in China: summary of a report of 72314 cases from the Chinese Center for Disease Control and Prevention. JAMA. **323**, 1239–1242 (2020). https://doi.org/10.1001/jama.2020.2648
3. Carabelli, A.M., Peacock, T.P., Thorne, L.G., Harvey, W.T., Hughes, J., COVID-19 Genomics UK Consortium, Peacock, S.J., Barclay, W.S., de Silva, T.I., Towers, G.J., et al.: SARS-CoV-2 Variant Biology: Immune Escape, Transmission and Fitness. Nat. Rev. Microbiol. **21**, 162–177 (2023). https://doi.org/10.1038/s41579-022-00841-7
4. Hsiang, S., Allen, D., Annan-Phan, S., Bell, K., Bolliger, I., Chong, T., Druckenmiller, H., Huang, L.Y., Hultgren, A., Krasovich, E., et al.: The effect of large-scale anti-contagion policies on the COVID-19 pandemic. Nature. **584**, 262–267 (2020). https://doi.org/10.1038/s41586-020-2404-8
5. Tuli, S., Tuli, S., Tuli, R., Gill, S.S.: Predicting the growth and trend of COVID-19 pandemic using machine learning and cloud computing, 100222. Internet Things. (2020). https://doi.org/10.1016/j.iot.2020.100222. ISSN 2542-6605. https://www.sciencedirect.com/science/article/pii/S254266052030055X
6. Zhao, S., Lin, Q., Ran, J., Musa, S.S., Yang, G., Wang, W., Lou, Y., Gao, D., Yang, L., He, D., Wang, M.H.: Preliminary estimation of the basic reproduction number of novel coronavirus (2019-nCoV) in China, from 2019 to 2020: a data-driven analysis in the early phase of the outbreak. Int. J. Infect. Dis. **92**, 214–217., ISSN 1201-9712 (2020). https://doi.org/10.1016/j.ijid.2020.01.050

7. Levitt, M., Scaiewicz, A., Zonta, F.: Predicting the trajectory of any COVID19 epidemic from the best straight line. medRxiv [Preprint], 2020.06.26.20140814 (2020). https://doi.org/10.1101/2020.06.26.20140814. PMID: 32607515; PMCID: PMC7325180
8. Zhou, P., Yang, X., Wang, X., Hu, B., Zhang, L., Zhang, W., Si, H., Zhu, Y., Li, B., Huang, C., Chen, H., Chen, J., Luo, Y., Guo, H., Jiang, R., Liu, M., Chen, Y., Shen, X., Wang, X., Zheng, X., Zhao, K., Chen, Q., Deng, F., Liu, L., Yan, B., Zhan, F., Wang, Y., Xiao, G., Shi, Z.: A pneumonia outbreak associated with a new coronavirus of probable bat origin. Nature. **579**, 270–273 (2020)
9. Tabish, S.A.: COVID-19 pandemic: emerging perspectives and future trends. J. Public Health Res. **9**(1), 1786 (2020). https://doi.org/10.4081/jphr.2020.1786. PMID: 32550223; PMCID: PMC7282311
10. Zhou, C.-M., Qin, X.-R., Yan, L.-N., Jiang, Y., Yu, X.-J.: Global trends in COVID-19. Inf. Med. **1**(1), 31–39., ISSN 2772-431X (2022). https://doi.org/10.1016/j.imj.2021.08.001

Students Evaluating Teachers Before, During and After the COVID19 Pandemics. Human Interaction in Education, a Psychological and Behavioral Perspective

Mihaela Cîrlugea(✉)

Technical University of Cluj-Napoca, Cluj-Napoca, Romania
Mihaela.Cirlugea@bel.utcluj.ro

Abstract. The paper presents a comparative survey regarding the student behavior and perception of the educational process, before, during and after the pandemics. The study was done on subjects in the second year of study, where the author and the team teaches. Student comments and evaluations are discussed over three periods of time: before pandemics (moderate), during pandemics (critical) and after pandemics (positive). The educational process was evaluated in terms of percentages of the qualifications reflecting the students satisfaction degree in respect to the teaching methods for different curricula. The conclusion showed that before the pandemics while the scholar process was a moderate classical face-to-face one with medium results, during pandemics the whole process was stressful, impersonal and negatively perceived, while after pandemics a collaborative human interaction was developed. A crisis situation determined the creation of a positive learning environment, where human interaction proved to be very important. The skills we obtained while online teaching were useful but complementary to the collaborative on-site learning process.

Keywords: Pandemics · Education · Students · Evaluation · Human

1 Introduction

Schools and universities lockdowns did affect students and teachers psychically. Both sides needed to adapt, requiring a great change in mentality and acceptance. Technology was adapted to educational needs but finally used as a complementary teaching instrument [5]. On long term, the crisis subsequently improved the collaboration between students and educators, directing them towards supporting each other. Stressors appeared on different levels [6], which made both sides to discover their human side, growing a need in compassionate and caring communication. Both sides needed to reach out, to prove accommodation willingness. Before pandemics, the educational process was taken for granted. Psycho-social aspects intervened, forcing the educators to change their role, from an obsolete intimidating factor of power, judging the students [7], to guides,

S. Vlad and N. M. Roman (Eds.): MEDITECH 2024, IFMBE Proceedings, 130, pp. 201–212, 2025.
https://doi.org/10.1007/978-3-031-95671-3_21

maybe friends or even "surrogate parents" [6, 8, 9]. The process of dictating the curriculum, making them listen and learn, translated into a communicative and collaborative process which was in the benefit of everyone. Coaching or even trying to solve personal problems became new activities that made both parts realize the human aspect of the learning process that now had to be rediscovered in a contemporary perspective. Surveys were made proving that stress factors in educational units caused depressions, anxiety, self-injury, suicidal intention or mental health problems [2, 4, 10]. Students were seeking for help in friends or family but also in faculty professors. Some of them didn't look for help, because they didn't dare to open, were not sure where to go or believing that people wouldn't understand them [10].

The nowadays student generations are fragile and shaken by the Covid-19 situations and their consequences. Besides acting weird and aggressive, students are actually in the need of help and understanding. The stress students face daily has sources like social-media pressure, personal relationships or high academic expectations [3]. These can anytime deviate in overwhelming anxiety crises or considering suicide. Therefore, teachers, as educators, should focus on the knowledge transfer while simultaneously providing a secure, supporting and healthy environment.

2 Survey Background

In research [4] the influence of the online and onsite teaching was studied and how that affected the students. Now, we present how that situation affected the student perception and the educational process. Many students were traumatized by the crisis, but also because of the lack of human interaction. This paper refers to the need of the human interaction in the educational process.

The survey was made observing the data and the students in the Electronics and Communications, English Section, second year of study, first semester, where the author and the team are teaching. The evaluations and corresponding comments were analyzed over academic years 2018–2024. In [4], the study indicated that the students can be divided in years 1 and 2 as young and less experienced, with more insecurities and less basic technical knowledge; years 3 and 4 of study were more self confident and could better deal by themselves with learning procedures and project design.

We may consider 100 the average number of students, because during time, their number differed somewhere between 79–120. The constant over these years were the identical evaluation forms together with the evaluation criteria. This is why, when studying the data in chapter "4 Evaluation Percentual Values" we refer to percent that are the "common denominator".

For the study and data representation, the time was divided into three parts: first semester in academic years **2018–2019** and **2019–2020** is pre-pandemic **P1**, first semester in **2020–2021** during pandemics is **P2**, the results "happening" in year 2020 being relevant for that period, and **2021–2024** is **P3** post-pandemic. The survey was done observing the students' evaluation material of classes and teachers. In all periods **P1**, **P2** and **P3**, the curricula had the same content and course-seminary-lab-project structure, similar exam structures and same teachers.

3 Student Evaluation Comments

3.1 The Time Period P1

During **P1**, before the pandemics, students took the traditional onsite teaching for granted. Part of the students were interested in studying while others were not showing openness to learning. The comments were combined, positive and negative, but with a distant attitude towards the educational system. It is possible that the easily reached confidence in social media made them become arrogant in real life [1].

Students' negative evaluation comments on the left are accompanied by our educator opinion on the right, in Table 1. These tables contain a parallel between their perception and the teacher's intentions and facts. Table 1 shows a cat and mouse relationship provoked by the students who looked for excuses not to attend the classes and blaming the teacher. Also because they were mostly absent, they couldn't have the perspective of the whole and didn't understand the main concepts that build the curricula.

Table 1. Negative Comments in **P1**.

Student Comments	Teachers Opinion & Point of View
We didn't like the vibe of the teacher, the first impression was not a nice one	It is possible the teacher had not a good day, though the focus should be on the curriculum
The teacher came angry to the first class I attended, so I never came to the classes again and studied by myself and search on Google; Bipolar, subjective and changing behavior during the course	The teacher could pass through a divorce or the death of a parent or other personal problems; but a student to decide never to attend a class because of the teacher's mood is a pretext, pushing the blame on the teacher The teacher could be mad because the students were impertinent
The exam contained exercises that were not studied during the classes	This is not compulsory and not efficient, the curricula is dense and students have also to study by themselves New problems in the exam help to see if the students understand the teaching material and can further develop it
Weak seminaries and too short	The time has to be 2 hours with breaks; therefore in our previous paper we proposed the modular teaching
Why are not taught exactly the same problems to all student groups during seminaries?	This is inefficient: When presenting different problems to different groups during seminaries, they can share the solved problems amongst them for more examples

On the bright side, there were also positive evaluations as presented in Table 2. Students attending classes, who understood the teacher's intention and appreciated the effort, received a whole perspective and understood the curricula and the teaching approach.

Table 2. Positive Comments in **P1**.

<table>
<tr><th>Student's comments</th><th>Our Point of View</th></tr>
<tr><td>Teaching very well organized during courses</td><td rowspan="7">Thank you for attending the classes and understanding the teachers and the curriculum!
Students who participate to at least 70–80% of the classes, understand the main ideas of the course and also create a bonding with the educators and gaining trust in them and in the teaching process
Also a pleasant feeling for the educators to see that their intentions, knowledge and methods are appreciated</td></tr>
<tr><td>I appreciate the creativity and openness during the teaching process, the original intuitive explanations and also the good evaluating process</td></tr>
<tr><td>Insists on teaching us to think, not work like robots</td></tr>
<tr><td>I am impressed that you tried to help us understand the curricula and tried to explain as good as you could</td></tr>
<tr><td>The teacher is so nice and kind, I like that spirit and the teaching way</td></tr>
<tr><td>A very interesting and amusing liberty in talking and teaching during classes, so the distance between student and teacher disappears</td></tr>
<tr><td>It is great classes are not boring and it makes me attend the classes, learn and keep up my moral</td></tr>
</table>

3.2 The Time Period P2

Unfortunately, in P2 there were only negative remarks. We tried to explain the cause of them, in Table 3. The pandemics shock and distancing, made human beings emotional, unsure and therefore aggressive or emotionally unstable. The non-responsible ones put the blame on others. The fear and aggresiveness stands out from the negative student comments intended to hurt the teachers as persons. The same negative comments were received by different teachers and not referring to curricula or teaching methods. Also many technical curricula are impossible to be taught online [4] without human interaction. Yes, formally they can be taught but the need of attention span over a log time period is too much to be asked from a student.

In **P2** were no nice, pleasant nor optimistic comments. They were personally addressed, not to the curriculum. Similar mean comments were found in many teachers' evaluations, not only for one particular teacher, signs that students were scared, frustrated because not able to participate in the online process, unhappy with the grades and

Table 3. Negative Comments in **P2**.

Student's comments	Our Point of View
I would encourage the teacher to teach more professional, we have gaps in learning	Students didn't turn on the cameras during online classes; they didn't answer our questions, they didn't say if they didn't understand One of us, addressed them a very simple question like what is the equivalent resistance of two serial resistances and they made the teacher repeat 3 times the question and still had no answer. This proves they didn't pay attention and had no intention in doing so. They were hiding behind virtual icons wearing their names The blame was put on the educator, very few students made efforts to learn and pay attention
Lessons were hard to follow online; we couldn't actively participate	Most of the students couldn't technically interact with the white board, it was not like they could write in the class at the blackboard. It was a most unilateral teaching process
It was not an objective evaluation	The evaluation was performed by the computer, u-sing same problems and the same criteria as for written exams But sadly, students didn't understand the curricula because most of them didn't pay attention during online classes
Some colleagues, who the teacher knew, got better marks	1.The evaluation was computer calculated 2.Teachers didn't know the students in person, in most cases they didn't ever seen them in real life
The course slides were chaotic, couldn't use them for learning	Same course support was used in the next years and nobody complained about
2 hours were not enough for the exam	The exam always(in the last 30 years) lasted 2 hours, even when written on paper Online was a grid with answers, the time should have been more than plenty

(*continued*)

Table 3. (*continued*)

Student's comments	Our Point of View
We would have been appreciated if the documentation would have been done for our student series	The materials were made in that sense!
The teacher was talking nonsense during the course, about coffees and planes	Students couldn't pay attention while online, the lessons were about creating circuit designs for elevators or coffee machines States were defined like: The machine takes out the cup, pours the water and coffee, indicates if the user pushes the button for sugar and for cream During another class they were told to look on the window to see the plane flying according to the schedule they could follow on radar in their laboratory lesson

situation. They blamed teachers for the grading, though the exams were online and marks were calculated by the computer. So, from the student's psychological perspective, the lack of social interaction, frustration, the lack of feeling the empathy of the colleagues or of the teacher, left a bad mark on the identity development of the students. They couldn't cognitively develop and lost their self-concept.

While online, it was hard to teach without student feedback, not seeing their faces, sometimes it was like talking to the monitor. Students also felt lonely in front of the screen, not part of a team, some of them maybe not even knowing the other students in the virtual class and therefore not being motivated to talk or answer questions.

It is easy to accuse someone you don't know or have never seen in person, than a friend or an acquaintance. New and absurd situations made people lose their temper by emotional stimuli so they couldn't focus on the real problems. The conflict with school and educators was a false one, created by fears and false perceptions affecting them emotionally, in grades and scholar results. Thus, the pandemics created stressful experiences, making people insecure and aggressive.

The impact on teachers was hard, so that, in the next years, teachers tried to come closer to students, to show them openness and help them understand lessons and self-improve, to be available when students needed it.

3.3 The Time Period P3

During **P3**, the first semester 2021–2022 was mixed online-onsite, so we could adapt to the schedule and the curricula typology, but also to the student needs. Then, in 2022–2023 and 2023–2024 normality returned, classes were onsite again, allowing the presence at class and the student-teacher interaction. Amazing how by actively participating in class, students became interested, courses and educators being appreciated. The human factor is thus essential, changing dramatically subjective perceptions and not any virtual environment can replace the warmth of persons when collaborating. In Table 4, the comments from **P3** period are listed, on both columns. No negative ideas, maybe some recommendations sometimes.

Table 4. Positive Comments in **P3**.

Student's Comments	
The best teacher Teaches with love and helps the students	A sympathetic teacher, sometimes too many details in class and hard to follow, but warm and dear
An exceptional way of teaching! Always considering the student's opinion.	The teacher that explained a university curricula as if it was for kindergarten, so easy and nicely explained
Thank you that you always tried to find a solution to each problem, it was a pleasure to participate to the classes!	The courses were interactive, fun but we learned a lot, I appreciated that everything was explained using the blackboard, not only slides
The problems were explained concisely, clearly, detailed	Sometimes the teacher is too nice, I would 10/10 recommend
Students could actively participate to the classes, I am sorry you are no longer our teacher. Can't imagine a better person teaching this subject!	The teacher is very communicative, direct and sincere, trying to understand what students understand or not, so supplementary explanations can be added when needed
I am happy you were my teacher, I always attended classes with pleasure; I wish you good luck in the future!	I always attended the courses with pleasure, the quality of information is very good, superior, always in good mood
I always went away from classes in a good mood, a very understanding teacher, one can discuss anything with!	The teacher helped us a lot during the semester, with every explanation we needed, trying the best so that we understood

After the pandemic shock, students were eager to learn, to attend classes. They did appreciate if someone was willing to explain them and they didn't even consider it was the teacher's fault when something didn't go right. These were generations that were obliged to learn in the 12th grade for the bachelor's degree online, during pandemics, and/or attend the first university year also online, so, afterwards, they highly appreciated the possibility of really meeting their teachers in person. They appreciated each intention and offer that came from the educators. Maybe the students in **P3** didn't necessarily know how to ask, but they surely did know how to receive!

Besides the psychological perspective that made students in P3 to socially and professionally develop, the human interaction made them more secure, more eager to socialize and to learn. From a behavioral point of view, the educator that showed empathy, understanding, willingness to help was successfully imitated by the students. The positive behavior was taken over, from teacher to students, the availability to help, to communicate, to collaborate, all this proved to create an encouraging learning environment. Human interaction has influenced students and their relationships with educators. When teachers were insecure and nervous, students were influenced to observe and criticize, not to focus on the learning process. When the educators opened themselves to the students, their engagement increased and improved the student's learning behaviors.

4 Evaluation Percentual Values

As presented in introduction, the average number of students can be considered 100 in each year, but we considered percentual values so that we can compare them over the years.

The evaluation criteria contained aspects like: teaching mode, educator-student relationship, quality of curricula information, existence of bibliography, evaluation quality, information accuracy. These aspects could be appreciated by each student giving a qualifier from 1 to 4, where: **1 = Very Good, 2 = Good, 3 = Satisfactory, 4 = Unsatisfactory.** If all the students in a year that completed the evaluation are considered 100%, for the first criteria in Fig. 1, the Teaching mode, in year 2018, 62.5% gave the qualification 1, 31.25% gave the qualification 2, and the rest of 6.25% gave qualifiers 3 and 4. In the Tables 5, 6, and 7 the values are indicated for the qualifiers presented in the graphical examples in Fig. 1. As stated in [4], some curricula are appropriate for online teaching but in case of electronic or electrical circuits, or even mathematics, online teaching without human interaction and collaboration cannot function well.

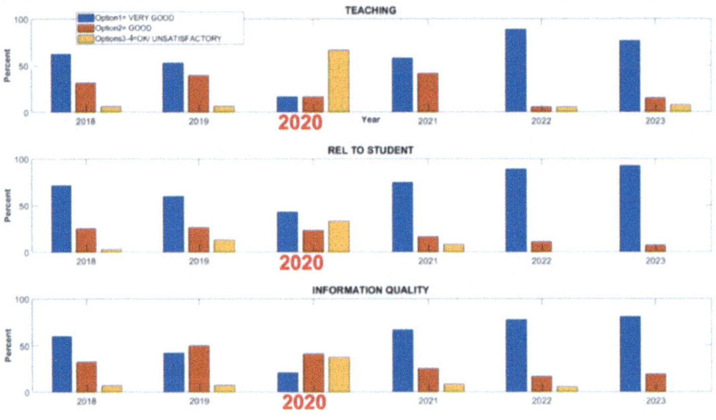

Fig. 1. Percentual student->teacher evaluations; a circuits design curriculum.

The charts in Figs. 1 and 2 contain data gathered from the evaluation tables, which were afterwards processed in Matlab procedures that transpose the tables into bar charts. Figure 1 presents the situation for a curricula dealing with circuits while Fig. 2 indicates the results for a programming curriculum. Both were taught by the author and the team. The first 2 years are **P1**, the one indicated with red is the pandemic **P2**, the last three columns are **P3**. The first blue column shows the percent for option 1 = very good, the second orange column is for option 2 = good, while the third column is the sum of percentages of options 3 + 4. In **P1** the first two columns are high and pretty much

Table 5. Evaluation Percent for Teaching Mode

Year	% for 1 = Very good	% for 2 = Good	% for 3 + 4 = Not good
2018	62.5	31.25	6.25
2019	53.33	40.00	6.67
2020	16.67	16.67	86.67
2021	58.33	41.67	–
2022	88.89	5.56	5.56
2023	76.92	15.38	7.69

Table 6. Evaluation Percent for Student-Teacher Relationship

Year	% for 1 = Very good	% for 2 = Good	% for 3 + 4 = Not good
2018	71.88	25.00	3.13
2019	60.00	26.67	13.33
2020	43.33	23.33	33.34
2021	75.00	16.67	8.33
2022	88.89	11.11	–
2023	92.31	7.69	–

Table 7. Evaluation Percent for Information Quality

Year	% for 1 = Very good	% for 2 = Good	% for 3 + 4 = Not good
2018	60.00	33.33	6.67
2019	42.86	50.00	7.14
2020	20.69	41.38	38.93
2021	66.67	25.00	8.33
2022	77.78	16.67	5.56
2023	80.77	19.23	–

equilibrated for both curricula. Then, the **P2** part is visibly low and options 3 and 4 prevail. In **P3**, especially the first column for option 1 = very good is visible. In **P3** 2021 when the mixed online-onsite mode was on, for the curricula with code and programming the blue column for "very good" was high. Students did appreciate the online programming classes [4] where all results were instantly obtained. The evaluation comments presented in chap. 3 of this article sustain the numbers, in **P1** were combined good and bad opinions, in **P2** the criticism and frustration was at maximum and in **P3**, students became more eager to learn, grateful for the chances they had to participate to real classes.

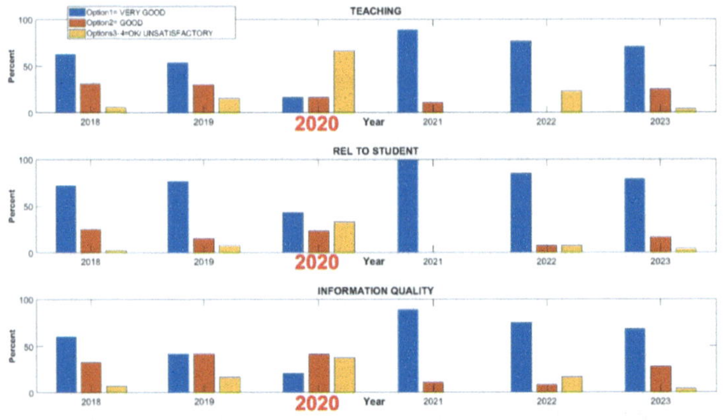

Fig. 2. Percentual student->teacher evaluations, a programming curriculum.

5 The Teaching Goal

The study presented here is a partial one, considering only aspects from 5 curricula studied by a student group over a semester. The students were chosen from the second year, because they are younger, more vulnerable and unexperienced, but the second year curricula is more technical and specialized than the generalities taught in the first year. The teachers are a team, the evaluation results are more complex but in this paper, because of the data confidentiality, only result excerpts regarding the author evaluation were made public. Comparing the three situations for the chosen time periods P1, P2 and P3, the differences are visible.

Yes, generations differ from year to year, we felt that while explaining, simplifying or detailing each lesson for adapting it to each generation. Though the analysis done here refers more to the influence of historical facts over people more than to generation differences.

After the shock of passing from traditional face-to-face to impersonal online teaching, the relation became virtual, impersonal, almost artificial. Students felt the need to hide, to oppose. Even for curricula appropriate for online, teaching was hard and guilt was entirely shifted to teachers. Therefore in P3 we tried to correct those aspects day by day, to work WITH the students, to collaborate, help them participate to the class, giving them the chance to express their opinion, choose subjects for their projects where they can use their creativity, understand the curriculum using their individual intuition.

It is important to help students to define their objectives, to develop real competencies for life, to admit if something is wrong so we can correct it. We try to educate them for the generative love for the learning product.

The teacher's goal is to determine learning needs, establish objectives, find strategies and push students towards open discussions. We tried to find creative ways of teaching and of student implication, to help them understand that failure is a step in the learning process. In case of low grades or if they don't understand, we need to help students understand, not humiliating them, help them to improve, being there in case of need.

6 Conclusions

The article presents an analysis of the student-teacher evaluations compared before, during and after the Covid19 pandemics. The subjects were students in the second year of a technical university where the author teaches. The analysis was made on groups of an average value of 100 students, because for each year their number differs. On one side the student comments were studied, on the other the qualifiers they gave for different teachers and curricula. Before the pandemics, the students were more self aware, had a more clear group identity, but were functioning more as a critical cell in relation to the teacher. Probably the traditional teacher behavior was more aggressive than assertive, not enough open to the young, more coercive and focused on imposing the knowledge. After the pandemics, after bad evaluation qualifiers from the students, teachers needed to reevaluate the educational process and to be more aware of the student's insecurities. By opening themselves and focusing more on collaboration, creativity, positive behavior and participative learning environments, teachers helped students to imitate the teachers, to be more receptive during classes and to dare to be themselves, to discover their needs, identity and self concept.

Learning and understanding should be the journey to the true reality, essential and full of mystery. The teaching process should include the human factor in the aspects of communication, collaboration, understanding and support, the satisfaction of curious students who want to learn and be surprised. We are trying on a regular basis to maintain a pleasant learning environment, so that students can focus on understanding things, on developing themselves and not to be frightened and stressed. We tried to get rid of old conceptions where educators were almighty and provide instead a product where students are the most important client we have to work with.

The author declares no conflict of interest.

References

1. American Council on Education: College Student Mental Health Proposal Talking Points. ACE (2024)
2. McDonough, P.: Helping Students at Risk of Self-Harm: Considerations for the New Academic Year. Higher Education Today (2019)
3. Bossman, A., Agyei, S.: Technology and instructor dimensions, e-learning satisfaction and academic performance of distance students in Ghana. Heliyon. **8**, e09200 (2022)
4. Cirlugea, M., Farago, P.: Post-Covid19 conclusion regarding the education in a technical University in Romania. In: How Does Stress Influence the Educational Process, International Conference on Advancement of Medicine and Health Care through Technology. Springer Nature, Switzerland (2022)
5. Yeigh, T., Lynch, D.: Is online teaching here to stay? Acad. Lett. Open Access, Article 24 (2021)
6. Dean, K.L.: Too much of a good thing: escalating developmental needs in the educator-student relationship. Acad. Lett. Open Access, Article 495 (2021)
7. Senge, P., McCabe, N., Lucas, T.: Schools that Learn, 2012. Crown Publications (2012)

8. Kruger, K., Jarrat, D.: The Pandemic Exposes Just how Much Support College Students Need. The Hechinger Report (2020)
9. Berdik, C.: What Aspects of Teaching Should Remain Human? The Hechinger Report, (2024)
10. Eisenberg, D., Ketchen Lipson, S., Heinze, J.: The Healthy Minds Study, 2022–2023 Data Report. American Council of Education

Miscellaneous Topics

13.56 MHz RFID Module – From Application to Process Modelling and Effects on Human Health

Claudia Constantinescu[1,2](✉), Claudia Pacurar[1], Adina Giurgiuman[1,2], Calin Munteanu[1], Sergiu Andreica[1,2], Marian Gliga[1,2], and Sabin Dusa[1]

[1] Department of Electrotechnics and Measurements, Technical University of Cluj-Napoca, Cluj-Napoca, Romania
Claudia.Constantinescu@ethm.utcluj.ro
[2] Academy of Romanian Scientists, Ilfov 3 Bucharest, Romania

Abstract. RFID (Radio Frequency Identification) is a technology very used in the last few years due to its reliability and ease of use. The applications are present in a large range of frequencies, but this paper is focused on a13.56 MHz module which is usually combined with an Arduino board. First, a simple application for student monitoring capable of storing information about each student and the time he/she checked in and out of class is presented. The advantages of such a system are highlighted and the components used to construct this system are described. The next step of this paper is to determine how the RFID module functions by numerically modelling it with and without the tags and cards near it, attempting also to show how the positioning of the tags influences the characteristics of the active tag's antenna. The last step of the study is centered on determining the influence of such a tag near human tissues and concludes by determining if it is safe or not to use an RFID for long periods of time.

Keywords: RFID tag · distance · electric and magnetic field · SAR · Arduino application

1 Introduction

RFID technology was first used in the '90 starting from a primitive espionage tool developed by the Soviet Union, reaching state-of-the-art technologies present in industry and our daily lives, RFID has become a standardized technology for a multitude of applications today. The growing interest and increasing demand for this technology have necessitated new approaches in the design and implementation of RFID systems [1]. Radio Frequency Identification (RFID) is a technology that has experienced significant growth in recent years and is used in a wide range of applications. From access control, inventory tracking, and animal monitoring to real-time object identification and location, RFID has revolutionized the way we communicate and interact with the world around us. RFID is a technology that allows the transmission of information through radio waves. It consists of an RFID tag or label, which can be attached to an object or person, and

S. Vlad and N. M. Roman (Eds.): MEDITECH 2024, IFMBE Proceedings, 130, pp. 215–223, 2025.
https://doi.org/10.1007/978-3-031-95671-3_22

an RFID reader, which can read the information stored on the tag. Generally, RFID tags consist of an electronic chip and an antenna that can be used to communicate with the RFID reader [2].

Generally, RFID tags operate in one of the three frequency bands: 125 kHz, 13,56 MHz or 860–960 MHz [3] The operating frequency influences the distance at which the tag can be read, with higher frequency tags being capable of communicating at greater distances than lower frequency ones. There are two main types of RFID tags: active tags and passive tags. Active tags contain an internal battery that allows them to actively communicate with the RFID reader, making them capable of communicating at greater distances than passive tags. Passive tags do not contain an internal battery and are activated by the radio waves emitted by the RFID reader. These are cheaper and have a longer lifespan than active tags. The tag is powered remotely by the reader using wireless energy transfer, where it reflects back a portion of the received electromagnetic wave to communicate with the outside world [4–6].

2 Description of the Application in Which the RFID Studied Tags Are Used

The researchers designed a monitoring circuit used for student attendance evidence. For this application the circuit was made of an ESP8266 NodeMCU board, a RFID RC 522 module, some tags for the module, an LCD display, all presented in Fig. 1. By using this system, the teachers have a lot of advantages like save time and effort, as they no longer have to manually record the attendance of each student. Additionally, the accuracy of the records is improved, as the information is transmitted directly from the circuit to Google Sheet, ensuring data integrity. Students can check their attendance in real time and receive immediate feedback if there are issues with registration or connectivity. The use of this circuit with Wi-Fi transmission functionality, integrated with Google Sheet, simplifies and streamlines the process of monitoring student attendance in labs, giving teachers more time for teaching activities and reducing the manual work associated with attendance recording.

Fig. 1. The circuit designed for the student monitoring application.

The ESP8266 board is one of the most popular ESP8266 development boards. It has 4 MB of flash memory, 11 GPIO pins, and a 10-bit resolution ADC pin. It is most commonly used in projects involving Wi-Fi and IoT. It is extremely versatile and perfect for beginners.

The board includes the following components: ESP-12E module, power section, peripherals, serial communication, on-board switches, and LED indicator. The RC522 RFID module, based on the MFRC522 integrated circuit from NXP, is one of the cheapest RFID options available online. It usually comes with an RFID card and a round tag with 1 KB of memory, allowing any message to be stored on the tag. The reader can communicate with a microcontroller through a 4-pin SPI interface with a maximum data rate of 10 Mbps. Its operating frequency is 13.56 MHz, the feeding voltage is 2.5–3.3 V and the reading distance is 5 cm. We used a number of 5 tags to ensure the system is appropriate for larger number of students. These RFID tags operate in the RF range of 13.56 MHz and have a non-reprogrammable unique chip. They are passive devices, meaning they do not contain a power source (battery). They consist of a microchip and an NFC coil (antenna). The microchip is for storing data, and the NFC coil (antenna) is for transmitting data to the RFID reader module. The LCD display is a simple one being able to display 16X2 characters. The code behind the circuit is composed of three different parts: the code used to write information to the tags, the code for reading the information from the tags in Arduino IDE and transmitting the information to Google Apps Script and the code from Google Apps Scripts which ensures that each information from the tag is written in the indicated column from the Google Sheets table (Figs. 2 and 3).

ID	Nume	Prenume	Numar telefon	Grupa
1	Fedusca	Marieta	0754271321	1341
2	Ghirean	Dorian	0771268735	1141
3	Marian	Florin	0741648523	1141
4	Popescu	Raisa	0756321487	1341
5	Marie	Mihai	0741987546	1111

Fig. 2. The values inserted on the 5 tags and cards.

	A	B	C	D	E	F	G	H	I
1	ID	Ora intrare	Ora iesire	Sala	Data	Nume	Prenume	Numar telefon	Grupa
2	1	12:06:24 PM		Sala 16A	06/30/2023	Fedusca	Marieta	0754271321	1341
3	1	11:57:32 AM	13:43:41 PM	Sala 16A	06/12/2022	Fedusca	Marieta	0754271321	1341
4	5	11:57:39 AM	13:43:45 PM	Sala 16A	06/12/2022	Marie	Mihai	0741987546	1111
5	2	11:58:02 AM	13:43:49 PM	Sala 16A	06/12/2022	Ghirean	Dorian	0771268735	1141
6	3	11:58:32 AM	13:43:54 PM	Sala 16A	06/12/2022	Marian	Florin	0741648523	1141
7	4	12:03:56 PM	13:44:00 PM	Sala 16A	06/12/2022	Popescu	Raisa	0756321487	1341
8	1	13:05:32 PM	14:50:11 PM	Sala 16A	07/12/2022	Fedusca	Marieta	0754271321	1341
9	5	13:05:35 PM	14:50:13 PM	Sala 16A	07/12/2022	Marie	Mihai	0741987546	1111
10	4	13:05:39 PM	14:50:20 PM	Sala 16A	07/12/2022	Popescu	Raisa	0756321487	1341
11	2	13:05:42 PM	14:50:23 PM	Sala 16A	07/12/2022	Ghirean	Dorian	0771268735	1141
12	3	13:05:44PM	14:50:41 PM	Sala 16A	07/12/2022	Marian	Florin	0741648523	1141
13	1	11:59:11 AM	13:43:41 PM	Sala 16A	13/12/2022	Fedusca	Marieta	0754271321	1341
14	5	11:57:32 AM	13:43:41 PM	Sala 16A	13/12/2022	Marie	Mihai	0741987546	1111
15	2	11:57:32 AM	13:43:41 PM	Sala 16A	13/12/2022	Ghirean	Dorian	0771268735	1141
16	4	11:59:11 AM	13:43:41 PM	Sala 16A	13/12/2022	Popescu	Raisa	0756321487	1341
17	3	11:57:32 AM	13:43:41 PM	Sala 16A	13/12/2022	Marian	Florin	0741648523	1141

Fig. 3. The information stored in Google Sheets when a tag is detected.

At first, the code was written in Arduino IDE software to be able to store some information on each tag. The information about the students who were considered were: name, surname, phone number and group.

3 Description of the Application in Which the RFID Studied Tags Are Used

The RFID module is constructed out of an antenna and some other electronic components, but this study is aimed at understanding the antenna functionality.

3.1 The Geometrical Parameters and Construction of the RFID Tag

The module's antenna was modelled first with the help of the numerical model from Ansys, Q3D extractor in order to determine the inductance, internal capacitance and resistance as it can be seen in Fig. 4.

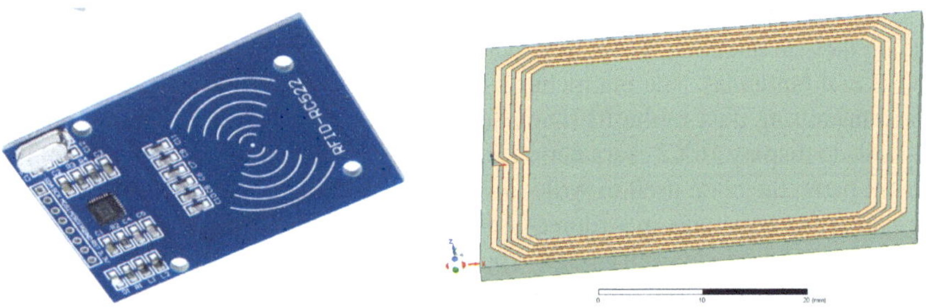

a) b)

Fig. 4. The structure to be analyzed: a) physical structure, b) structured modelled with Q3D extractor

The dimensions of the antenna are the length of the coil L = 41 mm, the width of the coil l = 38 mm, with a copper thickness of 0.03 mm. The dielectric is FR4 epoxy and has a thickness of 1.51 mm. After the analysis it was determined that the inductance of the coil is 1.34182 uH, while the resistance is 1.420 pF, thus negligible [7].

The initial structure was then inserted in the Ansys HFSS program and at its ends a port was connected. It was observed that it doesn't resonate at 13.56 MHz, thus there is a need to be matched. As it can be observed also on the initial circuit from the Arduino project, the way to do it is to insert capacitors, resistors and inductors in the circuit. Thus, knowing the inductance of the coil and the needed resonance frequency of 13.56 MHz, formula (1) is applied, and the capacitance needed is calculated as 104 pF. The results, although the resonance frequency is not exactly at 13.56 MHz are in accordance and with a little tuning the frequency can be reached. For these results a resistance of 50 Ω and a capacitor of 115 pF were inserted in parallel with the antenna. The connections of the source and the RLC port to the antenna from the program can be observed in Fig. 4, while in Fig. 5 the S parameters for the structure can be seen.

The electric and magnetic fields on the spiral inductor are presented in Fig. 6. The gain of the antenna was determined, and it can be stated that it has a very small value, fact that once more shows that this kind of structures resonating at lower frequency values

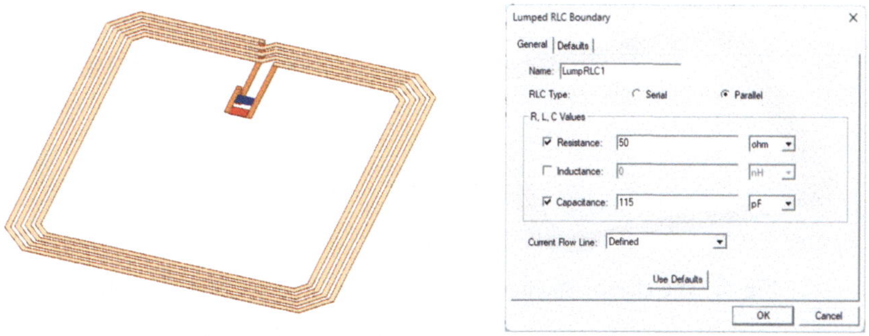

a) b)

Fig. 5. Matched structure from HFSS: a) matched structure, b) value of the resistance and capacitance for the matching.

are inductors, and they are not able to function when a tag or a card are further from them.

$$C = \frac{1}{(2\pi f)^1 \bullet L} \quad (1)$$

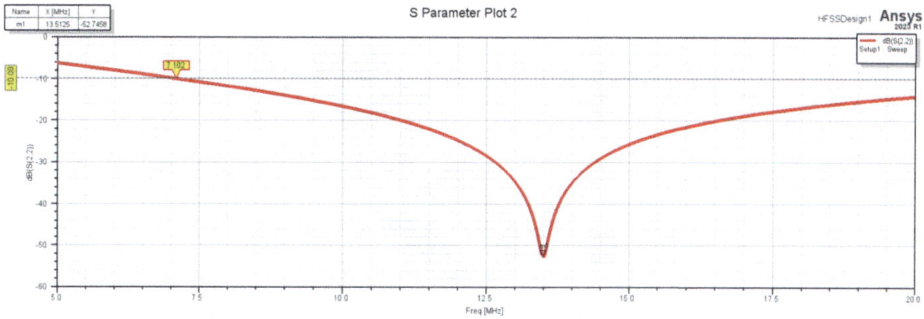

Fig. 6. The S parameters obtained for the RFID antenna.

3.2 Influence of the Distance Between the Tags

The electric and magnetic fields around the tag are represented in Fig. 7.

The limits for the electric and magnetic field according to ICNIRP 2020 [8] for a period higher than 30 min are calculated with the formula (2) and (3) for general public where E_{inc} is the incident E field strength and H_{inc} is the incident H field strength. According to this calculation the values for this device are E_{inc} = 6.2027 V/m and H_{inc} = 0.16224 A/m, which are in accordance, while for an exposure between 0 and 6 minutes the limit values are higher. The SAR values are much lower than the limits imposed by the standards as it can be seen in Fig. 8 as a representation in the human

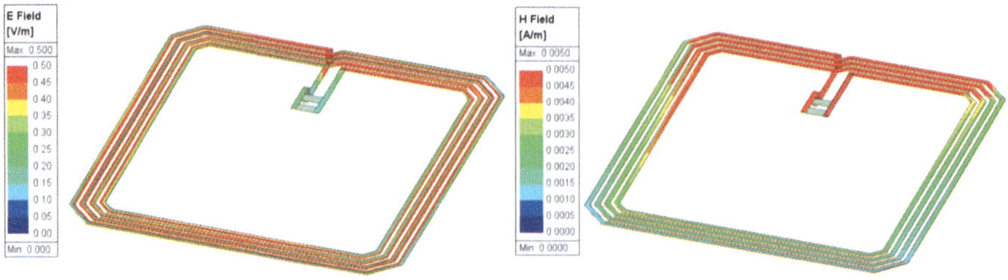

Fig. 7. The electric and magnetic field values for the RFID antenna.

hand.

$$E_{inc} = 300/f_M^{0.7} \tag{2}$$

$$H_{inc} = 2.2/f_M \tag{3}$$

It is known that the radio waves are the ones activating the passive tag, so a circular tag like the ones given with the module was modelled and the distance between the tags was varied in 3 different positions: distance of 2 mm, 20 mm and 40 mm as it can be seen in Fig. 9 [9–12].

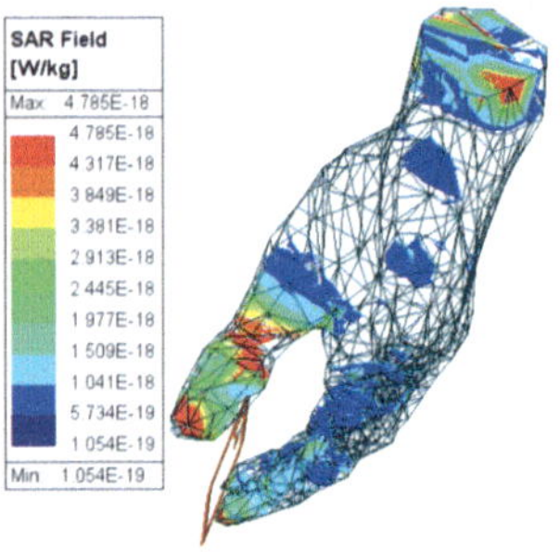

Fig. 8. The SAR values for the RFID antenna near a human hand.

The electric and magnetic field for the two RFID tags at different distances from each other can be observed in Fig. 10. It can be seen how a larger distance between the tags makes the electric and magnetic field not to be present, thus the signals are not sent from one to another. The higher values for a bigger area are present when the tags are closer, while at 40 mm the electric and magnetic field do not reach the second coil considered to be the second RFID tag.

In practice, the passive tag is placed next to an active tag, thus a numerical modeling of this situation is presented in Fig. 11, where the values of the electric, magnetic and SAR are presented with a distance of 20 mm between the tags. It can be stated that all the values are much lower than the standard values, thus using RFID tags at a frequency of 13.56 MHz is safe for the user.

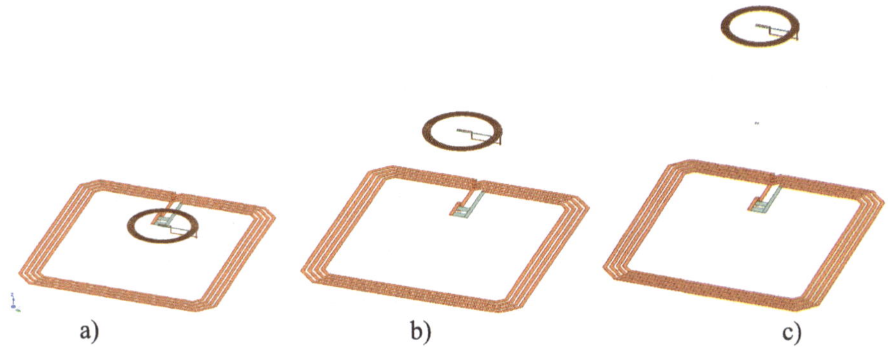

Fig. 9. Positioning of the two tags considering the distance between them a) 2 mm, b) 20 mm and c) 40 mm.

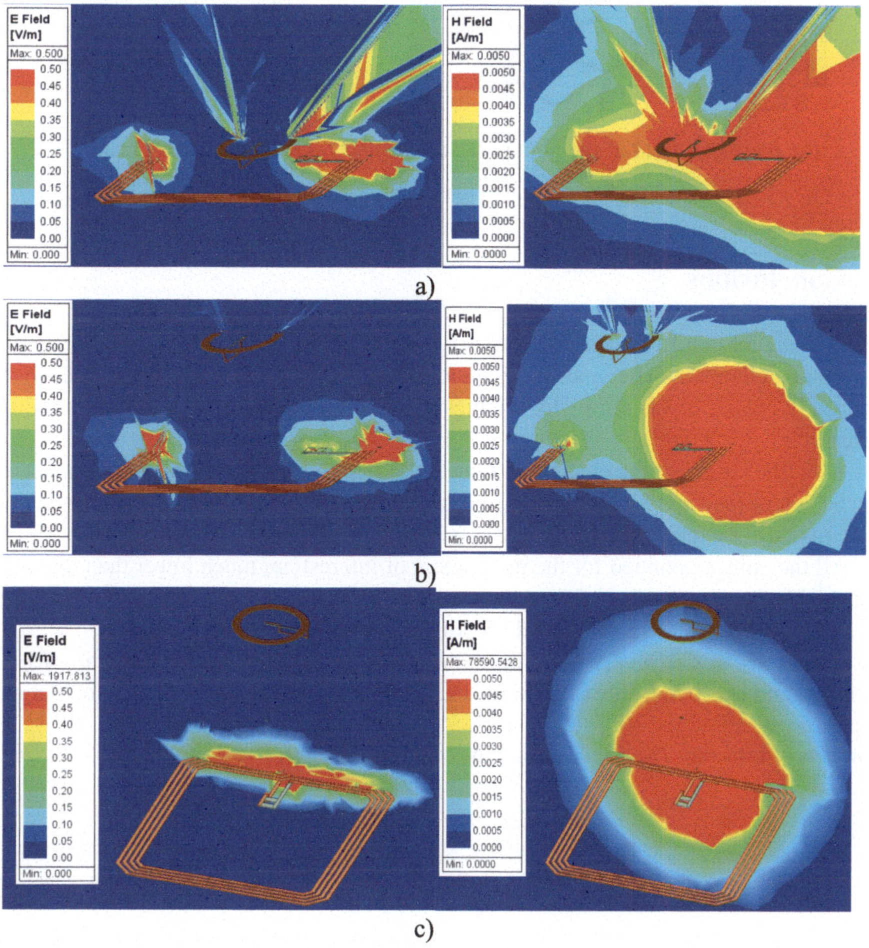

Fig. 10. Electric and magnetic field between two RFID tags at different distances from each other: a) 2 mm, b) 20 mm and c) 40 mm.

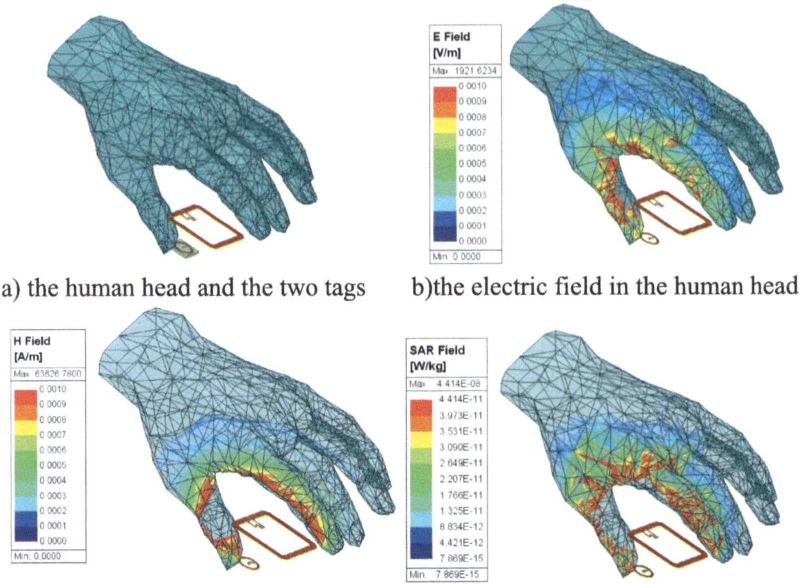

a) the human head and the two tags b)the electric field in the human head

c)the magnetic field in the human head d)the SAR in the human head

Fig. 11. The parameters of interest in the human head when two RFID tags interact.

4 Conclusions

When designing an RFID system, it is very important to understand its operating conditions and elements. After analyzing the RFID tags used in the application with which this study began, it was seen why the tags must be so close for the information to be transmitted.

Also, the values of the electric and magnetic field generated by the spiral inductors can be determined and for all the different types of coils the maximum distance at which the coils must be one from the other can be easily determined.

All the values obtained for the parameters of interest are much lower than the ones imposed by the standards in force, thus we can conclude that the use of RFID tags at lower frequencies like 13.56 MHz does not affect human health.

This study is also a very good multidisciplinary teaching method regarding RFID concepts and use, being able to combine the practical part with the numerical modeling.

Conflicts of Interest. The authors declare no conflict of interest.

References

1. Al-Qahtani, A., Khan, F.: RFID technology in healthcare: a review. J. Infect. Public Health. **11**(6), 797–804 (2023)
2. Finkenzeller, K.: RFID Handbook: Fundamentals and Applications in Contactless Smart Cards, Radio Frequency Identification and Near-Field Communication, pp. 208–316. John Wiley & Sons
3. Gao, Y., Wang, S., Li, X., Tao, R.: A survey on RFID technology and its applications. Mob. Inf. Syst., 1–14 (2016)
4. Ezzeddine, H., Merakeb, Y., Huillery, J., Breard, A., Duroc, Y.: Vollaire C.: simulation framework for studying UHF RFID systems in pulse wave mode. In: IEEE International Conference on RFID Technology and Applicationns (RFID-TA), Pisa, Italy, pp. 120–124 (2019)
5. Mayordomo, I., Heyd, N.V.D., Dräger, T., Bernhard, J.: Development of a wireless power transmission system for embedded passive sensors using LF RFID technology. In: 2012 Fourth International EURASIP Workshop on RFID Technology, Turin, Italy, pp. 65–71 (2012)
6. Andreica, S., Gliga, M., Racasan, A., Munteanu, C., Pacurar, C., Constantinescu, C.: Study of conducted electromagnetic emissions of a wireless power system. In: International Conference on Electromechanical and Power Systems (Sielmen), pp. 191–195 (2017)
7. Claudia, P., Topa, V., Giurgiuman, A., Munteanu, C., Constantinescu, C., Gliga, M., Andreica, S.: The construction of a wireless power supply system using planar spiral inductors. In: 8th International Conference on Modern Power Systems (MPS), Cluj-Napoca, Cluj, Romania, pp. 1–6 (2019)
8. ICNIRP: Guidelines for limiting exposure to electromagnetic fields (100 KHz to 300 GHz), published. Health Phys. **118**(5), 483–524 (2020)
9. Mattei, E., Lucano, E., Censi, F., Triventi, M., Calcagnini, G.: Provocative testing for the assessment of the electromagnetic interference of RFID and NFC readers on implantable pacemaker. IEEE Trans. Electromagn. Compat. **58**(1), 314–322 (2016)
10. Jomaa, J.K., Ndagijimana F., Ayad H., Fadlallah M. and Jomaah, J.: Near-field characterization for 13.56 MHz RFID antenna, 2017 International Symposium on Electromagnetic Compatibility - EMC EUROPE, Angers, France, 1-4 (2017)
11. Benassi, F., Masotti, D., Costanzo, A.: Engineered and miniaturized 13.56 MHz omnidirectional WPT system for medical applications. In: 2019 IEEE International Conference on RFID Technology and Applications (RFID-TA), Pisa, Italy, pp. 306–309 (2019)
12. Giurgiuman, A., Gliga, M., Bojita, A., Andreica, S., Munteanu, C., Topa, V., Constantinescu, C., Pacurar, C.: Software program for the evaluation of human exposure to electric and magnetic fields. Technologies. **11**, 159 (2023). https://doi.org/10.3390/technologies11060159

YourDiagnoseFriend - Integration of Machine Learning in Medicine

Laura Marina Nemeş, Emilia Şipoş(✉), and Laura-Nicoleta Ivanciu

Bases of Electronics Department, Technical University of Cluj-Napoca, 26-28 G. Bariţiu Street, 400027 Cluj-Napoca, Romania
emilia.sipos@bel.utcluj.ro

Abstract. Artificial Intelligence has significantly transformed numerous aspects of daily lives, particularly in the healthcare sector where it performs tasks that traditionally requires human intelligence. *YourDiagnoseFriend* is a groundbreaking web-based tool designed to classify and provide potential medical diagnosis based on user-reported symptoms. Implemented within the Django framework to ensure a robust and scalable platform, *YourDiagnoseFriend* is crafted to be user-friendly, accessible, offering accurate and reliable diagnostics. Access is restricted to authenticated users to ensure data privacy and security. Users can select from various symptom categories and input their specific symptoms, which the system then analyzes to predict possible diagnoses and recommend specialized doctors. The application facilitates early disease detection and guides users toward timely and appropriate healthcare steps. This thoughtful design and functionality aim to provide users with valuable health insights, enhancing their overall health management experience and promoting proactive healthcare practices.

Keywords: Disease Prediction · Decision Tree Algorithm · Django Framework

1 Introduction

While Artificial Intelligence (AI) was hardly used in medicine until the 2000s, it has since become popular, including disease diagnosis facilities, disease management information, selecting the most effective course of treatment for a specific set of symptoms, and spotting anomalies that are difficult for the human eye to detect. Nowadays, AI and Machine Learning (ML) are able to assist doctors to diagnose, treat, monitor, evaluate and follow-up of patients. Compared to traditional methods, modern AI and ML systems have the ability to identify complex patterns and analyze large amounts of medical data to provide more accurate diagnoses [1].

Decision tree (DT) is a supervised ML algorithm, with a variety of forms, each with unique properties. The algorithm used in the proposed application is CART – Classification and Regression Tree, which can be used to solve both classification and regression problems [2].

Several AI-based health applications with notable accuracy utilize machine learning to predict diseases, but also facing challenges such as high computational requirements and dataset limitations.

S. Vlad and N. M. Roman (Eds.): MEDITECH 2024, IFMBE Proceedings, 130, pp. 224–234, 2025.
https://doi.org/10.1007/978-3-031-95671-3_23

An AI-based medical chatbot for infectious disease prediction is proposed [3]. The chatbot is implemented using Deep Feedforward Multilayer Perceptron (MLP) algorithm, achieving the best accuracy value of 94.32%. High computational requirements and only one group of diseases (infectious) are two major drawbacks of the paper.

DT algorithms are used in [4], where an Alzheimer's disease prediction application using speech analysis is proposed. Neural networks (NN), Random Forest and Extreme Gradient Boosting (XGB) classifier are other algorithm used to predict the disease, with an accuracy value up to 92.05%. The application is limited by dataset size, high computational resources being also needed.

Other DT prediction applications are presented in [5, 6]. The heart disease prediction is made using, in addition to DT, other ML techniques as Naive Bayes, K-Nearest Neighbor (K-NN) and Random Forest (RF) in [5] and Support Vector Machine (SVM), Logistic Regression (LR), and RF in [6]. A combination of DT, KNN, SVM and XGB is used in [7] to diagnose diabetes with an accuracy range of 76% to 80%. Since large datasets are required to achieve good results, computational efficiency is the main drawback of the papers.

DT, LR, Random Forest, SVM, and Adaptive Boosting (AdaBoost) algorithms are applied in [8] to predict several diseases. The best results for accuracy are: 98.57% for breast cancer, 85.71% for diabetes, and 87.1% for heart diseases. High computational are required and the model accuracy varies across used datasets.

A symptom-based health checker that uses machine learning algorithms to rank potential diseases is proposed in [9]. The dataset includes 75 symptoms and risk factors tailored for 22 cough-related diagnoses. The accuracy score ranges from 78,69% to 100%, underlining the need for further refinement of the dataset.

The systematic review in [10] shows that for online symptom checkers, the highest diagnostic accuracy (out of nine studies) is around 58%, which paves the way for approaches with higher accuracy.

In the current Digital Decade, one important objective for 2030 in digitalization of public services area is to implement e-Health, meaning 100% of citizens must have access to medical records online [11]. Moreover, the importance of an early diagnosis is well known, significantly improving patient outcomes, streamlining healthcare costs, and enabling personalized medicine [12].

The paper aims to contribute to these objectives, proposing an accessible and trustable application for early diseases detection, that can be used both by those with basic digital knowledge and experts. A comprehensive dataset containing various symptoms and corresponding diseases, supervised by the specialist physicians, is used to develop the online application.

To enhance the user experience, the application integrates a robust database managed through Structured Query Language (SQL) Server Management Studio (SSMS), which categorizes symptoms and associates them with specific diseases and specialists. The trained model is seamlessly integrated into the Django platform, providing users with possible diagnoses and personalized recommendations to specialists, with the help of an intuitive interface.

The paper is organized as follows: Section 2 presents the ML model development; Section 3 is dedicated to the user interface and authentication process; Section 4 provides the results and Sect. 5 concludes the paper.

2 Machine Learning Model Development

The development of a classification model using DT algorithm involves several steps. Together with the open-source Kaggle dataset [13], suitable training, validation and testing percents of data impact the classification outputs.

For all diagnoses, 131 different symptoms were used, organized in columns, and a final column representing the medical conditions. All values in the symptom columns are binary, with 0/1 indicating the absence/presence of a symptom in the disease. The initial dataset was preprocessed, to eliminate incomplete or double records, and to prevent overfitting. Finally, the dictionary of considered symptoms organized in categories and the relationships between symptoms and medical conditions are supervised by a specialist. All 131 symptoms are grouped in 11 categories, and 42 representative possible diagnostics were considered.

Each category includes unique identity document (ID), unique name, and the path to the representative image, in order to implement an intuitive category interface – Fig. 1.

ID_Category	Category_Name	ImagePath
1	Dermatological Symptoms	images/dermatologic.jpeg
2	General Symptoms	images/general.jpeg
3	Gastrointestinal Symptoms	images/gastrointestinal.jpeg
4	Respiratory Symptoms	images/respiratorii.jpeg
5	Excretory Symptoms	images/excretorii.jpeg
6	Neurological and Behavioral Symptoms	images/neuro.png
7	Cardiovascular Symptoms	images/cardiovascular.jpeg
8	Endocrine Symptoms	images/Endocrine.jpg
9	Ocular Symptoms	images/Ocular.jpg
10	Hematological Symptoms	images/Hematologic.jpg
11	Musculoskeletal Symptoms	images/musc.jpg

Fig. 1. Disease categories.

To further streamline data management, an SSMS database was developed, comprising three key tables: Symptom, Disease, and Categories. All medical conditions are recorded in the Disease table having associated specialist physicians, thus facilitating accurate and relevant diagnostic recommendations – Fig. 2.

The training subset is 64% of the database, 16% for validation and 20% of data is reserved for testing.

Using the training subset, CART is employed to learn patterns and relationships between symptoms and medical conditions. Once the model has been thoroughly trained and tested, the trained model is saved, integrated into the web application and used for real-time predictions – Fig. 3.

ID_Disease	Disease	Doctor
1	Fungal infection	Dermatologist or Infectious Disease Specialist
2	Allergy	Allergist/Immunologist
3	Chronic cholestasis	Hepatologist or Gastroenterologist
4	Drug Reaction	Allergist/Immunologist or Dermatologist
5	Peptic ulcer disease	Gastroenterologist
6	AIDS	Infectious Disease Specialist
7	Diabetes	Endocrinologist
8	Gastroenteritis	Gastroenterologist or General Practitioner

Fig. 2. Example of physician-disease association from Disease table

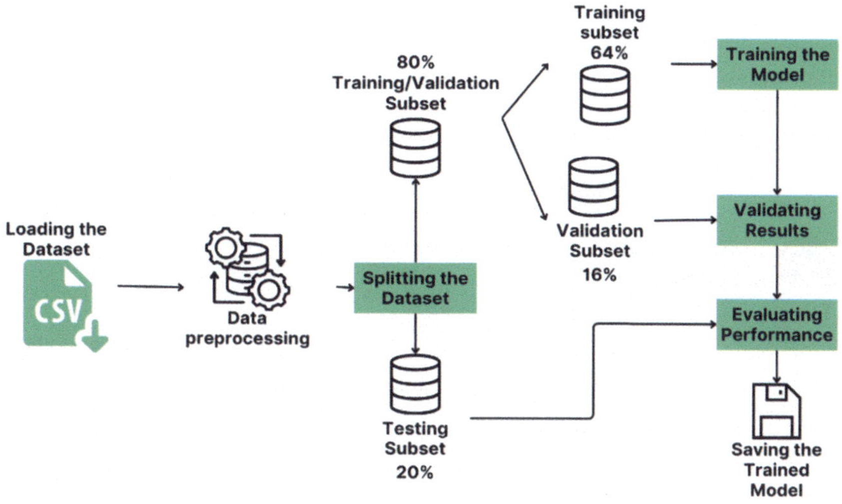

Fig. 3. The process of training and evaluation of the classification model.

3 User Interface and Authentication

The back-end development of *YourDiagnoseFriend* is critical for its functionality and security. Django's Model-View-Template (MVT) architectural pattern facilitates the separation of business logic from the user interface, ensuring a clean and maintainable codebase.

Creating a user-friendly and intuitive interface was another main objective of the application. The design principles, color scheme, and accessibility features are all tailored to provide users with a seamless experience that encourages frequent use and trust in the application. By prioritizing user-centric design, the application is structured to make symptom input and navigation straightforward. Key design principles include:

- simplicity - helps users focus on the task at hand without getting overwhelmed;
- consistency –users can easily predict where to find information and features, making the application easier to navigate.

The layout includes clear headings, logical groupings of related elements, and a fixed navigation bar to enhance usability.

User registration and login are secured using Django's built-in authentication system, ensuring that only authenticated users can access the application, thereby protecting sensitive health data. To guarantee data security, multiple measures were implemented, including Cross-Site Request Forgery (CSRF) protection, SQL injection prevention, and secure session management. By combining these technologies and approaches, the back-end development of *YourDiagnoseFriend* delivers a reliable, scalable, and secure platform, enabling users to receive medical condition predictions based on their reported symptoms.

The users first register and authenticate the application. Next, by following an intuitive and straightforward navigation flow, they get to input their symptoms and initiate a diagnosis process. The system analyzes the input symptoms and processes the data, and then provides possible medical conditions and recommends appropriate specialists based on Disease table – Fig. 4.

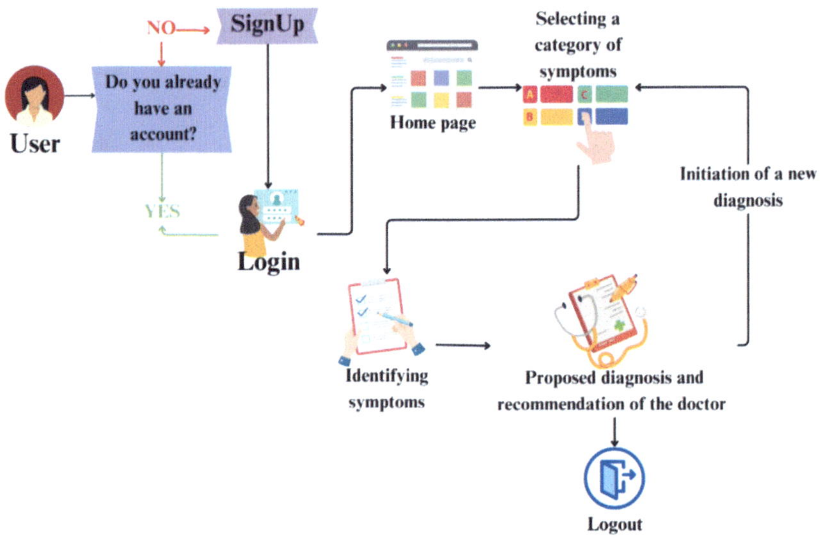

Fig. 4. Block diagram of the *YourDiagnoseFriend* application.

To manage and organize user data efficiently, two models were created in the Django admin interface, accessible only by the application administrator through a secure username and password. The *Users* model stores essential user information such names, phone numbers, email addresses, and encrypted passwords. Additionally, the *Predictions* model was implemented to track application usage. This model records each user's interaction, including the diagnosis and the date and time when a diagnosis was received.

4 Results and Discussions

The results follow users' criteria, as accessibility, easy-to-use especially for the people with basic digital knowledge, accuracy, trustability, data privacy and security. To interact with the application, users follow the flowchart in Fig. 5.

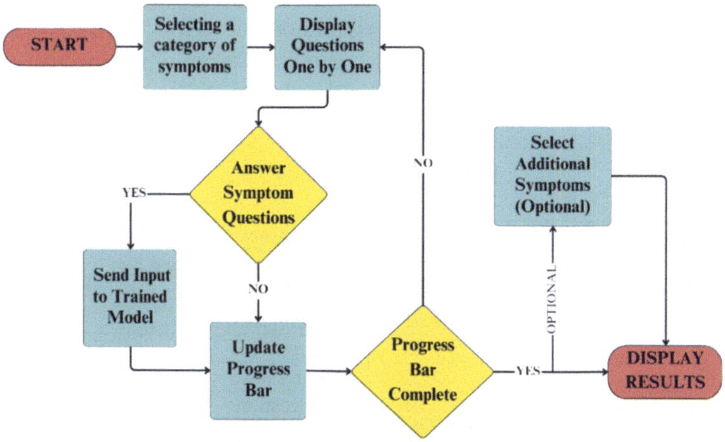

Fig. 5. The users' flowchart.

Simplicity in use is provided through the interface. A page with instructions is provided, where step-by-step guidance is delivered – Fig. 6a. Suggestive icons, some of them AI-generated, are used for each category of symptoms – Fig. 6b.

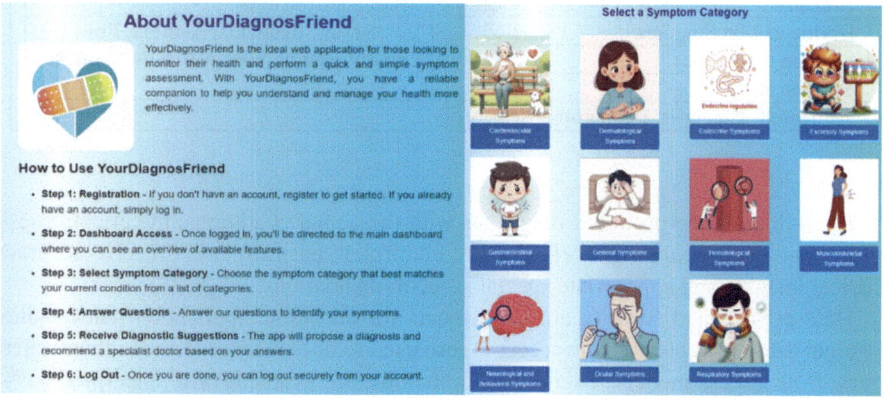

a) b)

Fig. 6. The user interface: a) Instructions page; b) Symptom categories.

Possible symptoms are presented as questionaries, with *Yes/No* buttons to confirm/deny their presence. The results are displayed after the symptoms' submission, alongside a specialist doctor's recommendation.

The application was tested with different scenarios, out of which two are presented here. The first test scenario is centered on dermatological symptoms, and for the second scenario, the user selected the category of respiratory symptoms and three additional symptoms from different categories: headache, fever, and tearing. Table 1 presents the user's responses to the presence of symptoms, for both scenarios. The last row shows the proposed diagnosis and the recommended specialist. The resulting medical condition confirms the accuracy of the diagnosis within the chosen category.

Table 1. Two case scenarios

Dermatological Symptoms		**Respiratory Symptoms**	
Skin rash	☑	Continuous sneezing	☑
Nodal skin eruptions	☑	Cough	☑
Skin peeling	☑	Breathlessness	☒
Dyschromic patches	☑	Phlegm	☑
Silver like dusting	☒	Runny nose	☑
Small dents in nails	☒	Congestion	☒
Blister	☒	Chest pain	☒
Inflammatory nails	☒	Throat irritation	☑
Yellow crust ooze	☒	Blood in sputum	☒
Red sore around nose	☑	Rusty sputum	☒
Red spots over body	☑	Mucoid sputum	☑
Yellowish skin	☒	Sinus Pressure	☑
Scarring	☒	Selecting other categories of symptoms	
Blackheads	☒	Headache	☑
Pus filled pimples	☑	High fever	☑
Bruising	☒	Watering from eyes	☑
Drying and tingling lips	☑	Result: **Bronchial Asthma, Pulmonologist or Allergist/Immunologist**	
Swollen extremities	☒		
Swollen legs	☒		
Result: **Psoriasis, Dermatologist**			

By validating the proposed diagnoses with the actual conditions, the effectiveness and reliability of *YourDiagnoseFriend* are confirmed, ensuring users receive accurate medical insights and appropriate specialist recommendations.

Data privacy and security are key aspects for an online application that handles personal data. To achieve it, the access is restricted to only registered and authenticated users. Both access models, Users and Predictions, are available with encrypted passwords, provided by the application at the registration step. Moreover, the Prediction access model offers information about application usage of the individual users. The administrator of the platform can check access to the application – Fig. 7 and can intervene if he notices suspicious activities.

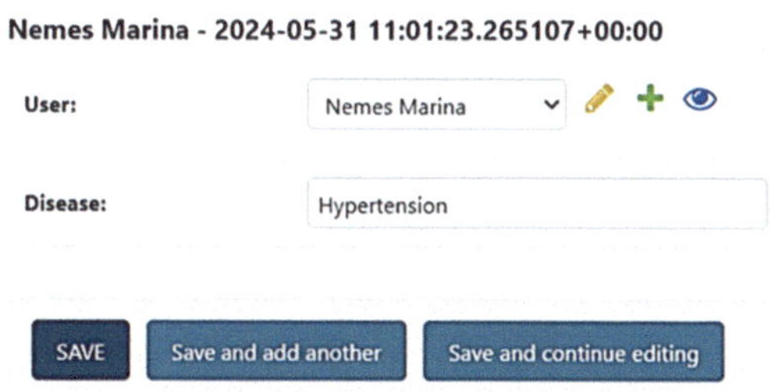

Fig. 7. Exemplification of *Prediction* access model.

To measure performances of the machine learning model, the confusion matrix was built, comparing predicted diagnosis against real diagnosis – Fig. 8.

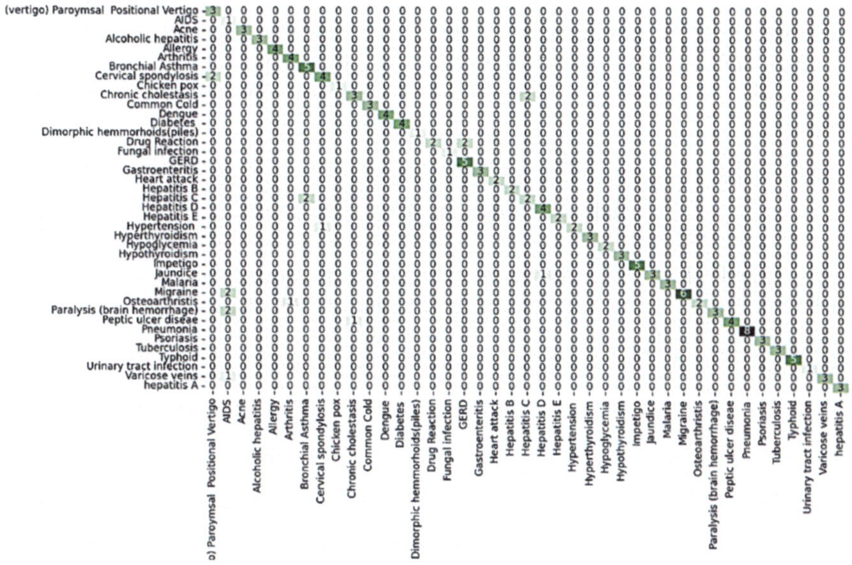

Fig. 8. The confusion matrix.

The accuracy of the classifier model is the ratio between correct predicted diagnosis and total number of diagnostics (Eq. 1). The main diagonal of the confusion matrix consists of correct estimates (126), the outlying values representing cases where the model misclassifies the medical conditions (17).

$$\text{Accuracy} = \frac{\text{Correct predictions}}{\text{Total number of predictions}} \tag{1}$$

The estimation errors are reduced, resulting an overall accuracy of 88.1%. Diseases as Migraine, Alcoholic Hepatitis and Bronchial Asthma are correctly classified, as observed from the confusion matrix. The classification errors (non-zero values outside the main diagonal) were obtained for Infectious Diseases (AIDS), misclassified twice as Migraine, or Osteoarthritis, misclassified as Arthritis. In both situations, the real diagnosis and the predicted one have multiple common symptoms, which leads to a limitation of the decision algorithm.

Between May and June 2024, three gender mixed focus groups with a total of 27 people were conducted to observe user perceptions and satisfaction. The participants were selected to provide a varied and balanced representation of the app's users. The first group consists of ten people between the ages of 18 and 25, undergraduate students. The second group consists of nine people, between 25 and 45 years old, students or employees. The third group included people over the age of 46, employees and retirees.

The questions were grouped into four categories: users' health-related behaviors, interest and use of the application, the user's experience with the application, and recommendations for future developments - Table 2. Additionally, for any symptom checker, the specific social context and the structural-social processes in which such applications are used, are necessary to be taken into consideration for a comprehensive ethical analysis [14].

Table 2. Evaluations of the focus groups

Focus groups FG	Useful	Easy-to-use	Functionalities	Would use	Strong points	Improvements
FG 1	90%	100%	80%	60%	User - friendly interface, quick results	a wider variety of diagnoses
FG 2	100%	100%	100%	100%	Clarity of information, ease of use	a wider variety of diagnoses, history of diagnoses
FG 3	75%	100%	100%	87.5%	User - friendly interface	a wider variety of diagnoses

The obtained 88.1% accuracy is considered good, but poor in comparison to other applications – Table 3. A possible cause is the reduced training dataset for the DT classification algorithm.

Table 3. Literature comparison

Application	Algorithm	Accuracy	Limitations
AI-Based Medical Chatbot [3]	MLP	94,32%	High computational, single disease
Alzheimer's Disease Prediction [4]	NN, DT, RF, XGB	Up to 92.05%	Small dataset size, high computational
Heart Disease Prediction [5]	Naive Bayes, DT, K-NN, RF	90.789% (K-NN)	High computational
Disease Prediction [8]	LR, DT, RF, SVM, AdaBoost	85.71%–98.57%	High computational, different accuracy
YourDiagnoseFriend	Decision tree	88.1%	Continuous model updates needed, small dataset

5 Conclusion

The paper presents the development and evaluation of *YourDiagnoseFriend*, a web application which primarily identifies potential diagnoses based on the symptoms experienced by users.

The application is accessible to a wide range of users, offering an intuitive interface and features, efficient user registration and authentication, and dynamic symptom identification. Its thoughtful design and functionality provide users with reliable and relevant information, enhancing their overall health management experience.

YourDiagnoseFriend stands out by providing a balanced combination of accuracy and user accessibility. Developed using the Django framework and Python programming language, it leverages the DT algorithm to offer users a quick and accurate evaluation of medical symptoms. The application integrates a robust database, categorizing symptoms and associating them with specific diseases and specialists.

While continuous updates to the model and maintaining high-quality data are necessary, *YourDiagnoseFriend* enhances the landscape of AI-driven health applications by addressing some of the limitations faced by other applications. This positions *YourDiagnoseFriend* as a valuable tool for personal health management and early disease detection, making advanced medical insights accessible to a broader audience.

Limitations of the proposed algorithm include misclassified diagnoses for diseases with high percentage of common symptoms, scalability, and non-customizable user interface.

Conflicts of Interest. The authors have no conflicts of interest to declare that are relevant to the content of this article.

References

1. Islam, S.: Artificial intelligence in healthcare. Int. J. Eng. Mater. Manuf. **6**(4), 319–323 (2021)
2. Timofeev, R.: Classification and Regression Trees (CART) Theory and Applications. Humboldt University, Berlin (2004)
3. Chakraborty, S., Paul, H., Ghatak, S., Pandey, S.K., Kumar, A., Singh, K.U., Shah, M.A.: An AI-based medical chatbot model for infectious disease prediction. IEEE Access. **10**, 128469–128483 (2022)
4. Kumar, Y., Maheshwari, P., Joshi, S., Baths, V.: ML-Based Analysis to Identify Speech Features Relevant in Predicting Alzheimer's Disease. In: Proceedings of the 8th International Conference on Computing and Artificial Intelligence, pp. 207–213 (2022)
5. Shah, D., Patel, S., Bharti, S.K.: Heart disease prediction using machine learning techniques. SN Comput. Sci. **1**(6), 345 (2020)
6. Vijaya Saraswathi, R., Gajavelly, K., Kousar Nikath, A., Vasavi, R., Reddy Anumasula, R.: Heart disease prediction using decision tree and SVM. In: Proceedings of Second International Conference on Advances in Computer Engineering and Communication Systems: ICACECS 2021, pp. 69–78. Springer Nature, Singapore
7. Dharmarathne, G., Jayasinghe, T.N., Bogahawaththa, M., Meddage, D.P.P., Rathnayake, U.: A novel machine learning approach for diagnosing diabetes with a self-explainable interface. Healthcare Analy. **5**, 100301 (2024)
8. Kohli, P.S., Arora, S.: Application of machine learning in disease prediction. In: 2018 4th International Conference on Computing Communication and Automation (ICCCA), pp. 1–4. IEEE (2018)
9. Aissaoui Ferhi, L., Ben Amar, M., Choubani, F., et al.: Empowering medical diagnosis: a machine learning approach for symptom-based health checker. Mobile Netw. Appl. **29**, 676–702 (2024)
10. Wallace, W., Chan, C., Chidambaram, S., et al.: The diagnostic and triage accuracy of digital and online symptom checker tools: a systematic review. npj Digit. Med. **5**, 118 (2022)
11. European Commission Homepage, https://commission.europa.eu/, last accessed 10 July 2024
12. Siddiq, M.: Use of machine learning to predict patient developing a disease or condition for early diagnose. Int. J. Multidiscip. Sci. Arts. **1**(1) (2022)
13. https://www.kaggle.com/datasets/kaushil268/disease-prediction-using-machine-learning
14. Müller, R., Klemmt, M., Koch, R., et al.: "That's just future medicine" - a qualitative study on users' experiences of symptom checker apps. BMC Med. Ethics. **25**, 17 (2024)

Foot Drop Orthosis, Low Cost, with Thermoplastic

Ionel Şerban(✉), Corneliu Nicolae Drugă, Barbu Cristian Braun, Alexandru Tulică, and Ileana Constanta Roşca

Faculty of Product Design and Environment/Product Design, Mechatronics and Environment Department, Transilvania University, Braşov, Romania
ionel.serban@unitbv.ro

Abstract. This paper aims to describe the realization of a proposed orthosis, made from an airy and lightweight low-temperature thermoplastic, used in the case of foot drop for supporting, stabilizing, and correcting the ankle joint in the process of locomotion. The described solution might be seen as accessible work with low-cost materials and little technical knowledge. Besides this, it can be personalized to the anatomy of every user, due to the thermoplastic's properties and it is composed of two parts easily connected through a quick-release band and Velcro. Thermoplastic can be modeled after being kept for a couple of minutes in water at a temperature of around 70 °C. The critical study indicates an abundance of available, on the market and not only, technical solutions that imply different materials, solutions, and methods including FES- Functional Electrical Stimulation. The paper also presents a bibliographic study made to understand the relevance of some keywords/terms, specific to this study, in a recognized database.

Keywords: foot drop · orthosis · thermoplastic

1 Introduction

Foot drop, also called drop foot, is defined by the difficulty in lifting the front part of the foot. In case of a foot drop, the front part of your foot might drag on the ground in a gait cycle [1].

Several different factors relate to the weakness or paralysis of the foot muscles that can cause foot drop pathology. Some of the factors are muscle or nerve disorders, brain or spinal cord disorders, and nerve injuries.

The symptoms of a foot drop are foot-dragging on the ground in a gait cycle, difficulty in lifting the foot, the thigh raising higher to avoid dragging (high stepping gait), pain and numbness in the foot while walking or lifting [2].

Regarding the material used for an orthosis, the first research paper recommended in Science Direct, concludes that neoprene might serve also as a substitute for conventional thermoplastic AFOs to improve the kinematics and kinetics of the gait of people with foot drop due to traumatic peroneal nerve injury. In our study, it is also used as a nonconventional solution [3, 4].

S. Vlad and N. M. Roman (Eds.): MEDITECH 2024, IFMBE Proceedings, 130, pp. 235–240, 2025.
https://doi.org/10.1007/978-3-031-95671-3_24

Searching Science Direct (in July 2024) using terms indicated in Table 1 it was found a great interest in this area of research regarding "foot drop" and "low cost" terms, and little interest, few results, regarding the terms "orthosis" and "thermoplastic" along with "foot drop". Although there is a greater impact when searching for "foot drop brace", it was found more suitable to use "foot drop orthosis" in the title of this paper.

Table 1. The critical study, in direct science database, regarding the number of studies, nowadays, considering some specific terms.

Terms used	No. of results	No. of review articles	No. of research articles
Foot drop orthosis low cost with thermoplastic	116	17	25
Foot drop	83,984	6243	48,996
Foot drop low-cost	25,820	2675	12,638
Foot drop brace	2812	299	1078
Foot drop orthosis	1862	274	778
Foot drop thermoplastic	1386	196	390

After filtering the results, for the terms "Foot drop orthosis low cost with thermoplastic" considering the research done in the last 10 years, it was found several (76 articles), which represent approx. 66% of the total. There are many studies on other databases that treat this subject, but it was more suitable for the authors to use Science Direct [5].

Regarding thermoplastics, the fabrication of custom-made fracture braces can be challenging because the client is normally in pain, and it requires the use of sizable thermoplastic parts that can be tricky to handle [6].

X-LITE® is for many considered the favorite choice when an airy, variable stiffness (see Fig. 1) and lightweight splint, cast, or brace is needed for a patient. X-LITE® is registered as a Medical Device Class 1 and is for human use [7].

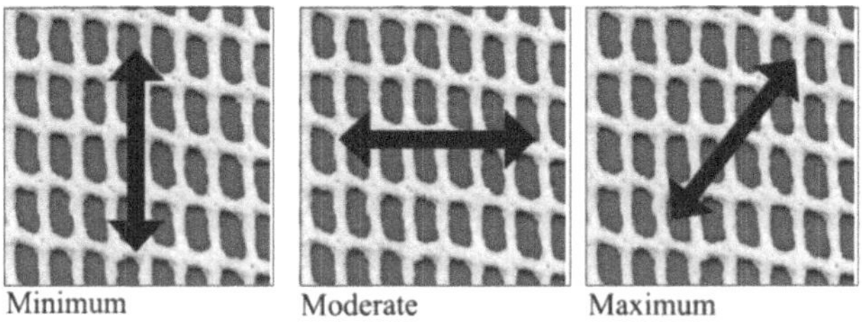

Fig. 1. The stiffness and elongation of the X-lite material in three directions. [8]

According to the ISO (International Standards Organization) definition, orthosis is a device for external use used to modify the structural or functional characteristics of the neuro-musculoskeletal system [9]. An orthosis (also called a brace or splint) is a device that sits around a part of the body. An orthosis can have various functions such as supporting, stabilizing, correcting, immobilizing, or relieving a body part [10].

Although most orthoses are made of light thermoplastic materials, many patients notice that they are heavy and uncomfortable, especially in the summer when the phenomenon of sweating between the two surfaces is accentuated. One solution is material reduction, following a finite element analysis, in areas that are not in high demand [11]. Another solution is the introduction of new lighter materials, such as x-lite, which have the property of reducing the phenomenon of sweating.

The realization of an orthosis made of thermoformable material, x-lite, involves going through the following stages, set out below [12]. The material is placed in a vessel of water which is around the temperature of 70 °C. The product must be immersed for at least 3 minutes to reach the condition required for application. After that, it can be placed on the foot and held there until it gets colder (a couple of minutes) so the material has the form of that anatomical part of the body. The thermoformable material consists of a cotton matrix impregnated with a copolymer solution. The main property of the copolymer is the possibility of repeating the thermoforming process at temperatures of 70 °C without losing its specific characteristics. Other important features of the product, x-lite, are non-toxic; biodegradable; incomparable ventilation; easy; reversible; unique bonding properties; variable stiffness (see Fig. 1); low price; and shape memory.

The original characteristics of our proposed orthosis consist of using low-cost, accessible materials that might be custom-made (thermoplastic but not only) and aid the foot drop pathology.

2 Realization of the Proposed Orthosis

A simple schematic of the forces, weight, and reaction force. This figure has the meaning to represent an approximation of the location of the center of mass of the foot along with the main forces that act on it (see Fig. 2). Considering the approximate location of the center of mass, the location of the quick-release band, from the plastic insert, for the force, in the quick release band, to act in that location.

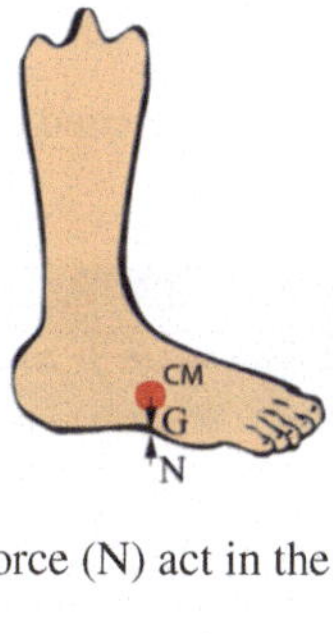

Fig. 2. The weight (G) and reaction force (N) act in the approximated center of mass location.

The foot drop orthosis (see Fig. 3) is composed of the following elements: ankle cuff, quick-release band, and plastic insert with lace fastening.

a) b)

Fig. 3. The main components of the orthosis: a) plastic insert with lace fastening, b) ankle cuff.

So that the ankle cuff does not cause discomfort, a 5 mm sponge is added in addition to the thermoformable material. A waterproof material fabric is used for the outer covering of the cuff, which will allow the wearer to use the orthosis even in rain or other activities that involve splashing water. As the chances of the wearer meeting water at temperatures above 70 °C are extremely low, the thermoformable material will retain its shape and still provide support. The plastic insert with lace fastening is made of thin, flexible plastic (it can also be made from thermoplastic) to facilitate the process of inserting it behind the laces. One end of the quick-release band will be sewn to the surface of the plastic piece, and the other end will be attached to the ankle cuff.

The steps to obtain the ankle cuff are: measure the needed material according to the perimeter of the subject foot; the thermoplastic (since it comes in roles- see Fig. 4a) will be inserted in hot water to straighten (see Fig. 4b) and left to cool in a straight/plane position (see Fig. 4c); a 5 mm sponge is added to the thermoplastic (see Fig. 5a) and the waterproof material (see Fig. 5b); at the end, it will be sewed/glued the quick release band and the Velcro (see Fig. 5c).

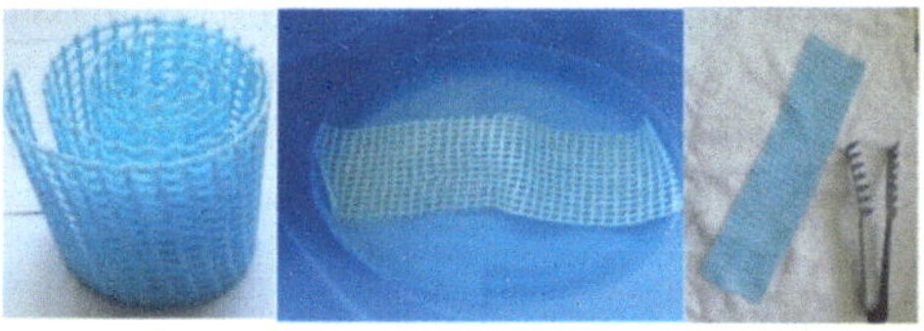

a) b) c)

Fig. 4. X-lite thermoplastic: a) as a roll; b) inserted in water at 70 °C; c) after extraction.

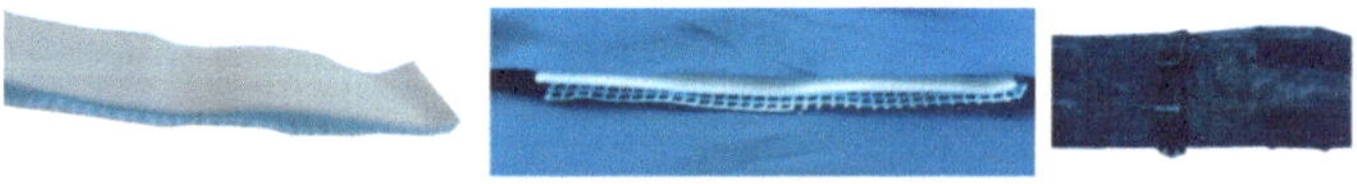

a) b) c)

Fig. 5. Ankle cuff: a) cooled thermoplastic with 5 mm sponge; b) added external waterproof material; c) final product with attached quick-release band and Velcro.

The final product, the ankle cuff, will again, be inserted in hot water and positioned on the leg to take its shape (see Fig. 6).

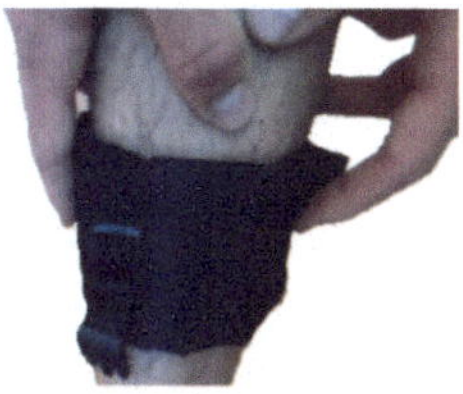

Fig. 6. Final step for the ankle cuff.

The steps to obtain the plastic insert with lace fastening are to measure and cut the needed plastic insert to fit the shoes and the holes for laces (see Fig. 7a) and then to sew/glue the quick-release band (see Fig. 7b).

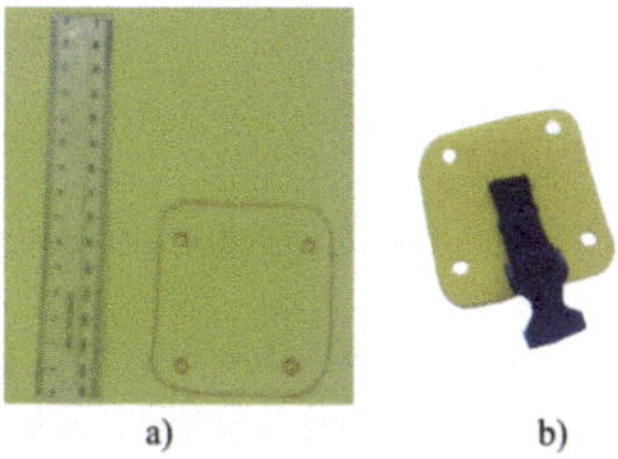

Fig. 7. a) Measuring and cutting the plastic insert; b) plastic insert with the quick-release band.

The final assembly (see Fig. 8) of the passive orthosis, made with thermoplastic, to assure a good anatomical fixation.

Fig. 8. Front view of the final product.

3 Conclusions

According to the critical study, there is much interest in this area. Nowadays solutions change following the advances in materials and techniques. This paper describes the steps to obtain a foot drop orthosis, a low cost, that might easily be done as a work with accessible materials and little technical knowledge. The polymer in X-LITE is fully reacted, which means it does not contain any of the residues in unreacted form, and therefore will not release such residues when applied or worn. There might be a further study regarding FEA, numerical, and biomechanical analysis, to complete and assess the presented orthosis but also other solutions available on the market, which is very diversified. The market for this type of orthosis consists of solutions that imply different types of materials but also mechatronic solutions that add sensors and actuators for aiding the measurement/analysis that might be done and for the advances in locomotion.

Acknowledgments. We would like to thank a former student, Eng. Mureşan Andreea Alexandra, for its involvement in the development of this paper.

Conflict of Interest. The authors declare they have no conflict of interest.

References

1. https://www.mayoclinic.org/diseases-conditions/foot-drop/symptoms-causes/syc-20372628 last accessed 30 Julu 2024
2. Shahar, F.S., Sultan, M.T.H., Lee, S.H., Jawaid, M., Shah, A.U.M., Safri, S.N.A., Sivasankaran, P.N.: A review on the orthotics and prosthetics and the potential of kenaf composites as alternative materials for ankle-foot orthosis. J. Mech. Behav. Biomed. Mater. **99**, 169–185 (2019)
3. Esfandiari, E., Saeedi, H., Ahmadi, A., Jalali, M., Pour, A.M., Abdollah, V.: Investigating the effects of conventional thermoplastic ankle-foot and the neoprene ankle-foot orthoses on the kinetics and kinematics of gait in people with foot drop following traumatic injury of the peroneal nerve: a pilot study. J. Foot. **50**, 101898 (2022)
4. Anbuvalanraj, A., Karuppudaiyan, S., Roy, S.: Effects of trimline cut in ankle foot orthosis: an experimental and finite element analysis. J. Orthop. **56**, 63–69 (2024)
5. https://www.sciencedirect.com/ last accessed 30 July 2024
6. Nouri, A., Wang, L., Li, Y., Wen, C.: Materials and Manufacturing for Ankle–Foot Orthoses: A Review. Adv. Eng. Mater. **25**, 1–24 (2023)
7. https://www.runlite.com/x-lite/ last accessed 30 July 2024
8. Şerban, I.: Studii şi cercetări privind influenţa mediului înconjurător asupra stabilităţii şi locomoţiei umane. PhD Thesis, Transilvania University of Brasov, Romania (2011)
9. Pratt, D.J.: Functional foot orthoses. J. Foot. **90**, 2–11 (1995)
10. https://www.spronken.com/en-en/categories/1/orthoses-and-braces last accessed 30 July 2024
11. Kutz, M.: Biomedical Engineering & Design Handbook, 2nd edn. The McGraw-Hill Companies., ISBN: 978-0-07-170472-4 (2009)
12. Monchaut, F., Şerban, I., Roşca, I.: On an orthesis for damaged ankle articulation. In: The 10th International Conference on Mechatronics and Precision Engineering, Bucureşti, România (2011)

Analysis of Human Exposure to Electric and Magnetic Fields While Charging and Driving an Electric Vehicle

Marian Gliga[1,2](✉), Adina Giurgiuman[1,2], Calin Munteanu[1], Sergiu Andreica[1,2], Claudia Pacurar[1], Claudia Constantinescu[1,2], Sabin Dusa[1], and Marius Botezatu[1]

[1] Department of Electrotechnics and Measurements, Technical University of Cluj-Napoca, Cluj-Napoca, Romania
Marian.Gliga@ethm.utcluj.ro

[2] Academy of Romanian Scientists, Ilfov 3 Bucharest, Romania

Abstract. The transition to electric vehicles is a promising strategy to reduce or eliminate harmful emissions, including carbon dioxide, nitrogen oxides, and particulate matter, from internal combustion vehicles. While electric vehicles offer environmental benefits, their operation involves the generation of electromagnetic fields (EMFs) in their immediate vicinity. Although the biological effects of EMFs have been extensively studied, the specific health risks to humans remain uncertain.

In electric vehicles, occupants are exposed to artificial EMFs due to their close proximity to high-capacity electrical systems operating for extended periods. Recognizing these potential concerns, electric vehicle manufacturers have incorporated various technological solutions during the design and manufacturing processes to mitigate EMF exposure. This study will conduct experimental measurements of the electromagnetic field generated by an electric vehicle under a range of operational conditions.

Keywords: electric cars · electromagnetic pollution · magnetic field exposure

1 Introduction

Electric vehicles and the hybrid ones have witnessed a significant surge in popularity in recent years. These vehicles are equipped with electric motors that operate at high speeds and produce zero emissions, thereby reducing environmental pollution caused by toxic chemicals [1].

While the majority of electric vehicles (EVs) currently operate at voltages below 400 volts, they often draw substantial currents, reaching several hundred amperes [2]. This high current flow leads to the generation of a magnetic field (MF) within the electric machinery that is considerably more intense than those produced by nearby electrical devices [3, 4].

Although magnetic fields diminish with distance, the proximity between the magnetic field source and the vehicle's passengers is minimal—approximately 1 cm—allowing the magnetic field to remain relatively intense [5]. Consequently, a substantial current

S. Vlad and N. M. Roman (Eds.): MEDITECH 2024, IFMBE Proceedings, 130, pp. 241–249, 2025.
https://doi.org/10.1007/978-3-031-95671-3_25

flows in close proximity to the passengers, generating an exceptionally strong magnetic field [6].

The electromagnetic fields (EMF) generated by these sources can potentially interfere with electronics such as mobile phones and smartwatches, and adversely affect humans within or in proximity of the vehicle. The susceptibility of electronic devices to EMF is a well-documented issue, often leading to electromagnetic interference (EMI). To mitigate these problems, electronic devices within the electromagnetic environment must adhere to strict electromagnetic compatibility (EMC) standards [7].

Bioelectromagnetics is the scientific field exploring the interactions between electromagnetic radiation and living organisms. This includes investigating the potential effects of such radiation on human health. Electromagnetic radiation can be categorized into two main types: ionizing and non-ionizing. The key distinction lies in their ability to disrupt the structure of atoms and molecules. Ultraviolet radiation serves as the dividing line between these two categories. It's important to note that electric vehicles emit electromagnetic radiation that falls into the non-ionizing category [8].

The biological effects of electromagnetic fields (EMF) vary depending on their specific characteristics. This variability has led to the European Union's establishment of distinct exposure limit values for low-frequency (0–10 MHz) and high-frequency (100 kHz–300 GHz) fields [9].

Long-term exposure to low-frequency electromagnetic fields (EMFs) can have harmful consequences for various human tissues, including the nervous system and muscles. Although high-frequency fields mainly cause tissue heating, excessive exposure to low-frequency EMFs can result in a variety of negative effects.

Long-term effects of EMF exposure are challenging to definitively document due to the extended time required for their manifestation and scientific evaluation. While long-term effects may be difficult to study, the short-term effects of non-ionizing radiation are well-understood. Biological effects have been observed from very low-frequency EMFs (up to several hundred kHz) and strong medium-frequency EMFs (radio and microwaves up to 300 GHz). It's important to note that thermal effects are minimal for field at approximately 100 kHz or below this frequency.

Electric vehicles employ high-powered electric motors, inverters, high-voltage cables, and batteries that operate within the low-frequency spectrum of 0–100 kilohertz. Consequently, exposure to magnetic fields is an unavoidable characteristic of their operation.

The International Agency for Research on Cancer (IARC) has classified exposure to low-frequency magnetic fields as possibly carcinogenic to humans, with a particular association to leukemia. The International Commission on Non-Ionizing Radiation Protection (ICNIRP) has developed guidelines to establish safe exposure limits for low-frequency magnetic fields [10].

Research, including the study referenced in [11], has investigated human exposure to magnetic fields emitted by electric vehicle motors. One particular study [12] found that magnetic field levels near the motor power cable inlet of a 70 kW drive were significantly more intense than the reference magnetic field level recommended by the International Commission on Non-Ionizing Radiation Protection (ICNIRP) in 1998. In fact, these levels were measured to be 170 times higher. To reduce the likelihood of excessive magnetic field exposure, it is essential to focus on the quality design and integration of high-voltage systems in hybrid or in electric vehicles.

Studies, such as those cited in [13, 14], have examined wireless charging technology for electric vehicles and concluded that it is a safe approach. The magnetic field levels generated by the high-voltage cable connecting the motor to the battery system have been analyzed in [15]. Electric vehicle manufacturers are actively working on technological innovations to decrease the magnetic field strength within the vehicle. One such solution involves combining the inverter with the electric motor and the energy management system with the battery system.

A model has been developed to evaluate the shielding effectiveness of the vehicle's frame and magnetic shielding, as discussed in [16]. This study also presents an equivalent model for batteries and cables. By analyzing various configurations, the magnetic field inside the car can be significantly reduced.

A study [17] examined the impact of electromagnetic fields on humans in electric vehicles by measuring the electric field intensity (6 Hz to 3 GHz) and magnetic field (6 Hz to 30 MHz) during vehicle operation. The study also analyzed the electric and magnetic field characteristics during charging and balancing modes. The results indicated that the electromagnetic radiation emitted by electric vehicles primarily consists of waves with frequencies ranging from 6 Hz to 500 Hz. Additionally, these levels may be higher during driving mode compared to charging or stationary modes.

Numerous studies on the electromagnetic fields generated by electric vehicles utilize computer modeling and simulation. Electromagnetic fields, which are invisible, are produced by both natural processes and human activities, particularly those involving electricity. These fields, measured in volts per meter (V/m) for electric fields and in Teslas (T) for magnetic fields, are most concentrated near their source and diminish with increasing distance.

Specialized equipment can be used to measure the electromagnetic pollution emitted by electric vehicles under various operating conditions. This study employed the ESM-100 H/E Fieldmeter from the German company Maschek Elektronik, as shown in Fig. 1. This device operates within a frequency range of 5 Hz to 400 kHz. It is capable of measuring magnetic fields in the range of 1 nT to 20 mT and electric fields from 0.1 V/m to 100 kV/m [18].

Fig. 1. Measuring instruments ESM-100 H/E Fieldmeter

This study seeks to quantify and analyze the electric and magnetic field intensities within electric vehicles across a range of operational conditions.

2 Human Exposure Recommendations

Figure 2 outlines the recommended electric and magnetic field levels for bedrooms, as specified in SBM-2015. According to these guidelines, magnetic field levels exceeding 500 ηT are deemed extremely high and potentially harmful to human health. While these strict recommendations are advised for bedrooms, higher electric and magnetic field values are permissible in office environments [19].

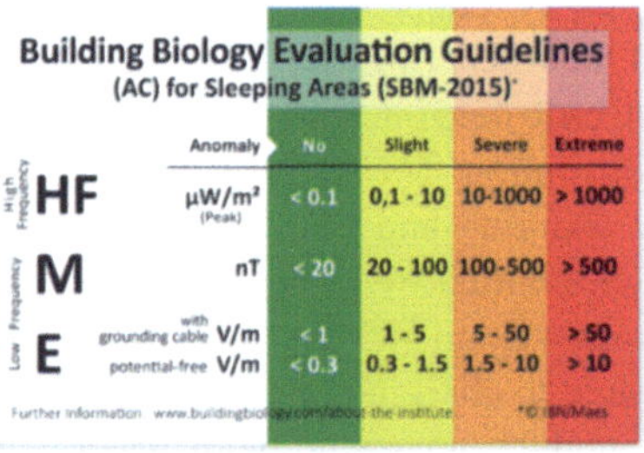

Fig. 2. Electromagnetic field levels

The International Commission on Non-Ionizing Radiation Protection (ICNIRP) suggests using induced electric field strength as the primary benchmark for exposure. Magnetic field strength (H) and magnetic flux density (B) are used as reference points for evaluating exposure. These reference points are based on fundamental limits using models that simulate a person standing upright and exposed to a consistent field in an open area.

The reference levels are averaged over the entire body of the exposed individual, assuming that local exposure limits are not exceeded. In general, reference levels for magnetic flux density decrease as frequency increases. For example, at frequencies between 8 and 25 Hz, the reference levels range from 0.625 to 0.2 mT; for frequencies between 25 and 400 Hz, the level is 0.2 mT; and for frequencies between 400 and 3000 Hz, the magnetic flux density ranges from 0.2 to 0.0267 mT. These values are mainly based on short-term exposure limits.

Magnetic field values can attain several hundred μT near particular electrical equipment. Under power lines, magnetic fields can be around 20 μT, and electric fields can reach several thousand volts per meter. As noted in [5], yearly exposure to magnetic fields at frequencies of 50 and 60 Hz and levels exceeding 0.3–0.4 μT may elevate the risk of childhood leukemia [20].

3 EMF Measurement and Results

Measurements of the electromagnetic field (EMF) generated by the electric vehicle were conducted using a Field Meter ESM-100 H/E. The EMF was measured outside the vehicle during charging with a CCS connector at a height of 1.8 meters. The four measurement points, as illustrated in Fig. 3, yielded varying values, which are detailed in Table 1. In charging mode, the electric field near the charging station reaches a maximum value of 1.2 V/m and the magnetic field near the charging station reaches a maximum value of 131 ηT.

Fig. 3. The vehicle during charging with connector CCS

Table 1. The electric and magnetic field values around the car during charging.

Position	Electric field V/m		Magnetic field ηT	
	Filter 'all'	Filter 'low'	Filter 'all'	Filter 'low'
In front (1)	1	0.7	131	129
On the left side (2)	1.2	0.5	46	43
On the right side (3)	0.9	0.3	26	19
In rear (4)	0.9	0.3	24	15

Additional measurements were conducted under various operating conditions and at different locations within the electric vehicle. Table 2 presents the electromagnetic field (EMF) readings for the driver's area, the front passenger seat, and the rear seat. These measurements were taken in different vehicle operating modes. During charging and ignition modes, the EMF levels are notably low. When the vehicle is stationary for charging, an electric field is recorded in the range of 0.9–0.4 V/m for the driver's seat, 1.12–0.6 V/m for the front passenger seat and 1.5–1.1 V/m for the rear seats. A magnetic field is also recorded in the range of 107–112 ηT for the driver's seat, 114–119 ηT for the front passenger seat and 197–225 ηT for the rear seats.

Table 2. The electric and magnetic field values inside the car during charging.

Position	Electric field V/m		Magnetic field ηT	
	Filter 'all'	Filter 'low'	Filter 'all'	Filter 'low'
The driver's area	0.9	0.4	107	112
The front passenger seat	1.1	0.6	114	119
The rear seat	1.5	1.1	197	225

The electromagnetic field measurements were conducted in three specific areas: the driver's seat, the front passenger's seat, and the rear seat (Fig. 4).

Fig. 4. Example of measuring the electric and magnetic field inside the car

When the electric vehicle is in motion at speeds of 50, 75, and 100 km/h, the electric field values were measured to be between 0.4 and 2.5 V/m, and the magnetic field values were measured to be between 43 and 987 ηT. These results are summarized in Table 3. The data in Table 3 show a clear correlation between vehicle speed and EMF values, with lower EMFs observed at lower speeds and higher EMFs at higher speeds.

Table 3. The electric and magnetic field values inside the car while moving at different speeds.

Speed	Position	Electric field V/m		Magnetic field ηT	
		Filter 'all'	Filter 'low'	Filter 'all'	Filter 'low'
0 km/h	The driver's area	1.2	0.7	46	43
	The front passenger seat	1.1	0.5	93	92
	The rear seat	1.3	0.7	26	21
50 km/h	The driver's area	1.4	0.5	202	170
	The front passenger seat	1.5	0.4	403	520
	The rear seat	0.9	1.1	182	224
75 km/h	The driver's area	1.8	1.6	280	344
	The front passenger seat	1.3	0.6	422	311
	The rear seat	1.2	0.9	304	396
100 km/h	The driver's area	1.7	1.6	241	296
	The front passenger seat	2.5	2.2	219	257
	The rear seat	1.3	1.6	425	987

4 Conclusions

During charging mode, the electric field near the charging station reaches a maximum value of 1.2 V/m, while the magnetic field in the same area attains a peak value of 131 ηT. When the vehicle is stationary for charging, measurements of the electric field within the vehicle reveal values ranging from 0.9 to 0.4 V/m at the driver's seat, 1.12 to 0.6 V/m at the front passenger seat, and 1.5 to 1.1 V/m at the rear seats. Simultaneously, the magnetic field inside the vehicle is recorded with values ranging from 107 to 112 ηT at the driver's seat, 114 to 119 ηT at the front passenger seat, and between 197 and 225 ηT at the rear seats. These measurements indicate that while the vehicle is stationary during charging, there is a significant presence of both electric and magnetic fields within the vehicle's interior.

When the electric vehicle is in motion at various speeds—specifically at 50, 75, and 100 km/h—the electric field values within the vehicle range from 0.4 to 2.5 V/m. Concurrently, the magnetic field values measured during vehicle motion vary between 43 and 987 ηT. The data presented in Table 3 clearly demonstrate a correlation between vehicle speed and electromagnetic field (EMF) levels, with lower EMF values recorded at reduced speeds and a noticeable increase in these values as the vehicle speed rises. This trend suggests that the operation of the vehicle at higher speeds results in greater electromagnetic emissions, which may have implications for the design and safety standards of electric vehicles.

Acknowledgments. This work was supported with a grant from the Academy of Romanian Scientists, project title: Digitalization of the Evaluation Process of Human Exposure to Electric and Magnetic Fields.

Conflicts of Interest. The authors declare no conflict of interest.

References

1. Garvanova, M., Garvanov, I., Ivanov, V., Borissova, D.: Measurement and estimation of the magnetic fields in electric vehicles. In: 2022 22nd International Symposium on Electrical Apparatus and Technologies (SIELA), Bourgas, Bulgaria, vol. 2022, pp. 1–4. https://doi.org/10.1109/SIELA54794.2022.9845773
2. Sztafrowski, D., Winiarz, J.: Electromagnetic field in electric and hybrid cars. J. Phys. Conf. Ser. **2408**, 012018 (2022)
3. Fotescu, R.-P., Svasta, P., Burciu, L.-M., Constantinescu, R.: Electromagnetic and electric field radiation inside electric vehicles and classic vehicles. In: 2022 IEEE 28th International Symposium for Design and Technology in Electronic Packaging (SIITME) (2022)
4. Vassilev, A., Ferber, A., Wehrmann, C., Pinaud, O., Schilling, M., Ruddle, A.R.: Magnetic field exposure assessment in electric vehicles. IEEE Trans. Electromag. Compatib. **57**, 35–43 (2015)
5. Moreno-Torres, P., Lafoz, M., Blanco, M., Arribas, J.R.: Passenger Exposure to Magnetic Fields in Electric Vehicles. InTech (2016)
6. R Hristov, S Stefanov and P Kostov, Investigation of the Electromagnetic Field in Electric and Hybrid Cars, 2010
7. IEEE Standard for Safety Levels with Respect to Human Exposure to Electromagnetic Fields, 0–3 kHz
8. IEEE Standard for Safety Levels with Respect to Human Exposure to Radio Frequency Electromagnetic Fields, 3 kHz to 300 GHz
9. SR EN 62110 – nivelurile câmpului electric şi magnetic generat de sistemele de alimentare în curent alternativ. Proceduri de măsurare a nivelurilor de expunere a publicului
10. ICNIRP GUIDELINES FOR LIMITING EXPOSURE TO TIME-VARYING ELECTRIC AND MAGNETIC FIELDS (1HZ – 100 kHZ)
11. Ruddle, R., Low, L.: Feasibility of estimating in-service vehicle occupant exposure to electrical powertrain magnetic fields using non-local magnetic field measurements. In: 2015 IEEE International Symposium on Electromagnetic Compatibility (EMC), vol. 2015, pp. 702–707. https://doi.org/10.1109/ISEMC.2015.7256249
12. Ruddle, R., Low, L., Armstrong, R., Dawson, L., Rowell, A.: Investigation of human exposure to magnetic fields from electrical powertrains: measured exposure levels and simulated impact on human body. In: 5th IET Hybrid and Electric Vehicles Conference (HEVC 2014), pp. 1–7 (2014). https://doi.org/10.1049/cp.2014.0956
13. Songcen, W., Xiaohua, H., Zishou, L., Jinxing, X., Fengxian, W., Lei, W.: Research on electromagnetic exposure measurement method for electric vehicle wireless charging system. In: 2018 IEEE 2nd International Electrical and Energy Conference (CIEEC), 2018, pp. 513–517. https://doi.org/10.1109/CIEEC.2018.8745925
14. Campi, T., Cruciani, S., De Santis, V., Maradei, F., Feliziani, M.: EMC and EMF safety issues in wireless charging system for an electric vehicle (EV). In: 2017 International Conference of Electrical and Electronic Technologies for Automotive, pp. 1–4 (2017). https://doi.org/10.23919/EETA.2017.7993214
15. Ionescu, V., Săpunaru, A., Popescu, M.O., Popescu, C.L.: Magnetic field constraints in the passenger compartment of electric vehicles. In: 2019 IEEE PES Innovative Smart Grid Technologies Europe (ISGTEurope), vol. 2019, pp. 1–5. https://doi.org/10.1109/ISGTEurope.2019.8905554

16. Pinaud, O., Chadebec, O., Rouve, L.L., Guichon, J.M., Vassilev, A.: Forward model computation of magnetostatic fields inside electric vehicles. IEEE Trans. Magn. **50**(2), 985–988 (2014). https://doi.org/10.1109/TMAG.2013.2283913
17. Wang, X., Li, H.F., Mei, N., Wang, X.Y., Liu, J., Fan, X., Wei, S.F., Shi, S.: The research and safety discussion of electromagnetic radiation in electric vehicles. In: 2021 International Applied Computational Electromagnetics Society Symposium (ACES), vol. 2021, pp. 1–3. https://doi.org/10.1109/ACES53325.2021.00038
18. MASCHEK Homepage, https://www.maschek.de/en/products/esm-100-h-e-fieldmeter
19. Standard of Building Biology Testing Methods SBM Homepage, https://buildingbiology.com/building-biology-standard/
20. Giurgiuman, A., Gliga, M., Bojita, A., Andreica, S., Munteanu, C., Topa, V., Constantinescu, C., Pacurar, C.: Software program for the evaluation of human exposure to electric and magnetic fields. Technologies. **11**, 159 (2023). https://doi.org/10.3390/technologies11060159

Product Strategy on Ventricular Assist Devices Market

Constantinescu-Dobra Anca(✉), Veliche Madalina, and Martis Claudia Steluta

Technical University of Cluj-Napoca, Cluj-Napoca, Romania
Anca.constantinescu@enm.utcluj.ro

Abstract. The global market for artificial heart and ventricular assist devices (VADs) has witnessed significant growth lately, driven by technological advancements and strategic initiatives by leading companies. Prominent manufacturers such as Abbott Laboratories and Abiomed carve out a substantial market share, mainly due to a diversified product strategy and strategic expansions. This article provides an overview of the key players, recent mergers and market trends, focusing on the products strategy of those companies throughout the last decade, considering the evolving environment of this critical segment of the medical devices industry. The technique used for product strategy assessment was the strategic positioning map. It was taken into consideration the diversity of the product range, the price, the promotion as well as the additional services provided.

While some companies focus on a narrowing their range of medical products with a particular emphasis on temporary artificial heart devices and hemodynamic support, others offer a broader and more diverse range of medical devices, including ventricular assist devices, services, drugs, and nutritional supplements that allow them a consolidated position on the global market.

Keywords: product strategy · ventricular assistance devices · strategic positioning map

1 Introduction

1.1 The Global Trends on Ventricular Assist Devices Global Market

Artificial hearts and VADs are critical technologies for patients suffering from severe heart diseases. These devices serve as life-saving solutions by either temporarily or permanently supporting heart function. The market for these devices is experiencing rapid growth, fueled by an aging population, increasing incidence of heart diseases, and continuous technological innovations.

The VAD market is expanding rapidly, with a market size valued at USD 1.25 billion in 2023 and expected to almost double by 2032, growing at a CAGR of 9.9% [1, 2]. This growth is driven by the rising occurrence of heart failure, particularly among the aging population, which has led to an increased need for advanced cardiac support systems. Also, along with the fact that the people aged up to 65 years reach the percentage of 10%, while the forecasts present an increase to16% in 2050 for this specific target segment

S. Vlad and N. M. Roman (Eds.): MEDITECH 2024, IFMBE Proceedings, 130, pp. 250–260, 2025.
https://doi.org/10.1007/978-3-031-95671-3_26

[3]. Because of the facts above, the World Health Organization recommends developing strategies for elderly people to remain independent and to maintain their health state.

VADs are vital not only for elderly people, but for young patients as well with severe heart conditions, assisting in pumping blood from the heart to the rest of the body. The increasing demand for these devices is further catalyzed by advancements in medical technology, including the integration of artificial intelligence to improve device efficiency and patient outcomes.

The *rising incidence of heart failure* is a key factor contributing to the increasing demand for VADs. According to the Journal of Cardiac Failure, approximately 6.7 million Americans over the age of 20 are affected by heart failure, a number projected to rise to 8.5 million by 2030 [4, 5]. Furthermore, as heart transplants are not always feasible due to the limited number of donors, VADs serve as an important long-term solution for patients with heart failure. Market players are focusing nowadays especially on research and development functions to create more efficient and effective VADs. For instance, Abbott's HeartMate 3 has shown improved survival rates in patients, indicating a promising effectiveness for such devices in managing heart failure.

Even though the market faced a setback during the COVID-19 pandemic, when non-emergency surgeries were postponed, after this period the demand for VAD implantations increased again. Additionally, disruptions in the supply chain affected the availability of these devices in pandemic period. However, the market witnesses an inflexion point for cardiac treatment after 2021 [6]. In terms of technological advancements, the development of product strategy in the following directions will maintain market growth [7–11]:

- efficient devices with lower rotational speeds,
- smaller sizes,
- enhanced anatomic designs.
- fully implantable VADs (such as Medtronic's HVAD system) that eliminate the need for external power sources.

This advancement reduces the risk of infection and enhances the quality of life for patients that need long-term cardiac support. The integration of AI and machine learning in these devices further enhances their performance, allowing for better monitoring of clinical data and more precise surgical outcomes.

These devices witnessed also some disadvantages, as well; mainly the high cost of VAD implantation remains a significant barrier to widespread adoption. Additionally, potential complications post-implantation, such as device malfunction or the need for replacement, represent further challenges. These factors may limit the widespread penetration of VADs in the global market.

The objective of this article is to assess the effectiveness of product strategy of the leading companies that run their business in VAT market, by emphasizing the specificity of their implemented strategies. This paper will examine the product portfolio of five leading companies in the last decade and several positioning maps are elaborated, considering the core lines of their offered products and services. The article serves as well for the managers from the companies that run their businesses in this field, to set up their product strategies and decide the best way to differentiate in this competitive landscape through an optimal strategic positioning.

1.2 The Competitive Landscape on VAD Market

Several major companies dominate the artificial heart and VAD market [12–16]. Abbott Laboratories and ABIOMED are among the top players, with Abbott holding a significant global market share in 2023, due to robust sales of its VAD products. The company's focus on expanding its product range globally has been a key driver of its market.

Other notable companies in the market include Berlin Heart GmbH, Jarvik Heart Inc., Medtronic, and Reliant Heart. These companies are actively working to obtain regulatory approvals to expand their product portfolios and strengthen their market positions. Additionally, CH Biomedical Inc., and Calon Cardio Technology Ltd. have their products in the research and development stage, with a prediction to enter on the market in the coming years and capture a significant share.

The artificial heart and VAD market has seen significant mergers and acquisitions, which have reshaped the competitive landscape. In 2022, Johnson & Johnson acquired ABIOMED, allowing it to operate as an independent business within its MedTech segment. In the same year, Abbott acquired Walk Vascular LLC, a company specializing in mechanical thrombectomy systems, further expanding its product portfolio.

Earlier, in 2021, ABIOMED acquired preCARDIA, a developer of a proprietary catheter and controller system, to enhance its offerings for patients with acute decompensated heart failure (ADHF). Meanwhile, CH Biomedical received approval for its CH-VAD system from the National Medical Products Administration (NMPA) in China, marking a significant milestone for the company.

The artificial heart and VAD market continued to growth, driven by technological innovations, strategic acquisitions, and the expansion of product portfolios by key players. As companies like Abbott and Abiomed continue to lead the market, emerging players and new entrants are expected to challenge the existent situation, further fueling competition and innovation in this vital sector.

2 Methodology

For analyzing the product strategies in VAD market, the positioning map was implemented as a strategic tool in market analysis.

A positioning map enables the comparison of a company's product portfolio comparatively with its competitors, helping to identify opportunities for new market entries. The key stages followed in the present analysis consisted of:

1. Defining the objective that could be identifying the market gaps, implementing a benchmarking analysis or market entry strategies
2. Identifying the key market players and their products launched on the market
3. Selecting key attributes that could position the company on the prospective customers mind e.g. safety vs. clinical performance on long term (such as thrombosis risk, stroke rate), durability of the device or hospitals adopting globally, regulatory approval, innovation degree etc.
4. Obtaining market data from markets reports, clinical studies, industry statistics, companies financial reports and product portfolio

5. Drawing up the map with two or three axes, with the chosen criteria and setting up the position for the most important players on VAD market on the obtained quadrant.
6. Identifying the clusters and the gaps in the markets
7. Exploring the insights: how emerging players might disrupt the market and shift strategic positioning, interpret the market gaps (e.g. high penetration vs. low innovation), examining the potential opportunities
8. Develop a competitive strategy e.g. adjust R&D focus to serve specific market niches, optimize price policies

The specific goal for the assessment was related to entering strategies in VAD market, while the key criteria refer to product policy or price.

The main advantages of using it in strategic planning are the following [18–21]:

1. Understanding the market dynamic;
2. Identifying market opportunities;
3. Differentiating from competitors;

There are also some disadvantages, as well:

1. Over-simplification.
2. Subjectivity of perceptions.
3. Rapid changes in consumer preferences.

In the case of VAD, several positioning maps have been created based on specific criteria to this field.

In the sample were included the most important companies in this field, that are leading this sector in the global market.

A short description of them and their mail products is as follows.

Abiomed: Pioneering Cardiovascular Devices [13]

Abiomed is renowned for developing the world's smallest heart pump, known as Impella, designed to provide temporary circulatory support to patients with severe heart failure. Impella is primarily used in cardiac surgery and intensive care units to support patients during critical periods of treatment. Additionally, the company markets other products, such as:

- Impella Connect: A real-time patient monitoring platform for those using Impella. It allows physicians to monitor patients' vital signs in real-time, offer remote medical assistance, and quickly identify any issues or complications.
- Breethe OXY-1 System: A portable blood oxygenation device that can be used outside of the hospital to support patients with respiratory failure.

Abiomed, Inc. also offers a range of other products and accessories, including catheters, tubes, and infusion sets, which are used to support patients with heart failure and other cardiovascular conditions. The company's product ranges are structured into five main categories, each with separate product lines. The wide range of products allows customers to choose the most suitable solutions for specific cases. Abiomed sells highly differentiated products with various features, making them unique compared to those of competitors.

Abbott Laboratories: A Comprehensive Product Strategy [12]

Abbott Laboratories implement a product strategy focused on developing and promoting a wide range of products and solutions for various medical conditions and healthcare needs. The company maintains a complex portfolio that includes different categories of products such as:

1. Medications: Abbott produces drugs such as Humira (for autoimmune diseases), Depakote (for epilepsy and bipolar disorders), Synthroid (for hypothyroidism), and Kaletra (for HIV infection).
2. Nutritional Supplements: Products like Ensure, Pedialyte, and Similac, used to support adequate nutrition for patients, including children and adults.
3. Medical Devices: Abbott produces a wide range of medical devices for various conditions, such as:
   - Ventricular assistance devices: HeartMate III, HeartMate II, CardioWest Total Artificial Heart.
   - Stents.
   - Heart failure monitoring systems.
   - Atrial fibrillation devices.
   - Blood pressure monitoring devices.

Abbott Laboratories also emphasizes addressing patient needs and preferences by developing customized solutions tailored to individual patient characteristics.

Berlin Heart GmbH: Focused on High-Quality Ventricular Assist Devices [14]

The primary objective of Berlin Heart GmbH's product strategy is the development and supply of ventricular assist devices for children and adults with severe heart failure. The company's product strategy focuses on creating high-quality ventricular assist devices that are safe, reliable, and easy to use. The products developed by this company include:

1. EXCOR Pediatric Device
2. EXCOR Adult Device
3. INCOR Device

Key elements of Berlin Heart GmbH's product strategy include:

- Innovation: The company invests in research and development to bring innovative products to the market that meet the needs of patients and physicians.
- Quality: Emphasis on product quality, with rigorous quality control processes to ensure the reliability and safety of products for patients.
- Adaptability: Product development is customized and tailored to the individual needs of patients.
- Accessibility: Berlin Heart GmbH offers high-quality medical devices at accessible prices, ensuring that patients with severe heart conditions can access the necessary treatment.

Jarvik Heart, Inc.: Specialized in Ventricular Assist Devices [15]

Jarvik Heart, Inc. focuses on the development and commercialization of the Jarvik 2000 ventricular assist device (VAD), a key component of its product strategy. The following components are part of this strategy:

1. Target Market Identification: Jarvik Heart, Inc. targets patients with end-stage heart failure who are awaiting heart transplants or are not eligible for them.
2. Product Design and Development: Jarvik 2000 is designed to be compact, lightweight, and fully implantable, enhancing patient comfort and convenience. The company invests in research and development to maintain the device's functionality, reliability, and safety.
3. Regulatory Compliance: Before marketing and selling Jarvik 2000, Jarvik Heart obtained the necessary approvals from the relevant authorities.

Medtronic: Broad Range of Medical Solutions [17]

Medtronic offers a wide range of medical products covering various medical specialties, including cardiology, neurology, orthopedics, surgery, and diabetes. These products include:

(a) Cardiac devices.
(b) Neurological systems.
(c) Orthopedic devices.
(d) Surgical systems.
(e) Diabetes therapy systems.
(f) Pain management products.

3 Results and Discussions

A series of positioning maps have been elaborated for VAD market, according to several creations specific to this field. The first map is a centralizing matrix that includes the types of products and services offered by the companies from the sample. This tool considers price and market niches addressed by the competitors.

According to Fig. 1, the analysis of competitiveness in this field highlights the fact that the market leaders Abbott and Abiomed are implementing as many strategies as possible to diversify their product portfolio. Thus, Abbott is present in the categories of ventricular assist devices, monitoring devices, drugs and supplements. Abiomed deals with monitoring devices, ventricular assist devices as well as other medical devices (catheters, tubes, infusion sets, orthopedic devices, surgical systems). While Abiomed and Abbott cover a broader category, Berlin Heart is limited to adult and pediatric ventricular assistance devices. In addition to ventricular assist devices, Medtronic also deals in drugs (pain killers) and medical supplements. Jarvik Heart only addresses VAD target segment.

In terms of pricing, Abiomed employs a premium pricing strategy, while Abbott Laboratories uses a diversified pricing strategy.

In terms of diversity of products and price, the positioning map is the following:

According to Fig. 2, Abbott offers the largest range of products along with Medtronic, while Abbot, Abiomed and Jarvik Heart are using a skimming strategy for pricing.

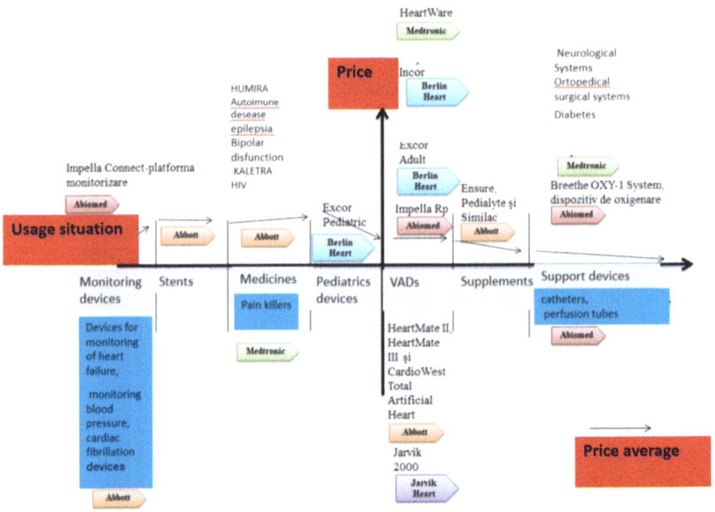

Fig. 1. A strategic map based upon prices and type of products for the VAD companies in the sample

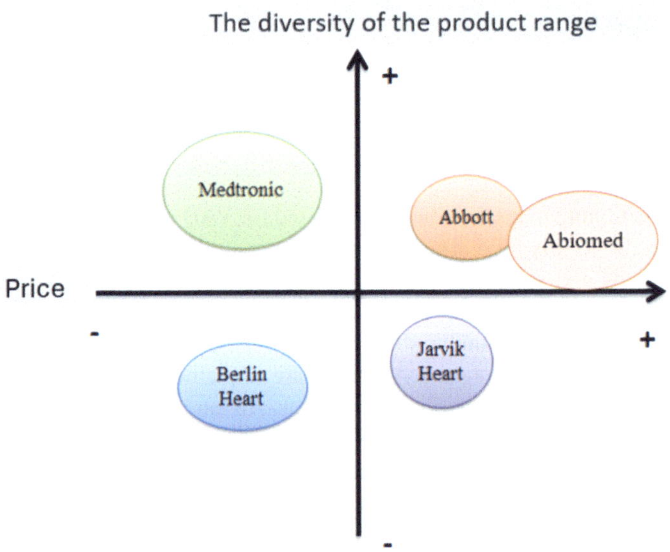

Fig. 2. A competitive map based upon price and diversity of product range

Analyzing based upon the specific range of product and additional offered services the following positioning map is obtained, displayed in Fig. 3.

Jarvik Heart and Berlin Heart have specific usage (permanent hearts and devices for children) and Abiomed with a narrower range of VAD products (or VAD related) for specific niches and has a sustained promotion strategy. Abbott chooses to offer

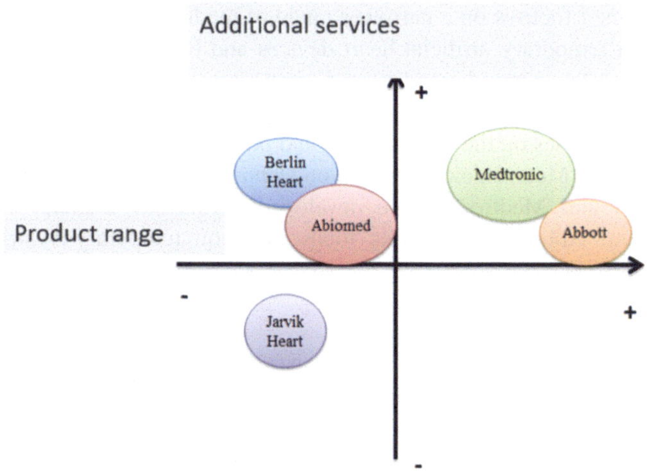

Fig. 3. A competitive map based upon diversity of product range and additional services

more products for a larger segment, with a wide range of VADs and sustained promotion. Instead, Medtronic offers products for specific use (cardiac devices, neurological systems, orthopedic devices).

There isn't a direct correlation between the number of VAD products a company offers and its market share or profitability. The extensive portfolio has contributed to Abbott's significant presence in the VAD market. Abiomed, specialized in the Impella series, one of the smallest heart pumps, despite a narrower focus compared to Abbott, has achieved substantial market penetration. Assessing the facts from a specialization vs. diversification perspective, we cannot conclude that profitability is determined by one of these. Companies like Abiomed demonstrate that specializing in a specific technology can lead to high profitability within that niche, even with a limited product line. Success depends on aligning products with market needs and continuous innovation. Abbott's diverse VAD solutions meet various patient requirements, contributing to its market leadership. While a diverse product range can enhance market presence, factors such as specialization, innovation, operational efficiency, and alignment with market demand play crucial roles in determining profitability and market share.

In Fig. 3, it can be seen also that most of the competing companies on the market choose to offer additional services in addition to the main product launched on the market. These consist of loyalty cards, financial assistance, volume and loyalty discounts, personalized financing plans, etc.

The only company that has not adopted these strategies is the Jarvik Heart company that does not offer the additional services available to the other companies. One explanation could be that Jarvik Heart focuses particularly on patients for specific procedures and less on medical companies.

Comparative conclusions among major manufacturers from the perspective of product strategy led to the following:

- While Abiomed focuses on a narrower range of medical products with a particular emphasis on temporary artificial heart devices and hemodynamic support, Abbott Laboratories offers a broader and more diverse range of medical products, including ventricular assist devices, drugs, and nutritional supplements.
- Berlin Heart focuses on the development and production of pediatric devices, while Jarvik Heart and Medtronic concentrate on technological solutions for adults.
- Jarvik Heart and Medtronic have developed permanent artificial hearts, while Abiomed and Abbott Laboratories focus more on temporary ventricular assist systems.

4 Conclusions

In conclusion, the VAD market witnessed substantial growth due to increasing heart failure cases and advancements in technology.

The ventricular assist device market analysis reveals the following trends.

A consistent increase in demand for VAD in recent years.

Advanced technology – the development of more efficient and smaller devices – provides better quality for patients.

High costs – due to the high degree of innovation, advanced technology, and the expensive process of production and implantation.

Competitors Analysis reveals a general dispersion in the positioning map based upon product range diversity and price quadrants.

- Product mix length analysis – existence of seven distinct categories in the VAD branch and related to this.
- Firms offer more and more additional services for support

While cost and potential complications raise challenges in this branch, the market's future emphasize for innovation and increasing awareness of the importance of cardiac support systems. The focus on enhancing efficient devices for patients in need through technological developments will continue to propel the market forward, making VADs an essential component in the management of severe heart conditions.

Even though the high barriers (especially capital ones) make difficult the entrance of emerging players into the VAD market along with the continuous innovations, there is the potential to disrupt the established market leaders, shifting the competitive landscape and strategic positioning of key companies.

Emerging companies are likely to release products with ***new VAD technologies*** incorporated, that could challenge the dominance of current leaders such as Abbott, Medtronic, and Abiomed. These disruptions could be related to miniaturization of VADs, aiming to create smaller, more durable alternatives to current devices. Another direction could be investing in research for biocompatible and wearable solutions. Companies like Carmat from France are investing in implantable artificial hearts with advanced biomaterials and autoregulation systems, offering a solution that could be more "acceptable" from physiological point of view. They gained approval in Europe and could directly compete with Jarvik Heart in the permanent heart replacement segment.

Another direction for disruption could be introduced by focusing on ***artificial intelligence and remote monitoring*** smart VAD by med-tech startups, integrating real-time predictive analytics for better patient outcomes. A lot of this innovative VAD are not yet largely adopted in emerging markets; hence the cost-effective strategies may open others developing directions, challenging the premium pricing strategies used by Abbott, Abiomed, and Medtronic.

The VAD market is poised for major shifts due to disruptive technological advancements, emerging competitors, and cost-effective solutions. Established companies must continuously innovate, integrate AI-driven solutions, and expand globally to maintain their leadership. As emerging players could reshape the landscape, established leaders may need to pivot their strategies.

Conflict of Interest. The authors have no conflicts of interest to declare.

References

1. Ventricular Assist Device (VAD) Market Size Envisioned at USD 3.08 Billion by 2032, https://www.towardshealthcare.com/insights/ventricular-assist-device-market-sizing
2. Implantable Ventricular Assist Devices (VAD) Market 2024 Growth and Size Predictions for Success, https://www.marketsizeandtrends.com/download-sample/487746/&utm_source=Pulse-Glob2&utm_medium=362
3. Gore A.: Ventricular Assist Devices VAD Market Report 2024 (Global Edition), https://www.cognitivemarketresearch.com/ventricular-assist-devices-vad-market-report?campaign_source=google_ads&campaign_name=CMR_EU_DSA&gad_source=1&gclid=CjwKCAiAjeW6BhBAEiwAdKltMpfeJMWnvJTj4JKtkEKVcL02nFgAEo0JhXiMSEdpvAcuCWBnPnFUnxoCONkQAvD_BwE
4. Iacomi, A.: Populaţia lumii îmbătrâneşte, Ziarul BURSA #Internaţional/1 iunie 2023., https://www.bursa.ro/populatia-lumii-imbatraneste-46909947
5. Bozkurt, B., et al.: HF STATS 2024: heart failure epidemiology and outcomes statistics an updated 2024 report from the Heart Failure Society of America. J. Card. Fail. **31**(1), 66
6. Heart Failure (HF) Rate Expected to Rise to 8.5 Million Americans by 2030. New HFSA Initiative Identifies Rapidly Changing Landscape of HF in the U.S., https://hfsa.org/heart-failure-hf-rate-expected-rise-85-million-americans-2030-new-hfsa-initiative-identifies
7. Swapna, S., Onkar, S.: Ventricular Assist Devices Market Research, 2031. Ventricular Assist Devices Market (2023) https://www.alliedmarketresearch.com/ventricular-assist-devices-market
8. Strategic Market Research: Industry Report and Statistics (Facts & Figures) - Sales Volume, ASP & Number of Procedure Conducted by Product Type & Application, (2022)
9. https://www.strategicmarketresearch.com/market-report/ventricular-assist-devices-market
10. Han, J., Trumble, D.R.: Cardiac assist devices: early concepts, current technologies, and future innovations. Bioengineering. **6**(1), 18 (2019) https://www.ncbi.nlm.nih.gov/pmc/articles/PMC6466092/
11. Strategic Market Research: Industry Report and Statistics (Facts & Figures) - Sales Volume, ASP & Number of Procedure Conducted by Product Type & Application (2022)
12. Fortune Business Insights, 2023, Ventricular Assist Device Market., https://www.fortunebusinessinsights.com/ventricular-assist-device-market-106377
13. Abbott Laboratories: Annual Report. Abbott Laboratories (2022)
14. Johnson & Johnson, ABIOMED Acquired by Johnson & Johnson. 2022
15. CH Biomedical, CH Biomedical Receives Approval for CH-VAD System in China. 2021

16. Jarvik Heart, Inc., Jarvik Heart, Inc. Presents New VAD Device in Kazakhstan. 2019
17. ABIOMED, ABIOMED Acquires preCARDIA. 2021
18. https://www.medtronic.com/ro-ro/index.html
19. Gigauri, I.: Perceptual mapping as a marketing research tool for brand positioning. SSRG Int. J. Econ. Manag. Stud. **6**(4), 73–79 (2019). https://doi.org/10.14445/23939125/IJEMS-V6I4P110
20. Moon, S., Kamakura, W.A.: A picture is worth a thousand words: translating product reviews into a product positioning map. Int. J. Res. Mark. **34**(1), 265–285, ISSN 0167-8116 (2017). https://doi.org/10.1016/j.ijresmar.2016.05.007
21. D'Aveni, R.A.: Mapping your competitive position. Harv. Bus. Rev. **85**(11), 110–20, 154 (2007) PMID: 1815979, https://europepmc.org/article/med/18159791
22. Lee Siu Ming, Industry Mapping and Value Chain Analysis of Medical Devices Companies in Penang, 03.02.2020, https://penanginstitute.org/wp-content/uploads/2020/02/Industry-Mapping-and-Value-Chain-Analysis-of-Medical-Devices-Companies-in-Penang.pdf

Efficient High-Voltage Probe Designs for EV Systems

Cirstinoiu Bogdan[1], Bonciog Dumitru-Daniel[1,2(✉)], Ionica Nicolae[1], Nyiredi Bogdan[1], Ordodi Valentin-Laurentiu[2], Marius Nicolae Roman[3], and Maranescu Valentin-Ioan[1,2]

[1] Vitesco Technologies Engineering, Timişoara, Romania
daniel.bonciog@upt.ro
[2] Politehnica University Timisoara, Timişoara, Romania
[3] Technical University, Cluj-Napoca, Romania

Abstract. This study investigates the performance of cost-effective high-voltage probes designed for electric vehicle projects (EV), with a particular emphasis on signal integrity across various frequencies and duty cycles. The paper addresses the challenges of measuring high-frequency signals in EV systems, where accuracy is essential but typically limited by the expensive cost of precise equipment. Significant errors were discovered at higher frequencies and duty cycles, which impacted measurement reliability. Both the dual opto-isolated probe, and inductive coupling probe shown a stable performance at lower frequencies, but their accuracy has decreased under more demanding conditions. These probes were deliberately intended to be cost-effective, acknowledging that high-voltage testing benches from the High Voltage Testing Laboratory cannot always be completely equipped for every project due to substantial investments and the high costs of precision instruments. The findings highlight the importance of optimizing affordable probe designs for high-voltage EV applications to ensure precise and reliable signal measurements while managing budget constraints.

Keywords: EV system diagnostics · high voltage probes · electric vehicle (EV) systems · cost-effective design · duty cycle errors · opto-isolated probes · measurement · debugging · automotive

1 Introduction

The electric vehicle industry (EV) is in continuous expansion, generating new challenges in diagnosing and testing critical components, such as DC-DC converters, battery inverters, and systems [1, 2].

These systems operate at various frequencies, from a few kHz to MHz, imposing strict requirements on the test equipment used to ensure optimum performance and safety. The reliability of these components is essential for the overall performance of electric vehicles so precise testing and monitoring of digital signals become critical factors [3, 4].

S. Vlad and N. M. Roman (Eds.): MEDITECH 2024, IFMBE Proceedings, 130, pp. 261–273, 2025.
https://doi.org/10.1007/978-3-031-95671-3_27

Currently, the costs of the test equipment are considerably high, considering that in a High Voltage Testing Laboratory specialized in testing high voltage systems, there are several test benches, each being designed to operate in parallel for various projects.

Thus, each test bench requires a complete set of similar equipment, which contributes significantly to increasing costs and, in some cases, makes it difficult to equip all benches.

In certain situations, testing requires the use of a system as economically as possible, given that many projects are compact and consist of several layers of interconnected printed circuit boards (PCBs). Such a system is essential to facilitate the quick diagnosis and identification of problems or defects that occurred in the device under test (DUT).

This system is designed to monitor the DUT digital signals, offering the test engine to the DUT when the DUT has no connection or has undergone a malfunction [5, 6].

In the context of this project, the emphasis is placed on digital signals, where the sampling frequency is not always the most critical parameter. However, the ability to quickly detect the growing and falling edges is essential to ensure that monitoring devices do not introduce delays that could affect the behavior of the tested system (DUT) [7]. Maintaining a minimum current extracted from the measured signal is important to prevent alteration of the original characteristics of the signal, thus ensuring the maximum integrity of the collected data.

The reduction of the extracted current is vital not only for the accuracy of the measurements but also to minimize the impact on the tested system (DUT). In designing measuring probes for EV systems, this objective becomes a priority. To address this challenge, the present work explores two distinct models of measurement probes, each designed to optimize the current consumption in the measured signal, without compromising the ability to quickly detect the signal edges.

The first proposed model is based on a simplified design, reducing the current extracted to less than 1.5 mA. It uses dual digital insulators, providing a balance between complexity and efficiency, which allows for easier implementation and increased compatibility with existing infrastructure. This model is ideal for applications where the small size and minimum energy consumption are essential factors while ensuring robust protection against external interference.

The second model, more advanced, integrates isolated pads and sources, providing a pre-buffer stage to minimize the impact on the measured signal. This approach allows more precise control of the current and a significant reduction in the potential electromagnetic interference, thus offering a higher solution for applications where maximum accuracy is critical.

In addition, advanced design contributes to reducing wiring complexity and optimizing the performance of the measurement system.

This holistic approach ensures that EV systems benefit from precise, essential diagnoses for their long-term reliable operation.

Therefore, the results of this study contribute significantly to the development of NDT test equipment more efficient and adapted to the increasingly demanding requirements of the electric vehicle industry.

2 Conceptual Framework

2.1 Design Concepts

Designing a compact and efficient measurement probe is a challenge, especially when the available space is limited, and the performance requirements are high.

In this context, we have developed a probe concept that optimizes both size and functionality, integrating two digital inputs on the same printed circuit (PCB).

Figure 1 illustrates the scheme of this design, where dual digital inputs are strategically placed to reduce the total size of the PCB.

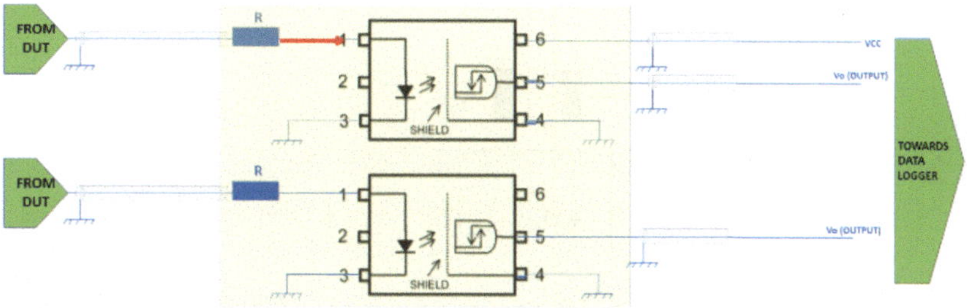

Fig. 1. Diagram of the dual opto-isolated probes.

An important aspect of this design is the reduction of the input current, indicated by the red arrow in the scheme. It was achieved by optimizing the components and selecting elements with a reduced polarization current, essential for the efficient functioning of the opto-isolator inputs. In addition, the two digital insulators share the same power supply and ground line, which not only simplifies the wiring but also reduces the number of wires to operate the probe.

Figure 2 presents the 3D visualization of the printed circuit plate, highlighting the compact way in which the components are arranged and how efficiently the space is used. This design ensures that the probe can be easily integrated into our EV projects, offering high performance in a small format.

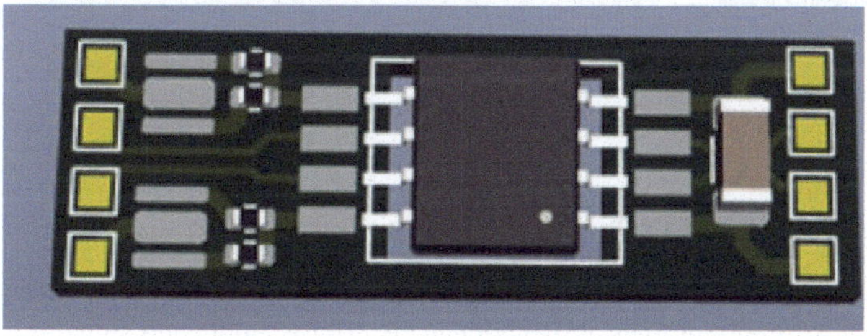

Fig. 2. 3D visualization for the dual opto-isolated probes.

This innovative approach in designing the measurement probe not only optimizes the size and consumption of energy but also ensures a reliable and simplified operation, which is essential in the context of testing and diagnosing high-voltage DUT.

The second proposed alternative is based on the use of an integrated circuit that incorporates both an inductive coupling insulator and an isolated power supply.

This solution allows the implementation of an additional buffer stage at the entry of the probe, aiming to reduce the current extracted from the signal, if the situation requires it. The objective for this buffer stage is to limit the current extracted to a maximum of 1 μA, while the integrated circuit estimates a maximum current extracted from the measured signal of approximately 150 μA, which represents a significant reduction from the initial design, with an order smaller size (Figs. 3 and 4).

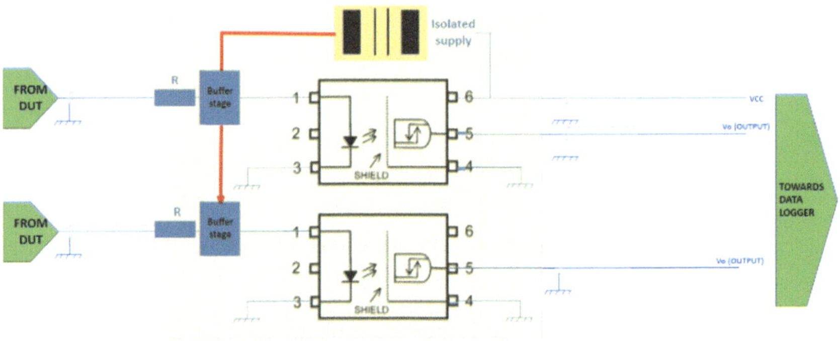

Fig. 3. Diagram of the dual digital isolator with buffer stage.

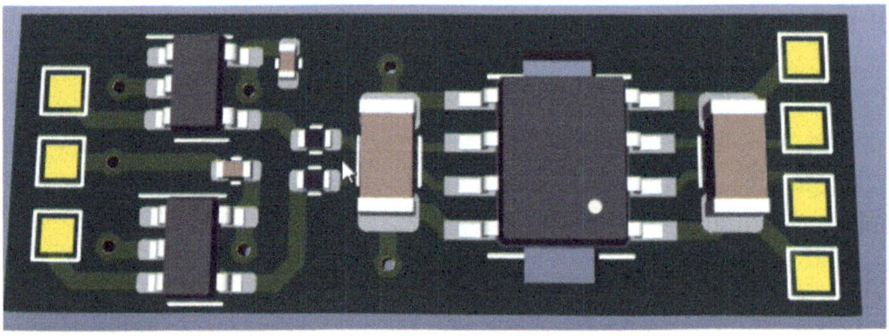

Fig. 4. 3D visualization for the dual digital isolator with buffer stage.

2.2 Power Supply and Probes Functionality

To ensure that the probe works efficiently, whether it uses optical or transformer-based insulation, it is important to establish both an electricity supply and signal transmission path.

An isolated power supply is required to provide the voltage corresponding to the integrated circuit (IC) in the probe. Since the device will predominantly utilize low-power logic gates in conjunction with the probe, the isolated power supply can be implemented using a traditional flyback converter or a push-pull converter, particularly when the supply voltage remains constant.

Figure 5 illustrates the block diagram of this configuration. The high-voltage test bench represents the environment in which the DUT and isolated probes are placed for testing. The main PCB is subdivided into several critical functional modules:

- the hysteresis comparator is used to stabilize the signal, eliminating minor fluctuations and maintaining a constant threshold.
- the evaluation board - signal injection allows both the signal monitoring and the injection of testing signals, thus facilitating a detailed analysis of the device response. This configuration guarantees the compatibility and optimal integration with various logical levels, thus increasing the adaptability and general performance of the measurement system.
- isolated power supply sources provide the voltage required for the operation of the associated circuits and circuits (5 V and 3.3 V), maintaining complete insulation to the external power source.

The external power source provides the voltages necessary for the entire system. The data acquired by the probes can be routed to either a standard data logger or a custom solution, depending on the user's requirements. This design strategy ensures that both the main printed circuit board (PCB) and the probes have a compact footprint, minimizing the space occupied near the device under test (DUT).

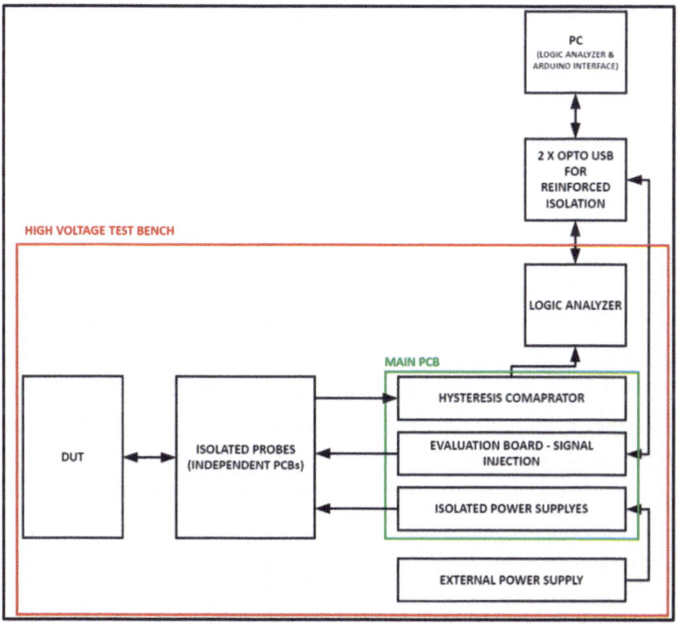

Fig. 5. Block diagram.

Given the potential variability of the length of the wire, a basic buffer circuit is required to maintain the integrity of the signal. Moreover, if signal injection is necessary, a level shifter and a logic board should be included to sequence the signals accordingly. As a result, the main board primarily consists of three key components: an isolated power supply, an array of hysteresis comparators (Schmitt triggers), and a programmable evaluation board.

3 Measurement Results

The measurement strategy involved injecting a predefined signal, generated by a signal generator, having a wave shape and a specified work cycle, at each frequency targeted, for various values of the duty cycle. This method was chosen to allow a systematic comparison of the results in a controlled environment, thus reducing the risk of human errors and ensuring the consistency of the collected data.

The analysis of the relative error was made in the low voltage domain, at room temperature (25 °C). For this evaluation, a signal generator was used to adjust the frequency and duty cycle of the rectangular signal.

The duty cycle of the signal applied to the optocoupler and the duty cycle of the signal at the output of the comparator with hysteresis was measured using an oscilloscope.

The relative error was calculated, and based on these data sets of curves were generated that illustrate the relative error for specific frequencies and duty cycles.

The formula used to calculate relative error is:

$$Relative\ Error = \frac{Duty\ Cycle_{output} - Duty\ Cycle_{input}}{Duty\ Cycle_{input}} \times 100\% \quad (1)$$

The results of the measurements are presented below for each probe, dual opto-isolated probe, and inductive coupling probe.

3.1 Dual Opto-Isolated Probe

Figure 6 illustrates the error of the duty cycle according to the frequency and the percentage of the duty cycle, highlighting the signal behavior under various testing conditions.

At higher frequencies (in the MHz range), errors are observed consistently across the range of duty cycles, particularly at lower duty cycles, with the error gradually diminishing as the duty cycle increases. Conversely, at lower frequencies, the error remains minimal across the entire analyzed range. Once the duty cycle exceeds 50%, the error significantly decreases for all frequencies, becoming nearly negligible at higher duty cycles.

The error of the duty cycle is considerable at high frequencies and small duty cycles, affecting the accuracy of the measurements. In applications that require high accuracy, solutions must be implemented to minimize these errors.

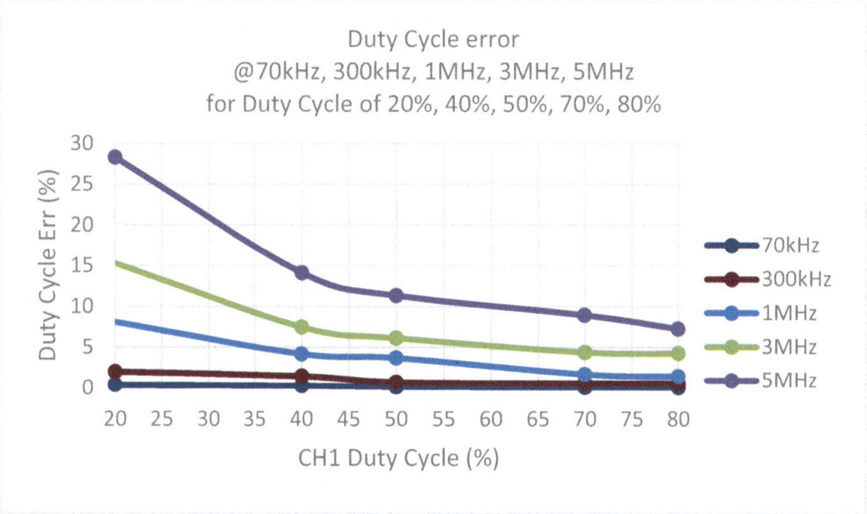

Fig. 6. Test results for dual opto-isolated probes.

At high frequencies (in the MHz range), substantial errors are observed in small duty cycles (20% and 40%), errors that decrease rapidly as the working cycle increases. This behavior is particularly evident at 5 MHz, where the error is most pronounced at lower duty cycles. Similarly, at 3 MHz, the error remains significant, but it is relatively lower. These results emphasize the challenges associated with maintaining the integrity of the signal at high frequencies, where opto-isolators require more precise synchronization and increased attention to the current extracted from the measured signal.

On the other hand, at lower frequencies, such as 70 kHz and 300 kHz, the error remains minimal throughout the analyzed range. At 300 kHz, for example, the error of the work cycle is constantly maintained at low levels, even in small duty cycles. This indicates that the double optical insulation probe operates more reliably in low frequencies, where the requirements related to synchronization and the extracted current are less stringent.

As the work cycle exceeds 50%, the error decreases significantly for all frequencies, becoming almost negligible at higher duty cycles. This reduction of errors in large work cycles suggests that the probe performance stabilizes as the signal occupies a larger portion of the cycle, reducing the relative impact of synchronization.

The error of the work cycle is considerable at high frequencies and small work cycles, affecting the accuracy of the measurements. In applications that require high accuracy, solutions must be implemented to minimize these errors. Potential improvements could include optimization of opto-isolators design for better performance at high frequencies or incorporating additional buffers to compensate for synchronization discrepancies.

In addition, the observed discrepancies emphasize the importance of choosing the proper frequency and duty cycle settings, depending on the specific requirements of the application. This aspect is particularly crucial in scenarios where high accuracy is required because high frequencies and small work cycles could lead to significant deviations from the expected results.

The analysis of the figure and data indicates that, although the double optical insulation probe is effective under certain conditions, there is room for improvements, especially at high frequencies, where maintaining the integrity of the signal becomes more difficult. Future improvements could focus on refining the probe design to minimize errors, especially in demanding applications, which involve high frequencies.

3.2 Inductive Coupling Probe

Figure 7 depicts the duty cycle error across varying frequencies and duty cycle percentages, providing insight into the signal's behavior under different testing conditions.

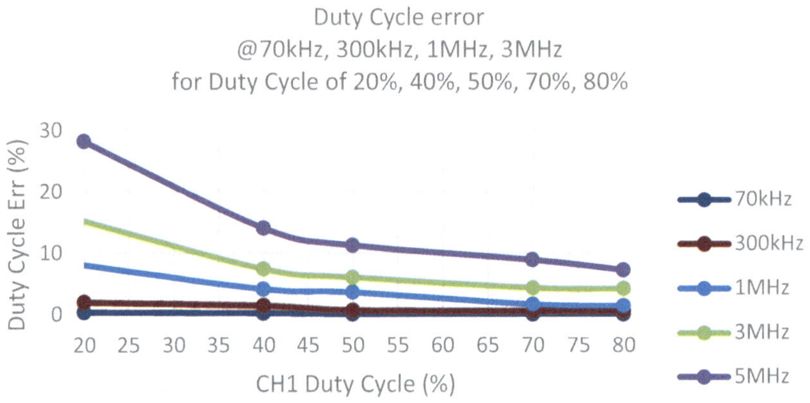

Fig. 7. Test results for inductive coupling probe.

At frequencies in the MHz range, significant errors are observed across the range of low-duty cycles, particularly at lower duty cycles. The error is most pronounced at higher frequencies, such as 5 MHz, while it remains somewhat less severe at 3 MHz.

For lower frequencies, in the range of 70 kHz and 300 kHz, the duty cycle error remains consistently low across the entire analyzed range.

Regardless of the frequency, the error decreases considerably as the duty cycle increases. At duty cycles above 50%, the error becomes almost negligible, approaching zero at 70% and 80% for all frequencies analyzed.

These results indicate that the inductive coupling probes exhibit higher errors at high frequencies and low-duty cycles. However, as the duty cycle increases, the error decreases significantly. Lower frequencies demonstrate greater stability, with minimal errors across the entire duty cycle range. These findings are crucial for optimizing performance in applications requiring high accuracy under varying operating conditions.

4 Future Improvements

The results of the comparison between these two models indicate that there is a significant potential for improvements. Based on the analyses performed, the general concept could be optimized by increasing the maximum operating frequency, which would allow us to obtain more accurate measurements, especially at high frequencies and duty small cycles, where current errors are more pronounced.

To reduce these errors at high frequencies and low-duty cycles, methods of improving the circuit response should be explored. Also, solutions could be studied to optimize the design of the inductive coupling sample, by using advanced components that minimize the losses and improve the general stability of the signal.

Another important aspect that requires attention is to reduce the number of wires needed to connect the wells. Fewer connections would bring considerable benefits, such as greater flexibility in the system configuration and a decrease in the risk of failure points.

4.1 Proposed Concept

The block diagram of the new probe concept is presented in Fig. 8.

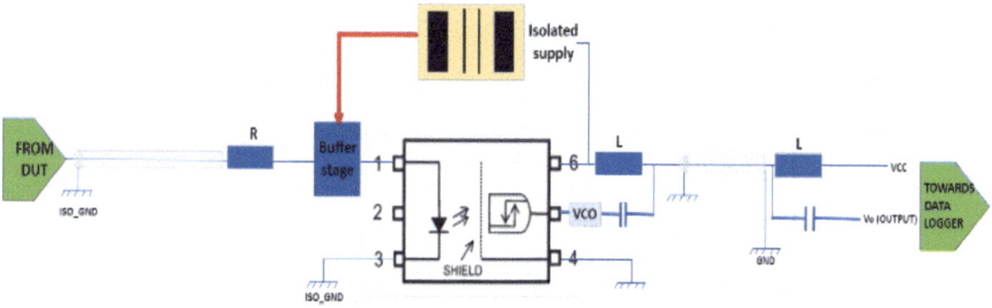

Fig. 8. Block diagram of the updated probe design.

The probe will incorporate, from left to right, a buffer stage supplied by a dedicated isolated supply. This configuration is crucial as it enables the integration of high-frequency rated opto-isolators, which typically draw more current from the measured signal. By using a buffer stage, the integrity of the signal is preserved, ensuring accurate transmission without introducing noise or distortion. The isolated supply that powers this stage can be designed using a simplified approach, such as a push-pull converter, which is efficient in minimizing the number of components required while still providing the necessary isolation and power stability (Fig. 9).

The buffer stage will amplify the measured signal and drive the isolator. The isolated supply will provide the necessary power to the buffer to ensure it can effectively drive the isolator. The isolated power supply is designed to ensure that the buffer receives a consistent power source, which is isolated from the rest of the circuit to prevent interference. This isolation is achieved through careful selection of components and circuit

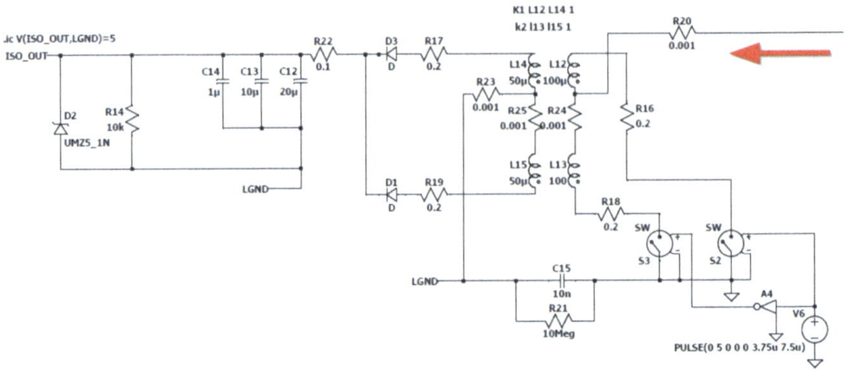

Fig. 9. Schematic diagram – isolated power supply.

design, ensuring that the power supplied to the buffer does not introduce noise or other forms of signal degradation.

Additionally, the use of high-frequency rated opto-isolators allows the probe to operate efficiently even in demanding environments where fast signal processing is required. The buffer stage, powered by the isolated supply, ensures that the opto-isolator receives a clean and stable signal, which is essential for maintaining accurate data transmission across the system (Fig. 10).

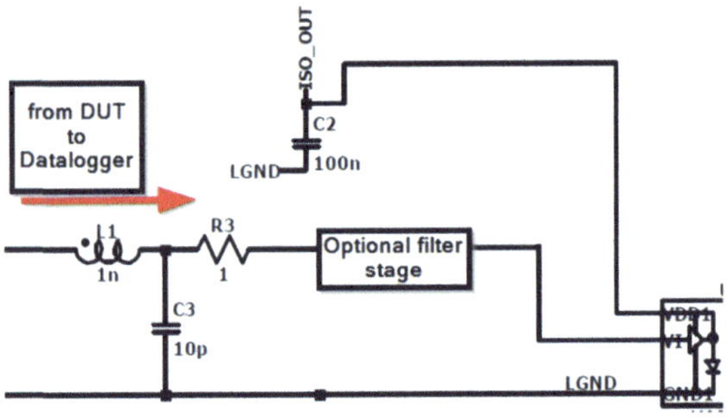

Fig. 10. Block diagram for the output stage towards the datalogger.

The output signal of the opto-isolator drives then a VCO which will generate a modulated signal. This signal is connected to the same wire as the power line using a bias-T configuration. This way, we can use the same coaxial cable to supply the probe and send the data to the data logger (Fig. 11).

Due to the injected signal, a series of low-pass filters need to be used on the supply path, prioritizing a small form factor for the probe. On the other side of the cable, towards the datalogger, the same bias T configuration is used to extract the signal from the power line. A decoder will recover the digital signal by demodulating the signal generated by the VCO.

After this, the signal can be interpreted by the datalogger as follows (Fig. 12).

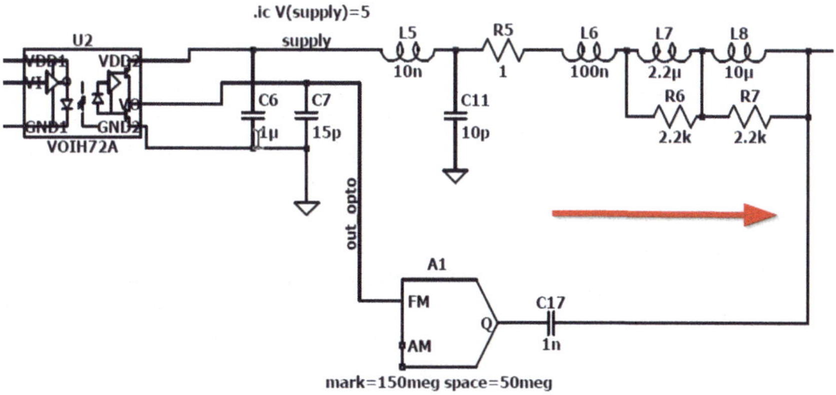

Fig. 11. Block diagram for the output stage towards the datalogger.

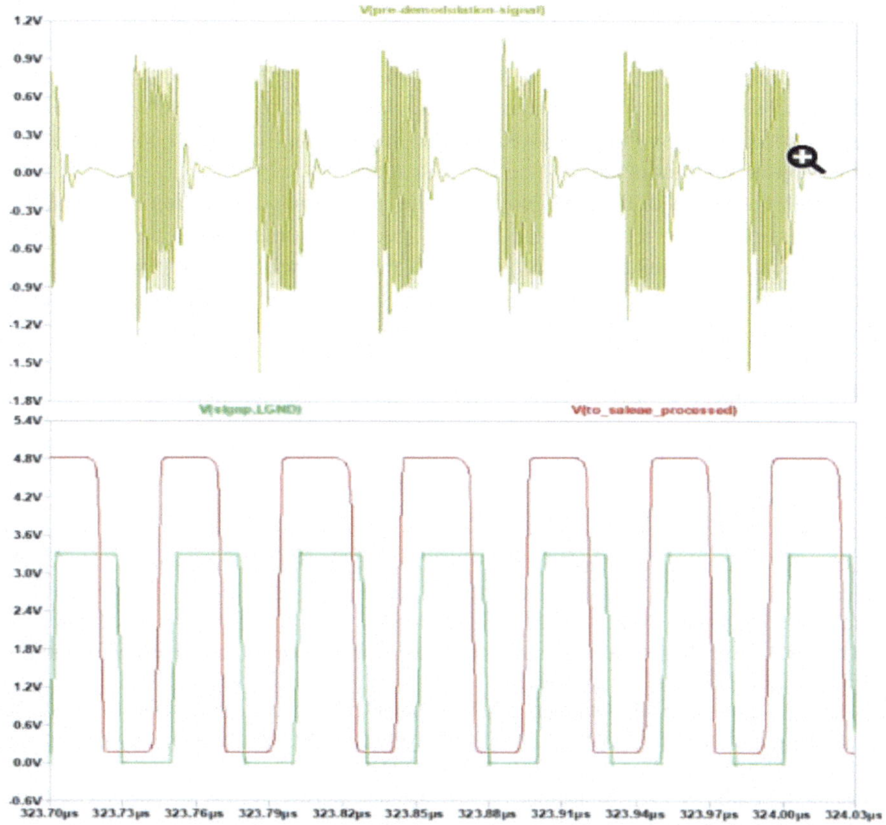

Fig. 12. Example of sampled signal 20 MHz, 50% duty cycle. Yellow - the preprocessed signal coming from the probe. Red - the signal captured by the datalogger. Green - initial measured signal.

5 Conclusions

The study evaluated the performance of the two designed probe types, dual opto-isolated probe, and inductive coupling probe, in the context of the measurement signals at different frequencies and duty cycles. The results highlighted the challenges associated with maintaining the integrity of the signal, especially at high frequencies (within the MHz range) and low-duty cycles. Under these conditions, the errors of the duty cycle were significant, affecting the accuracy of the measurements.

The dual opto-isolated probe demonstrated stable performance at low frequencies, where the synchronization requirements and the drawn are less stringent. However, at high frequencies, errors have been more pronounced, suggesting for applications that require high accuracy at high frequencies, additional design optimization is required.

The inductive coupling probe exhibited higher stability at low frequencies, with minimal errors throughout the duty cycle. However, at high frequencies, the performance of this probe was similarly affected by errors, especially at low-duty cycles.

In conclusion, although both probes are effective under certain conditions, there is a significant potential for improvement. Future improvements should focus on reducing high frequencies and low duty cycles, without compromising the size and cost of probes. In addition, it should focus on the duty cycle and achieve more precise results in frequency ranges up to 10 MHz, with an emphasis on the "high" time that the probe requires and can be accurately reproduced.

Overall, the results of the study underline the importance of selecting the frequency and the appropriate service cycle based on the specific requirements of the application. By implementing adequate technical solutions, the performance of the probes can be optimized, ensuring greater precision in the measurements at high frequencies and in demanding applications.

Conflict of Interest. The authors declare that they have no conflict of interest.

References

1. Reddy, K.J., Natarajan, S.: Energy sources and multi-input DC-DC converters used in hybrid electric vehicle applications–a review. Int. J. Hydrog. Energy. **43**(36), 17387–17408 (2018)
2. Acharige, S.S., Haque, M.E., Arif, M.T., Hosseinzadeh, N., Hasan, K.N., Oo, A.M.T.: Review of electric vehicle charging technologies, standards, architectures, and converter configurations. IEEE Access. **11**, 41218–41255 (2023)
3. Affam, A., Buswig, Y.M., Othman, A.K.B.H., Julai, N.B., Qays, O.: A review of multiple input DC-DC converter topologies linked with hybrid electric vehicles and renewable energy systems. Renew. Sust. Energ. Rev. **135**, 110186 (2021)
4. Islam, R., Rafin, S.S.H., Mohammed, O.A.: Comprehensive review of power electronic converters in electric vehicle applications. Forecasting. **5**(1), 22–80 (2022)
5. Chesi, A., Ferrara, G., Ferrari, L., Tarani, F.: Setup and characterisation of a multi-purpose test rig for R744 refrigerating cycles and equipment. Int. J. Refrig. **35**(7), 1848–1859 (2012)
6. Ranasinghe, W. R. A. A.: Universal Data Acquisition System for Structural & Environmental Monitoring Applications

7. Paul, C., Chatterjee, D., Kapat, S., Mallik, R., Jain, A.: Large-signal based design approach in boost PFC under constant on-time digital CMC with load feedforward for ultra-fast transient and stability. In: 2023 IEEE International Conference on Power Electronics, Smart Grid, and Renewable Energy (PESGRE), pp. 1–6. IEEE (2023, December)

Solutions for Heating in Healthcare Facilities: A Case Study on the Use of Heat Pumps and Energy Consumption Reduction

Constantin Cilibiu(✉) and Ancuța Coca Abrudan

Faculty of Building Services Engineering, Building Services Department, Technical University of Cluj-Napoca, Cluj-Napoca, Romania
constantin.cilibiu@insta.utcluj.ro

Abstract. In the context of rising energy costs and increasing emphasis on environmental sustainability, hospitals face significant challenges in optimizing their heating systems while maintaining patient comfort and reducing operational expenses. This paper explores the implementation of heat pump technology as a sustainable solution for a hospital heating system, focusing on its impact on energy consumption and carbon footprint reduction. The study presents a detailed case analysis of a hospital that transitioned to a heat pump-based heating system, highlighting the technical, economic, and environmental benefits achieved. Key findings demonstrate a substantial decrease in energy consumption, leading to lower operational costs and a significant reduction in CO_2 emissions. Additionally, the paper discusses the challenges encountered during the implementation process, including system integration with existing infrastructure and initial investment considerations. The results underscore the potential of heat pumps as a viable alternative to traditional heating methods in the healthcare sector, contributing to both economic savings and environmental stewardship. This case study provides valuable insights for engineers, facility managers, and decision-makers in the medical field who are considering the adoption of green technologies to enhance the sustainability of healthcare facilities.

Keywords: green energy · thermal comfort · energy efficiency

1 Introduction

Energy consumption in healthcare facilities is substantial due to their continuous operation and stringent climate control requirements. Integrating heat pumps as a renewable energy solution has shown significant potential in reducing energy usage and carbon emissions. Case studies from European hospitals have demonstrated that switching to heat pump systems can reduce operational costs and carbon footprints, while maintaining thermal comfort for patients and staff [1].

Decarbonizing healthcare facilities with heat pump technology has been identified as an essential step in reducing carbon emissions and increasing energy efficiency. Research highlights that heat pumps, when integrated into healthcare infrastructure, provide a

S. Vlad and N. M. Roman (Eds.): MEDITECH 2024, IFMBE Proceedings, 130, pp. 274–283, 2025.
https://doi.org/10.1007/978-3-031-95671-3_28

sustainable solution for climate control while lowering operating costs. Studies from healthcare facilities in Europe and the U.S. demonstrate the positive impact of replacing fossil-fuel-based systems with heat pumps, contributing to substantial energy savings and emission reductions [2].

A 2022 report by the European Commission highlights the significant role of heat pumps in reducing gas demand for heating across Europe, with expectations of lowering consumption by over 21 billion cubic meters by 2030. Additionally, the REPowerEU plan calls for doubling the deployment rate of heat pumps in buildings, with an emphasis on district heating and cooling systems in healthcare and other large-scale facilities. This is crucial for reducing reliance on fossil fuels and achieving the EU Green Deal goals [3].

A study by IRENA (2022) shows that electrically driven heat pumps, when applied in European buildings, have the potential to significantly cut both energy consumption and carbon emissions. Specifically, the report estimates that these systems can reduce direct carbon emissions by up to 10 gigatons annually in the building sector. This is particularly relevant for healthcare facilities, where high energy demands make heat pumps an attractive solution for decarbonizing heating systems [4].

In European hospitals, a key barrier to wider adoption of renewable energy is the integration with existing infrastructure. Many hospitals are old and have complex layouts, making it difficult to retrofit them with modern renewable technologies like photovoltaic panels or heat pumps. Additionally, stringent regulations for hospital buildings in the EU require compliance with safety and operational standards, which can slow down green energy projects. Some hospitals have successfully navigated these obstacles, but it requires substantial planning and investment [5].

Lastly, the intermittent nature of renewable energy sources like solar and wind presents another challenge. Critical facilities like operating rooms or ICUs demand a stable power supply, which renewables alone may not always provide. Hospitals often need to combine renewable systems with conventional backup power solutions, such as diesel generators, to ensure reliability, which can limit the environmental benefits of switching to green energy [6].

By addressing these issues through coordinated efforts, such as government subsidies, partnerships with green energy vendors, and advancements in energy storage, hospitals can overcome these critical barriers and move toward greener, more sustainable operations.

2 Methods

The paper presents a case study regarding the modernization of heating plant for a private medical care unit located in Cluj-Napoca, to improve energy efficiency while the thermal comfort for building occupants it is not affected by any means.

The analyzed building has a total area of 825 m^2 and three levels, namely basement, ground floor and first floor, as it is presented in the Fig. 1 (Fig. 1a, b and c).

In the year 2012, when this building was put into operation, it was equipped with two boilers for heating and preparation of domestic hot water, located in the basement of the building. One of the boilers had a power of 70 kW and the other boiler, provided

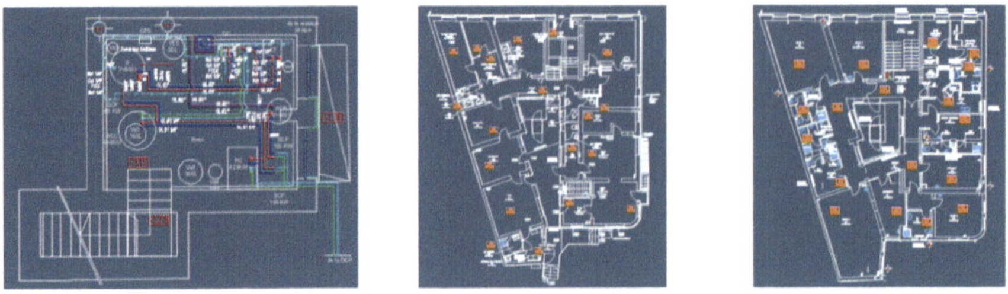

a. b. c.

Fig. 1. The three levels of the building: **a.** Basement, **b.** Ground floor, **c.** 1st Floor

for hot water and consumption peaks, had a power of 30 kW. Both boilers were operated on natural gas.

In the year 2023, building owners decided to rehabilitate the building heating plant in accordance with the new norms entered into force regarding the energy efficiency of buildings and related building services [7].

With the intention of implementing a heat pump and use green energy there were performed a series of simulations with the software application nPro [8] and the best results are presented below.

To determine the average heating demand, the measured consumption data were subjected to a meteorological adjustment according to regional number of degree-days. Annual energy demand for building heating purpose is 150 MWh with a peak power of 93 kW and for domestic hot water production, the building require 28.9 MWh with a peak power of 8 kW, the annual profile being presented in Fig. 2.

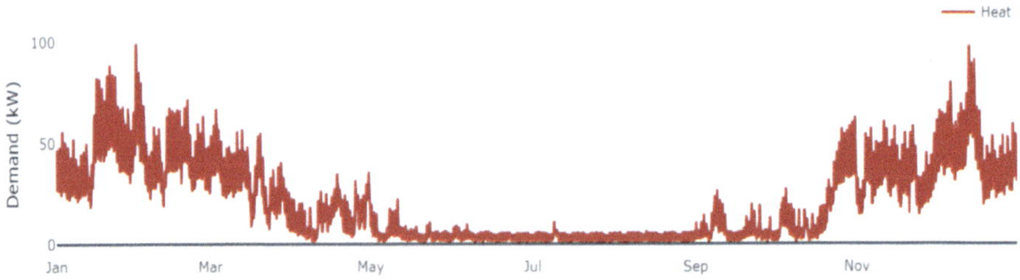

Fig. 2. Annual profile of energy demand for heating and hot water

Annual energy demand for electricity is 6.6 MWh with a peak power of 1.3 kW.

2.1 Initial Heating Plant Simulation

With the aim of substantiate the decisions related to the implementation of a heat pump and the use of photovoltaic panels to be able to provide part of the energy needed to feed the heat generation system, the simulations started from the existing situation, based on the two boilers operating on natural gas, this being necessary as a reference point for comparison with the proposed scenarios.

All simulations were carried out to optimize investment and operating costs as well as to reduce CO_2 emissions.

The simulation results calculated with nPro [8], for 1 year of operation, based on annual profile of energy demand for heating and hot water, for the smallest gas boiler are presented in Table 1.

Table 1. Results for 30 kW gas boiler operation [8]

Gas boiler 1	Unit	Value
Nominal heating power	kW_{th}	30
Generated heat	MWh	55
Fuel demand	MWh	61
Full load hours	h/year	1833

The simulation results calculated with nPro [8], for 1 year of operation, based on annual profile of energy demand for heating and hot water, for the largest gas boiler are presented in Table 2.

Table 2. Results for 70 kW gas boiler operation [8]

Gas boiler 2	Unit	Value
Nominal heating power	kW_{th}	71
Generated heat	MWh	124
Fuel demand	MWh	138
Full load hours	h/year	1747

The simulation results calculated with nPro [8], for 1 year of operation, based on annual profile of energy demand for heating and hot water, regarding energy import are presented in Table 3.

Table 3. Energy import for heating plant operation [8]

Energy import	Unit	Value
Electricity import from grid	MWh	7
Natural gas import	MWh	199

The simulation results calculated with nPro [8], for 1 year of operation, based on annual profile of energy demand for heating and hot water, regarding the CO_2 emissions and primary energy used for heating and hot water preparation are presented in Table 4.

Table 4. Emissions and primary energy for heating plant operation [8]

Emissions and primary energy	Unit	Value
CO_2 emissions	t	52
Primary energy	MWh	231

2.2 Air-Source Heat Pump with Local Photovoltaic Panels

First option analyzed for the heating plant was the installation of an air-source heat pump coupled with photovoltaics mounted on the building roof.

Total investment for the proposed heating plant modernization is 159,500 (€).

The simulation results calculated with nPro [8], for 1 year of operation, based on annual profile of energy demand for heating and hot water, for the photovoltaic panels are presented in Table 5.

Table 5. Results for photovoltaic panels operation [8]

Photovoltaics	Unit	Value
Installed capacity	kW_p	9
Collector area	m^2	53
Generated electricity	MWh	10
Full load hours	h/year	1150

The simulation results calculated with nPro [8], for 1 year of operation, based on annual profile of energy demand for heating and hot water, of the air-source heat pump are presented in Table 6.

Table 6. Results for air-source heat pump operation [8]

Air-source heat pump	Unit	Value
Generated heat	MWh	179
Electricity demand	MWh	41
Full load hours	h/year	1219
Nominal heating power	kW_{th}	136

The simulation results calculated with nPro [8], for 1 year of operation, based on annual profile of energy demand for heating and hot water, for air-source heat pump operation related to electricity import and green energy generation are presented in Table 7.

Table 7. Electricity import and renewable energy generation [8]

Energy import	Unit	Value
Electricity import from grid	MWh	44
Renewable energy generation	MWh	10

The simulation results calculated with nPro [8], for 1 year of operation, based on annual profile of energy demand for heating and hot water, for air-source heat pump operation regarding the CO_2 emissions and primary energy used for heating and hot water preparation are presented in Table 8.

Table 8. Emissions and primary energy for heating plant operation [8]

Emissions and primary energy	Unit	Value
CO_2 emissions	t	15
Primary energy	MWh	80

The optimal nominal capacity of the air-source heat pump was determined in the calculation.

The results presented for the heat pump are based on its coefficient of performance (COP), which reflects the ratio between the thermal power produced and the electrical power consumed. The higher this COP value, the higher the potential of the heat pump.

For the nominal operating point, a COP of 4 was assumed. Minimum COP was calculated to be 3 and maximum COP was adopted 5, while mean COP was 4.7, annual results being presented in the Fig. 3.

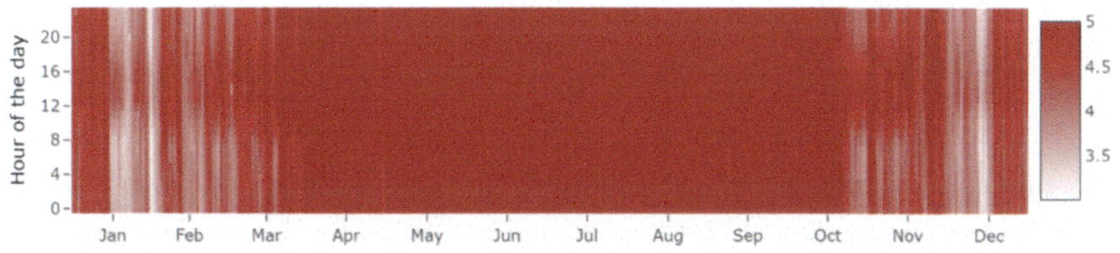

Fig. 3. Annual COP profile of air-source heat pump [8]

In Fig. 3, during the entire year this coefficient of performance varies, its calculated values being between 3 and 5, according to regional number of degree-days, the biggest variations being even during the heating season. This aspect is not at all unexpected, considering that this type of heat pump extracts its energy from the outside air, and during the heating season the temperature values reach negative minimums.

Annual maintenance costs for air-source heat pump system operation were estimated to 3839 (€).

2.3 Ground-Source Heat Pump with Local Photovoltaic Panels

Second option analyzed for the heating plant was the installation of a ground-source heat pump coupled with photovoltaics mounted also on the building roof.

Total investment for the proposed heating plant rehabilitation is 99,900 (€).

The simulation results calculated with nPro [8], for 1 year of operation, based on annual profile of energy demand for heating and hot water, for the ground-source heat pump operation are presented in Table 9.

Table 9. Results for ground-source heat pump operation [8]

Geothermal probes	Unit	Value
Extraction rate	kW	75
Generated heat	MWh	179
Electricity demand	MWh	36

The simulation results calculated with nPro [8], for 1 year of operation, based on annual profile of energy demand for heating and hot water, for the ground-source heat pump operation related to electricity import and green energy generation are presented in Table 10.

Table 10. Electricity import and renewable energy generation [8]

Energy import	Unit	Value
Electricity import from grid	MWh	39
Renewable energy generation	MWh	10

The simulation results calculated with nPro [8], for 1 year of operation, based on annual profile of energy demand for heating and hot water, for the ground-source heat pump operation regarding the CO_2 emissions and primary energy used for heating and hot water preparation are presented in Table 11.

The optimal nominal capacity of the ground-source heat pump was determined in the calculation. Minimum COP was calculated to be 5 and maximum COP was adopted 5, while mean COP was also 5, annual results being presented in the Fig. 4.

In the previously presented figure, throughout the whole year, the performance coefficient (COP) is constant, being calculated and having the same value throughout the heating season. This aspect is very important for the performance of the heat pump, which records the best values given the fact that heat is extracted from the ground, from

Table 11. Emissions and primary energy for heating plant operation [8]

Emissions and primary energy	Unit	Value
CO_2 emissions	t	14
Primary energy	MWh	70

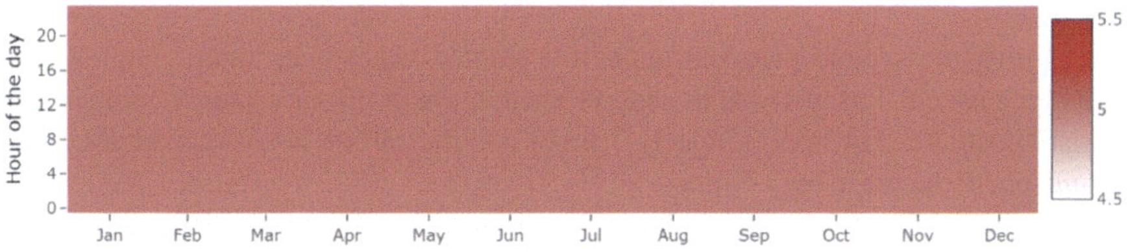

Fig. 4. Annual COP profile of ground-source heat pump [8]

a depth with a constant temperature regardless of the outside air temperature. Thermal energy is extracted with the help of deep vertical drillings (boreholes).

Annual maintenance costs for ground-source heat pump system operation were estimated to 2799 (€).

The calculation of boreholes for ground-source heat pump was based on the GeoHandLight method [9] as well as an open-source simulation core for geothermal borehole fields, based on the following parameters:

- Borehole resistance: $R_b = 0.075$ (m·K)/W
- Borehole radius: $r_b = 0.08$ m
- Borehole distance placement:8 m
- Borehole arrangement: $4 \times 7 = 28$ boreholes
- Probe length proposed: 150 m, resulting 150 m × 28 = 4200 m
- Specific heat extraction: 17.9 W/m.
- Temperature difference (probe): 3 K
- Thermal conductivity of the ground: $\lambda = 2.5$ W/(m·K)
- Volumetric heat capacity: $c_p \times \rho = 2.4$ MJ/(m^3·K)
- Geothermal heat flux density: $q_{geo} = 0.065$ W/m^2

Resulting annual heat extraction 143 MWh with a peak power of 75 kW.

Calculated borehole field area: 32 m × 56 m = 1792 m^2.

3 Discussion

Given the results previously presented, a brief analysis of the actual proposals is required.

Thus, for the initial situation, in which the thermal plant is composed only of gas boilers, the value of primary energy used for heating and domestic hot water preparation is 231 MWh annually from the national grid and 52 tons of CO_2 emissions are released in the atmosphere.

For the first proposal, an air-source heat pump with local photovoltaic panels, there are 80 MWh of primary energy used for heating and domestic hot water preparation. In regard of CO_2 emissions, only 15 tons are released annually into the atmosphere.

It is very easy to see that primary energy used decreases by about 2.5 times and carbon emissions are reduced by about 3.5 times. At the same time, green energy is also used at local site through photovoltaic panels, totaling a quantity of 10 MWh annually.

Another important benefit is related to the energy imported from the national grid, which is also reduced by 4.7 times, all this translating into a considerable decrease in operating costs.

Regarding the second proposal, namely a ground-water heat pump coupled with photovoltaic panels, it is possible to observe a reduction of primary energy compared to the existing situation by approximately 3 times and the carbon emissions released into the atmosphere decrease by 3.7 times.

The quantity of green energy generated is the same as in the first proposal, the proposed photovoltaic panels being provided to be mounted on the roof of the building, where the existing surface does not allow the placement of more than 53 m^2 of photovoltaics. The investment costs for this proposal are, on the other hand, lower by about 38% compared to the first proposal.

Thus, at first glance, the scale is favorably inclined to the second option, the results of the simulations recording better values.

However, considering that the proposed solutions will be applied to an existing building, it is imperative to consider the existing situation, more precisely the considerations related to the area required for the location of the boreholes. The existing building has an inner courtyard with an area of 650 m^2, made up of green spaces and car parks and buried under them are the building networks, respectively water, sewage and electrical low voltage lines. This is an area which in no form or variant can successfully meet the required distances for the boreholes layout.

Therefore, even if the results of the simulations lead to a variant of rehabilitating the thermal plant with a ground-water heat pump, the reality on the ground required the use of an air-water heat pump, despite the high costs, this being the only possible solution according to the actual situation.

4 Conclusions

Considering the National and European technical regulations regarding energy efficiency and the reduction of carbon emissions, it is required that most buildings to be thermally rehabilitated, both for the building envelope and for the building HVAC systems.

Moreover, all newly built buildings after 2023 or renovated must also have a local renewable primary energy source to comply with the requirements imposed to become Near Zero Energy Buildings.

The analyzed healthcare building is no exception to these regulations.

The proposed solutions successfully fit into the imposed energy trends, the calculations and simulations performed being detailed both on a technical and economic level.

The use of heat pumps coupled with photovoltaic panels located on the building or in its vicinity represents now the only viable solutions for meeting the requirements to reduce the primary energy consumed and the carbon emissions released into the atmosphere. However, in the case the of the existing buildings, their particularities decide which combination is more suitable for this purpose.

Each building is unique; therefore, any rehabilitation or modernization process requires a careful and customized design and planning process.

The case study presented above demonstrates that although the results of the simulations for a particular building indicate the adoption of a more effective solution from a technical and economic point of view, this theoretical solution cannot be every time applied, being necessary to find alternative adequate solution.

Conflict of Interest. The authors declare that there is no conflict of interest regarding the publication of this paper.

References

1. Sala, M., Alcamo, G., Nelli, L.C.: Energy-saving solutions for five hospitals in Europe. In: Sayigh, A. (ed.) Mediterranean Green Buildings & Renewable Energy. Springer, Cham (2017). https://doi.org/10.1007/978-3-319-30746-6_1
2. Cox, J., Belding, S., Campos, G., Lowder, T.: High Temperature Heat Pump Model Documentation and Case Studies. National Renewable Energy Laboratory, Golden, CO (2023) NREL/TP-7A40-84560. https://www.nrel.gov/docs/fy23osti/84560.pdf
3. https://energy.ec.europa.eu/topics/energy-efficiency/heat-pumps_en, last accessed 09 Sept 2024
4. IRENA: Renewable Solutions in End-Uses: Heat Pump Costs and Markets. International Renewable Energy Agency, Abu Dhabi (2022) https://www.irena.org/-/media/Files/IRENA/Agency/Publication/2022/Nov/IRENA_Heat_Pumps_Costs_Markets_2022.pdf
5. Sanyal, T., Rakshit, I., Bhattacharjee, P.: Green healthcare: initiatives and adaptations for sustainable future. Environ. Dev. Sustain. (2024). https://doi.org/10.1007/s10668-024-04957-z
6. https://www.wri.org/insights/powering-remote-hospitals-solar-overcoming-remaining-challenges, last accessed 09 Sept 2024
7. MC001-2022 Methodology for calculating the energy performance of buildings
8. nPro Software tool for planning building and district energy systems. https://acad.npro.energy/
9. Koenigsdorff et al.: GEO-HANDlight (Version 5.0) and user manual, 2022: https://innosued.de/energie/geothermie-software-2/
10. Directive 2010/31/EU of the European Parliament and of the Council of 19 May 2010 on the Energy performance of buildings; Official Journal of the European Union
11. LAW no. 372 of 13 December 2005 on the Energy performance of buildings republished on OFFICIAL GAZETTE No.868 of 23 September 2020

Author Index

S. Vlad and N. M. Roman (Eds.): MEDITECH 2024, IFMBE Proceedings, 130, pp. 285–286, 2025.
https://doi.org/10.1007/978-3-031-95671-3

The manufacturer's authorised representative in the EU is Springer Nature Customer Service Centre GmbH, Europaplatz 3, 69115 Heidelberg, Germany. If you have any concerns regarding our products, please contact ProductSafety@springernature.com

Printed and bound by CPI Group (UK) Ltd, Croydon, CR0 4YY
01/10/2026
02225911-0008